INSECT DIETARY

LONDON : GEOFFREY CUMBERLEGE
OXFORD UNIVERSITY PRESS

INSECT DIETARY

An Account of the Food Habits of Insects

BY

CHARLES T. BRUES

Professor of Entomology, Harvard University

HARVARD UNIVERSITY PRESS

Cambridge, Massachusetts

1946

PRINTED AT THE HARVARD UNIVERSITY PRINTING OFFICE
CAMBRIDGE, MASSACHUSETTS, U.S.A.

Library
UNIVERSITY OF MIAMI

TO

BEIRNE BARRETT BRUES

Whose keen and stimulating interest has furthered this and many
other entomological ventures undertaken by the author of this volume

FOREWORD

OVER A LONG PERIOD of years the writer has been deeply interested in the special phases of entomology that are dealt with in the several chapters of the present book. Although all relate to matters directly concerned in the acquisition and utilization of food, it may seem at first blush that some of the included material is slightly, or even highly, irrelevant to the main theme, and that a good deal of trimming might have been advisable. I shall not hide behind the cloak of platitude and recall the fact that all biological phenomena are interdependent among themselves and in turn dependent upon the more readily definable physical and chemical attributes of all matter. Modern science sees in the consummation of such a program of understanding a millennium when all the guesswork and superstition will be knocked out of biology. Thus purified she may hold her head aloft, balanced with mathematical precision and propped up on a framework of unquestioned strength and integrity. Such an unassailable position has so far been a will-o'-the-wisp, and as time goes on, it is increasingly evident that this goal lies in the misty future. No isolated discoveries, however grandiose, can be expected to answer the innumerable questions that engage the attention of biologists everywhere. Each and every application of the more exact sciences to biology marks some advance in our understanding of organic nature, but likewise it seems invariably to raise new questions that are still just beyond the reach of our present tools.

From this standpoint, the progress of biology during the last few decades has been disappointing. It has continually failed to simplify its problems and is introducing a spirit of frustration among those who must probe ever deeper to follow divergent pathways that isolate them more and more from their fellows. This is most noticeable in the purely experimental fields, the very ones that seek guidance most consistently from physics and chemistry.

It also involves more covertly the descriptive and observational fields, for they also yield vastly curtailed accomplishments when pursued as detached disciplines. Until recently they were cultivated independently, and a return to this status threatens unless they are able to maintain contact with experimental biology on the basis of mutual understanding. This can be fostered only by the free interchange of ideas.

Even in the classification of organisms, later dubbed taxonomy, the oldest and most extensively worked branch of zoölogy, things are not so serene as they were in the days of the cabinet naturalist. This is painfully evident in the efforts being made to renovate taxonomy under the alias of systematics and to invigorate it, mainly through the infusion of genetic principles. Such a transformation has been developing among the taxonomists themselves for several decades, in coöperation with paleontologists and other students of phylogenetics, but has only lately been seized upon by others who now see in it a fertile source of information concerning speciation and the broader problems of evolution. Many of the subjects presented on the following pages bear directly on these matters, and this is one of the chief reasons that they should be of general interest to biologists at the present time.

Entomologists can speak with fervor of the intricacies of taxonomic investigation. Among all biologists, they seem to have made the worst mess of it, characterizing so many families, genera, and species that they have far outstripped the whole field of their taxonomist brethren. This is not really their fault; it is merely a feeble attempt to sort out the avalanche of insects that Nature has lavished on the Earth. Taxonomic considerations cannot be neglected in dealing with the topics discussed in the present volume, but I shall try to hold them to a minimum, even in referring to the numerous genera, species, and higher categories that must needs be cited by name. Any discussion of questions concerning phylogeny, race history, or natural affinities can be undertaken only on the basis of genetic relationships, and classification is so far the only practical, universally valid guide.

The innate characteristics of structure and behavior, or form and function, show an array of mutualistic modifications in all organisms and reach their most elaborate configuration in highly organized animals like the insects. It is thus an idle pastime to speak of habits or behavior without reference to the structures that implement them. Consequently, some morphological material has been included. Likewise, adjustment to the inanimate, and particularly the animate, environment is the *sine qua non* of all gastronomic activity. The acquisition of enough food of the right kind at the right time is the goal, and barring paternally administered rationing, Nature has adhered strictly to this standard which she set as one of the first requisites of life on our planet. This is the struggle for

food and is the basis from which has sprung the merry-go-round of
hungry insects that is about to engage our attention. It is necessary, therefore, to make frequent references to varied environmental
relationships between insects and the flora or fauna upon which
they depend for food.

To facilitate the consideration of these matters, it will be most
convenient to treat the insects in several categories with reference
to the types of food they require, and the devious ways by which
they utilize their food plants and food animals. These categories
are not sharply limited, and in the nature of things we could not
expect to find them so, for they represent the momentary state of
a mass in flux. This is particularly true of the higher groups of insects which are in the more active stages of evolution and differentiation, both structurally and behavioristically. Practically, this
means that the relegation of certain materials to specific chapters
has necessarily been to some extent dependent upon convenience.

Any treatment of the habits of insects cannot evade frequent
reference to the fixed behavior pattern or instinctive actions that
are so innately characteristic of these animals. Insects are commonly
cited as furnishing the most perfect examples of the highly coördinated and purposeful instincts that lie on a plane that far transcends
the purely tropistic reactions. Much of this behavior, perhaps the
major part, is concerned with securing food, and frequently also even
in furthering the nourishment of their immediate offspring. Instinct
is amenable to experimental study in only a very minor way, and it
still appears to be one of the most clearly "vitalistic" phenomena.
Certain general concepts concerning the part it has played in the
evolution of food habits have emerged from the great mass of observations made of its varied manifestations in diverse insects.
These seem to show that we may be able to extract it from the
metaphysical shroud which hides its physical basis. We usually
think of the instincts of insects in connection with the social forms
like ants, bees, wasps, or termites. In these groups they attain
their greatest complexity and most startling character, but they
pervade more subtly all of the activities presently to be outlined
on the following pages. It is to the simpler, more widespread picture of instinctive behavior that we must turn if we are to tear it to
pieces for examination.

As with other branches of zoology, the literature relating to insects has reached unwieldy proportions. On this account I have
supplemented the present volume by the inclusion of a list of
relevant publications which will enable the reader to follow up many

of the matters which can be mentioned only briefly in the text. Some of these are cited by name in the text, and the content of others is generally indicated by their titles. They have been classified by topics and appended to the separate chapters. A few have been repeated where they are important in more than one connection, and numerous cross-references will be found under the several headings. It is hoped that these bibliographical interludes will not prove too annoying to the casual reader, who will wish to pass them over. They are intended for the professional biologist, or for the amateur who may desire to find out just how far entomologists have gone in dealing with some very special problem. Such a guide to the subject matter here included has not previously been available, and although the present one is far from complete, I trust that it will be helpful at least, and ask the indulgence of readers who may note the omissions of many references that might have been included.

The illustrations will, I hope, enliven the rather lengthy text and serve to clarify certain matters which do not lend themselves too readily to written description. They are derived from many sources. Most of the photographs are the handiwork of the author, mainly from negatives made over the course of many years for teaching purposes. The drawings are the work of Mrs. Anna S. O'Connor and Miss Florence Dunn. They have been redrawn from a great variety of technical publications and often simplified to adapt them to the more generalized purpose of the present book. The source of such illustrations is noted in each case. Mrs. O'Connor also typed a considerable part of the manuscript, and finally, Miss Ruth C. Dunn has aided greatly in the tedious process of seeing the book through the press.

I am chiefly indebted to several kind friends and colleagues who have read and criticized the manuscript. First to my wife and daughter, who unfalteringly pursued their reading to the bitter end — omitting the bibliographical interludes; to my colleagues, Professors F. M. Carpenter and F. L. Hisaw, for a critical reading of the biological contents; to Dr. C. M. Williams for a similarly indulgent perusal. Dr. A. Orville Dahl furnished some very helpful criticisms and suggestions on the botanical aspects of the insect galls that are discussed in Chapter IV, and my son, Dr. Austin M. Brues, reviewed the material in this chapter also, in connection with his own work on animal neoplasms.

C. T. B.

Cambridge, Massachusetts
March 1945.

CONTENTS

CONTENTS

ILLUSTRATIONS IN TEXT

LIST OF PLATES

INTRODUCTION

THE THREE PRIMARY INSTINCTS; to secure food, to protect itself from destructive agents and to provide for reproduction of its kind are necessarily common to all animals. The impress of each of these necessities is to be found in the behavior and structural constitution of every species, although by no means evident to the same extent, nor alike at all periods of ontogeny. Reproduction is confined to a limited part of the life span even in the most primitive types of animals, and the period of reproductive activity is still more obviously restricted in higher forms like the insects where it is almost invariably an attribute of only the final, unchanging imaginal stage. In these insects this stage must thus suddenly develop in final form all the structural and behavioristic adaptations of the sexually mature animal. Protection from destructive agencies is on the other hand a constant necessity, common to all periods of the life span preceding and including the reproductive stage. In the higher insects with complete metamorphosis that culminates in reproductive activity, protection is at all times equally important to the species. It is, however, in part passive in that it may depend to a great extent upon bodily characteristics, not necessarily associated with specialized behavior.

The instinct to secure food is equally a constant necessity during postembryonic growth, for only by its gratification can growth and development take place. Even more than reproductive activity, the taking of food by animals is usually rhythmic, cyclical, or interrupted by periods of varying length. In the case of species that normally hibernate or aestivate, a prolonged cessation of feeding occurs, coincident with a temporary stoppage of growth and a slowing down of metabolic processes. In all the higher insects the pupal stage represents a similar condition during which no food is taken, and in a few insects such as certain flies and moths the succeeding imaginal stage even lacks functional mouthparts, cannot feed and is consequently doomed to rapid dissolution. Furthermore, food requirements in insects are proportionately much greater during the preparatory stages, since it is during this period that large reserves of fat are produced and stored within the body. These are subsequently utilized, either in adult life by the more primitive types, or during both the pupal and imaginal stages by holometabolous insects. Feeding is thus by no means a uniform or continuous process, but its association with the several active stages is far more extended than is reproduction. Also, it is a purely active, never a passive process, and in consequence we find that it has modified structure and behavior even more profoundly than have the needs for protection.

A discussion of the food habits of insects must therefore include a consideration of the many peculiarities of structure and development which are involved in the quest and utilization of food. The interest which attaches to the food habits of these small animals is further intensified by the frequent association of many insects with highly specific kinds of food and by their entrance into definite, more complex, relations with other animals and plants to form circumscribed biological units of interdependent organisms, or biocoenoses. Such biocoenoses commonly involve several kinds of insects together with various plants and other animals. Some are quite discretely delimited, others less so, and finally it becomes difficult to distinguish the more loosely interwoven kinds from the interdependence which is everywhere evident between the several organisms that inhabit any environment. Although biocoenotic associations are manifestly not entirely trophic in origin, they are intimately dependent upon the food relations of the participating organisms. When any biocoenosis includes man, an economic situation is created. Many serious problems have arisen in this way, involving particularly the relations of insects to agriculture, forestry and public health. The innumerable kinds of insects that depend on plants for food come into direct competition with other partially or entirely herbivorous animals, including the human species, and in this way they exert a powerful influence upon agricultural production. Parasitic insects likewise affect man in common with many other animals and these likewise assume great economic importance especially as carriers of several important diseases of man and his domesticated animals. Other insects of various kinds affect the human food supply directly, serving as nourishment, delicacies or condiments for some of the savage, less completely denatured aborigines in various parts of the world. Indirectly, others produce or elaborate edible materials. At least one of these may be mentioned in polite society, and many tons of it are consumed annually. This is our familiar honey, produced by the hive-bee. Again many less fastidious vertebrates feed extensively on insects, transforming them into fish, flesh and fowl, fit for human consumption. The insect food of these food-animals is consequently related to the food supply of man.

Such practical considerations are beyond the scope of our present treatment, but it will be evident later that they have emerged from a biological background that forms the primary basis for an understanding of certain economic matters which are of vital importance to civilized man.

The competition for food between insects and all other animals combined, including man, is overshadowed by the struggle for sustenance among insects themselves. This internal struggle has been the most potent factor in guiding the evolution of food habits in insects and in moulding the correlated structural modifications that are everywhere to be seen in this group of animals. Also, man may learn much concerning the potential utilization of his own food supplies from a study of the insects.

Down through the ages insects have filled the open spaces and swarmed into every nook and cranny on our earth, discovering food of a kind everywhere, until it is practically impossible to find any materials capable of furnishing nutriment that have not already become the preferred diet of some member of this versatile group of animals. Corks, cayenne pepper, spinach, kiln-dried lumber, fur coats, cigarettes, even bologna or the Congressional Record form no exception as they all have their special devotees in the insect world.[1]

Many thousands of papers have been written dealing with various phases of the food habits of insects and others are appearing in print at such a rapidly increasing rate that it has become a tedious task even to list their titles on library cards, without attempting to read them and much less to digest their contents. They form an interesting and important phase of entomology which should be better understood by the intelligent layman as well as by the professional zoologist who is immersed in the study of other biological problems. If he has patience to read further, the former should at least lay down this book with the satisfaction that man is still free to enjoy the luxuries of the table, unhampered by the intricate adap-

---

[1] Corks, the moth (*Tinea cloacella*); cayenne pepper, the drug-store beetle (*Sitodrepa panicea*); spinach, the spinach flea-beetle (*Disonycha xanthomelaena*); kiln-dried lumber, the powder-post beetle (*Lyctus planicollis*); fur coats, the clothes-moth (*Tineola biselliella*); cigarettes, the cigarette beetle (*Lasioderma serricorne*); the Congressional Record, certain Termites (as *Reticulitermes flavipes*); bologna, the ham beetle (*Necrobia rufipes*).

tations and mechanized reactions to food that prevail among the insects. He will also have seen to what ends Nature may go in her search for the bizarre and unexpected, and learn to treat with greater tolerance his faddist friends whose food preferences he cannot personally appreciate. The biologist will find, as he may already have suspected, that entomologists have accumulated a vast amount of information on the food habits of insects but that they have thus far been unable to arrange these to their satisfaction in an orderly system. He will find also a great variety of material bearing on general biological problems; on evolution, instinct, adaptation, variation, inheritance and what not, much of which offers opportunities for fruitful experimental research. Insects have already become important laboratory animals; they deserve to be even more extensively utilized for such purposes.

INSECT DIETARY

CHAPTER I

THE ABUNDANCE AND DIVERSITY
OF INSECTS

> Indeed, biologists, Your Excellency, are beginning to doubt
> whether man can maintain his foothold upon this earth against
> his supremest enemy, the insects, without an application of
> science to life and conduct upon an unprecedented scale. The
> very insects may force man to an intelligent social and political
> ethics, or else, upon this planet at least, they may become his
> successors.
>
> ALBERT EDWARD WIGGAM, *The New Decalogue of Science*

NUMERICALLY, insects stand apart from all other highly organized
animals in two respects. Their evolution has progressed at such a
rapid rate that the number of existing species has come far to exceed
that of all other kinds of animals together. This by no means implies that diversity has suffered at the hands of mass production for
such is obviously not the case. This group of animals excels not
only in numbers of species and in populations of individuals, but
also in diversity of both structure and behavior. All these characteristics go hand in hand and the action of evolutionary forces has
been such that each factor has aided in the development of the others
until insects have dominated the earth. Rather we might better
say that the struggle among insects has been so intensified by great
variety and large numbers that it has necessarily produced extreme
adaptations to avoid the extinction of many types which must
otherwise have succumbed.

Aside from the microscopic Protozoa the only other animals
which in any way approach insects in numbers of individuals are
the earthworms and those pelagic crustaceans representing the
group Copepoda. Any one of us who has ever spaded a garden or
dug worms for fish-bait has a pretty good idea concerning the
enormous size of the earthworm population and it was many years
ago carefully investigated in connection with the formation of vegetable mould by Darwin. According to his carefully made estimates
there are in an acre of British soil on the average 53,000 earthworms

and this number may rise to half a million where the soil is particularly fertile. Recent workers in Europe have found much larger numbers of earthworms in both cultivated and forest land. Morris ('22) in England found slightly over a million individuals per acre in highly manured arable land and 458,000 in unfertilized soil. In Denmark, Bornebusch ('30) demonstrated the presence of large, although usually lesser numbers in forest soil, ranging from 73,000 to 716,000 per acre, although on one exceptionally favorable plot the earthworm population rose to 1,450,000 per acre. In numbers therefore earthworms vie with the insects of the soil and in actual bulk they far exceed these. The number of species is, of course, very small and they show neither the structural diversity nor the manifold behavioristic adaptations of insects.

Inhabiting the vast reaches of the open sea, certain small free-living copepods are everywhere abundant where they may be seen at night as myriad specks of light following in the wake of a vessel or bathing its sides in phosphorescence, for a few of these minute creatures flash like fireflies when the water is disturbed. Although the number of species is small a census would probably show some of them to be represented by more individuals than any insect. Thus, in the plankton of the offshore waters of the Gulf of Maine, Bigelow ('26) has found as many as 4,000 individuals of a common Atlantic species, *Calanus finmarchicus* (Fig. 1) present in one cubic meter of water. On this basis it is useless to speculate upon the almost limitless population of this widespread crustacean. Another indication of the abundance of this species is the recovery of two tons from a whale's stomach, representing more than one-half billion individuals serving as a single day's food for this particular whale. These crustaceans are marine, however, except for a few less abundant forms in fresh water, and consequently do not come into contact with insects as these avoid a marine environment almost entirely. As the copepods are much like insects in many respects, they really represent the insects of the sea.

The number of living species of insects which have so far been distinguished and described by entomologists is very great and far exceeds that of all other living animals, including the lowly Protozoa and man himself together with all the diverse types that bridge the gap between these two extremes. Even the number of described species of insects may be stated only in round numbers as there is by no means any complete catalogue or series of smaller lists that may be easily collated for this purpose. Handlirsch ('24) has most recently summarized the living insect fauna and estimated that 470,000 species had so far been described; since then probably 50,000 have been added. These are included in 33 orders, comprising 905 families (Brues and Melander, 1932). Of these orders the largest is the Coleoptera of which 195,000 species are known, although there are three others, the Hymenoptera, Lepidoptera

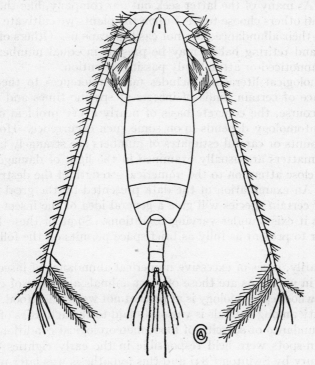

FIG. 1. The abundant and widespread copepod crustacean, *Calanus finmarchicus*. After Sars.

and Diptera, each with over 50,000 species accredited to them. Most of the others are much smaller, ranging from 2 to 21,000 species.

The insect fauna of many parts of the world is very incompletely known, especially that of the tropics where the variety and exuberance of life is greatest, and entomologists are continually increasing their estimates of the number of living species. That there exists an almost unbelievable number, probably about 10,000,000, is generally accepted to be near the truth, and as this enormous complex

is in a fair way to dominate the world or at least seriously to interfere with the earthly dominance of man, its mere abundance becomes a really important matter.

The actual and apparent abundance of several kinds varies within
extremely wide limits. Some which persist successfully are represented by very small populations, others appear at intervals in vast
hordes. As many of the latter seek out our company, like the mosquito, and others choose to feast upon the plants we cultivate in our
gardens, their abundance does not easily escape us. Others of more
modest and retiring habits may be present in equal numbers and
remain unnoticed or attract only passing attention.

Entomological literature includes many references to the great
abundance of certain kinds of insects at specific times and places
and, of course, the concrete basis of nearly every problem of economic entomology depends upon some such occurrence. However
actual counts or careful estimates of numbers are strangely lacking
as such matters are usually examined in the light of damage done
without close attention to the numerical strength of the destructive
insects. An examination of the data presented by the great abundance of certain species will give a general idea of the insect population as it exists under varying conditions. Some of these I shall
endeavor to present as fully as brief space permits on the following
pages.

Ordinarily, cases of excessive numerical abundance of insects are
periodic in nature as are those of most animals and even of certain
diseases whose epidemiology is at present not well understood. Such
periodicity among animals is variously laid to the activities of parasites, abundance or scarcity of food, meteorological conditions, etc.
Even sun-spots were held responsible in the early eighties of the
last century by Swinton ('83) and this hypothesis was later revived
by Elton ('24) who adduced further evidence. Very recently
Uichanco ('36) has demonstrated a close coincidence between outbreaks of the Philippine locust (*Locusta migratoria manilensis*) and
the sunspot cycle. Temperature and humidity are also correlated
with the cycle and these are undoubtedly the factors affecting the
locusts. Most recently, Elton has shown in the case of the Canada
lynx that the abundance of these animals in all of Canada has followed a cycle averaging 9.6 years over a period of more than a century. The lynx is practically monophagous, depending on rabbits
for food, so that its cycle is probably secondary to a similar one
involving its prey.

PLATE I

Two communal webs or "nests" of the American tent-caterpillar, *Malacosoma americanum*. Original photographs.

Plate II

A. Brine-fly (*Ephydra subopaca*). The larvae develop in salt pools, commonly at salinities greater than that of sea water. Photograph by the author.

B. Nest of the Texan fungus-growing, leaf-cutter ant (*Atta texana*), showing the multiple craters of excavated soil and the immense size of the colony. Photograph by Melander and Brues.

However complicated the basic causes of this periodicity may be, such a condition undoubtedly exists and it is frequently very pronounced. Outbreaks of forest insects like the larch saw-fly or the *Dendroctonus* bark-beetles in North America and the gipsy or nun moth in Europe are typically of this type exhibiting long quiescent periods. The North American tent-caterpillar (*Malacosoma americanum*) (Pl. I) and white marked tussock moth (*Orgyia leucostigma*) are similar but the outbreaks are usually separated by much shorter periods. Among such insects the outbreaks or epidemics occur at more or less widely separated intervals dependent probably to a great extent upon the abundance of parasitic insects and the rapidity of growth and recovery of their host trees. It may be noted also that they often cover large areas or that the successive appearances need not involve the same locality.

Other cases of abundance are primarily associated with migration or dispersion. These have been most commonly observed in Lepidoptera like the monarch butterfly (*Danais archippus*) swarms of which have been observed either before, during or after autumnal migration at one time or another by many American entomologists. I well remember one in the fall of 1899 near Austin, Texas, which so completely enveloped a large pecan tree that its branches drooped and the color of its foliage was replaced by the dull tint of the myriad butterflies which rested with wings folded upon its branches and twigs. Kenyon ('98) described a similar occurrence in Nebraska. No attempt was made to estimate the number of butterflies in this case, but it undoubtedly ran well into the thousands. Williams ('23, '25) in several papers has made observations on the migration of other butterflies in the American and African tropics, later incorporated into a comprehensive book (Williams '30). Among the various types, it appears that the swarms are usually smaller than those of *Anosia*, although sometimes mounting into thousands and possibly into millions. The large size and brilliant colors of butterflies, of course, render such swarms much more impressive than those of smaller, inconspicuous insects, and many earlier observers have dealt with such occurrences. Tothill ('22) believes in the case of the fall webworm (*Hyphantria cunea*) that some outbreaks may be initiated by migrations or flights of the moths.

Insects which are not readily disseminated by flight or otherwise frequently appear regularly in great abundance within circumscribed areas. One of this sort is the periodical cicada or 17-year locust (see Fig. 17), which appears in countless numbers, not every year, it is true, but when the adults of each generation mature. As is well known, the distribution of each of these broods is very definite and the appearance of each swarm may be accurately predicted. In the case of the periodical cicada the distribution does not appear to be dependent upon the type of vegetation or soil, although determined by these to some extent. The suddenness of the appearance of the cicada is naturally greatly heightened by the fact that its immature stages are hypogaeic and that the swarms are not of annual occurrence. Nevertheless their actual numbers are very great. Marlatt ('07) who has studied this species intensively states that the ground beneath a fair-sized tree (20 × 20 feet) may bring forth from 30,000 to 40,000 insects as he has counted 84 emergence holes on a superficial soil area of one square foot. Considering the very large size of the cicada this is really a stupendous quantity.

Insects of amphibiotic habits like the stone-flies and may-flies frequently appear in great swarms, transforming simultaneously in large numbers from aquatic nymphs into winged imagines. The swarming of the winged sexual phases of ants and termites is a similar phenomenon since it involves the simultaneous movement of a large population from a subterranean to an aërial environment.

The congregation of insects in large masses is frequently a feature of hibernation and has been most often noticed in coccinellid beetles. These Coleoptera overwinter in the imaginal stage and may either congregate near the scene of their summer's activities or they may migrate into adjacent mountainous country. There are many published notes on the abundance of these beetles on mountains in both Europe and North America and even in parts of the world as remote as New Zealand. The most remarkable account I have seen relates to a common North American species (*Hippodamia convergens*) in California (Carnes '12). In parts of California melon vines are severely damaged by aphids which are kept in check by aphidophagous coccinellids. It has been found by investigation that the mountainous regions of northern California harbor untold millions of *Hippodamia* during the winter. The beetles associate themselves into masses each containing thousands of individuals, hidden in the soil under the shelter of accumulated pine needles. According to Carnes ('12) two men working together may collect from 50 to 100 pounds of beetles in a single day. As each beetle weighs on an average of 20 milligrams, such a catch includes from 1,200,000 to 2,400,000 beetles! As the hibernating colonies always select certain situations on sunny well-drained slopes near running water, long experience enables the collector to locate the "lodes" of beetles with considerable ease.

Vallot ('12) records a still more remarkable case of another coccinellid, *Desoria glacialis*, which he found present in incredible numbers on the surface of a glacier at Chamonix in the Alps. The beetles were distributed over an area of 20 by 200 metres at the rate of 40,000,000 per 4,000 square metres, or 10,000 per square metre.

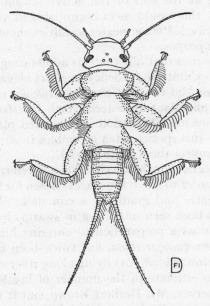

Fig. 2. Aquatic nymph of a stone-fly (*Perla media*). Stone-fly and May-fly nymphs often occur in great abundance in fresh water ponds and streams. Redrawn from Claassen.

Werner ('13) also has reported coccinellids congregated on mountains in Europe, Dobrzhansky ('25) in Turkestan, Hudson ('05) in New Zealand and many others have incidentally published similar observations. With the members of this family the aggregations are always associated with hibernation and with a desire to seek out an elevated spot. This is evinced even in the smaller overwintering colonies of *Adalia bipunctata* which I have frequently observed beneath shutters and windows of my own house near Boston. Here they seem to prefer places of concealment on the upper floors, although they are not entirely absent on the lower floor. This might, of course, be interpreted as a purely tropic reaction (negative geotropism) but may be conditioned by moisture, satiety of appetite, lack of food, temperature or more probably by a combination of all these, and no real explanation of the autumnal change in behavior can be reached by the simple postulation of an inhibited, intensified or reversed tropism. Poulton ('04) has suggested in connection with the gathering of coccinellids on mountain tops that this is an adaptation for mating at the end of the active season, preparatory to hibernation, but this would seem improbable in view of the great distance often traveled when nearer localities should provide equal facilities for this purpose.

Many other beetles that hibernate as adults congregate in masses but seem not to exhibit the tendency to select elevated places. The European chrysomelid beetle, *Galerucella luteola*, now very widely naturalized and abundant in eastern North America, regularly congregates in masses for hibernation in protected places. According to Marlatt ('08) this species often assembles in such numbers that it is possible to collect them by the quart.

Neither this beetle, nor the coccinellids just referred to, appear to form swarms in order to reach the places chosen for hibernation, but move independently and gradually accumulate. Another chrysomelid beetle has been seen migrating in swarms in the autumn, I think most likely as a preparation to entering hibernation. This species, *Disonycha quinquevittata*, has twice been observed in immense swarms (Howard '98, '01) flying along river courses in southern Arizona. No estimate of the number of beetles involved was made by the observer, Mr. Herbert Brown, but it must have been enormous as in one case the insects formed a belt 20–25 feet in thickness and as wide as the Colorado River at this point (doubtless several hundred feet at this season of the year). The second flight observed the following year was descending the course of the Gila River in a belt about fifteen feet thick and 100 yards wide. This swarm continued during the course of two days early in November, while the previous flight took place during October.

An interesting swarm of small tenebrionid beetles has been described by Hartman ('23). It appeared over one of the buildings of the University of Texas on a day in late February. From the ground this was mistaken for smoke and caused the ringing of a fire alarm. Hartman believed this to be a mating swarm, similar to the nuptial flight of social insects, issuing from a bat-infested, unoccupied tower. This beetle is known to develop in bat guano and the swarm appeared later to retreat like the genii into the recesses of the tower.

Among Hemiptera, many of which hibernate in the adult condition, the chinch bug, *Blissus leucopterus* (see Fig. 30), is a notable example of gregarious habits. These bugs live upon various Gramineae and are very destructive to wheat. In the more arid sections of the middle western states they overwinter in colonies concealed in tufts of bunch-grass, *Andropogon scoparius*, commonly as many as 1,000 individuals huddled in the shelter of a tuft only three inches in diameter (Headlee '10).

Similar associations, although usually less marked, are so common among insects of all sorts that a gregarious tendency at the time of hibernation may be regarded as a very prevalent phenomenon. It is typical in a small way, for example, of many Diptera, e.g., *Pollenia rudis, Muscina pascuorum*; of certain Hymenoptera, as *Polistes*; of other Coleoptera besides those mentioned, e.g., the carabid, *Brachinus*; and of certain Hemiptera, as *Zelus socius*, etc.

In one of the cases cited, that of *Polistes*, there is no change at hibernation as these wasps are social during the warm season before the colony disintegrates in autumn, so winter association means only a delay of the instinct to separate in the autumn long before undertaking the independent founding of new colonies. Indeed, the polistes wasps of tropical regions periodically leave the parental nest as swarms to establish new colonies since there is no cold period to inhibit their activities. This instinct may rarely persist even in temperate climates as has been noted by Rau ('41). In social bees like *Bombus*, however, autumn brings about a break in the social regime as the female bumblebee tends to become solitary at this time, in anticipation of her impending return to solitary life during the coming spring when she will rear her first offspring.

Some years ago (Brues '26) I had the good fortune to observe a very remarkable aggregation of small alleculid beetles, *Hymenorus densus*. These were swarming on the panicles of the Spanish dagger, *Yucca aloifolia* in southeastern Florida during the latter part of June. Over 9,800 beetles were shaken into an insect net from one of the large showy panicles of this *Yucca*, and as about half escaped, the swarm was estimated to include some 15,000 individuals. Within sight were dozens of Yuccas all covered in the same way and indicating a population of beetles mounting into the millions. The larvæ of this species live probably in the soil where they feed upon vegetable material of one kind or another. How the sandy sterile soil in the neighborhood of these Yuccas can serve to rear such immense numbers of these beetles is almost inconceivable. In this case it will be noted that the aggregation is for the purpose of feeding and that it is not associated with hibernation.

The most notable cases of abundance among insects occur in species which live at the expense of agricultural crops. As has frequently been pointed out, these outbreaks are due to changes in environment whereby a great abundance of food is supplied through the extensive growing of plants which serve as specific food for certain insects. When repeated from season to season, the opportunities for these insects to multiply increase until the situation becomes economically intolerable. Such cases of excessive abundance do not represent natural biological phenomena, although they serve as huge experiments that elucidate the factors which regulate the normal balance of nature among animals. Observations have frequently been recorded which show to what extent the actual insect population may increase under such conditions of almost unlimited opportunity, but a very few will suffice to illustrate the point. The vast, uniform areas where sugar cane is cultivated in the tropics offer ideal opportunities for the multiplication of cane pests. A cane-boring scarabæid beetle, *Dyscinetus bidentatus*, has been described as so abundant in British Guiana by Hutson ('17) that over 4,000,-000 were captured with nets in a single season on one estate. In Java, Bernard ('20) has described the great abundance of a slug-caterpillar, *Thosea cervina*, in tea plantations. Over an area of 120 acres as many as 7,500,000 larvæ and pupæ were destroyed and 200,000 individuals were collected in a single day.

Throughout the great corn-belt of the United States there is a common caterpillar, the corn ear-worm (*Heliothis obsoleta*) which is almost everywhere so uniformly present that practically every ear in the field is infested, almost always by a single caterpillar or at most by two (Quaintance & Brues '04). It is thus easy to calculate that the population per acre of this insect commonly rises to about 35,000 individuals.

Barber ('25) has made accurate observations on the number of larvæ of the European corn-borer (*Pyrausta nubilialis*) that may be present in corn-fields in Massachusetts. During the year of its greatest abundance in 1922, counts made by Barber indicate that the population of larvæ per acre may reach as high as 1,000,000 in sweet corn and 1,174,000 in field corn. When present in weeds of various kinds, including barnyard grass, on very fertile soil their numbers may reach 400,000 per acre.

Among insects affecting the cotton plant similar conditions prevail. A very abundant moth, *Alabama argillacea*, known as the cotton leaf-worm feeds in the larval stages upon the foliage of cotton. This species is widespread in the cotton growing districts of the United States, but appears never or at least very rarely to overwinter there. It is abundant in the West Indies and each season's crop is thought to result from an early migration of the moths which undergo several generations each summer. In the autumn a northward migration of the last brood of moths continues beyond the cotton belt and often as far as New England and eastern Canada (Gerould '15). During a residence in Boston, extending over many years, I have personally observed numerous autumnal flights of the cotton moth, which occur with considerable regularity. Such flights are very extensive and lead to no advantage as they remove the moths completely from their normal habitat. This is a very patent case of an instinct manifested in an utterly purposeless way at one season while its adaptive nature in the spring is equally evident.

The leaf-hoppers form a large group of inconspicuous though abundant insects. The apple leaf-hopper, *Empoasca fabae*, a common pest of potatoes, which acts also as vector of a destructive disease of this important crop plant, is a typical example of the abundance of this type of insects. Hartzell ('21) found that a census of the *Empoasca* population in a potato patch indicates the presence of five to six million nymphs hatching in a single acre of potatoes at the time the second generation is developing during July.

So far as unusual opportunities for securing food are concerned, there is another insect which has become abundant much as have agricultural pests. This is the housefly. In his compendium on this species Howard ('11) indicates the numbers of housefly larvæ that may develop in specific quantities of food. Thus over 10,000 were counted in fifteen pounds of stable manure in California by Herms who estimated that the population of the manure pile was about 455,000. Many years ago, before the advent of automotive fire-fighting machinery, I secured in a trap almost 100,000 adult flies which emerged from the basement of a fire-engine house where manure was stored, despite the fact that the place had been cleaned as thoroughly as possible by men with shovels previous to the installation of the trap. The housefly is, of course, one of the most abundant insects in the vicinity of human communities, but appears to be more numerous than it really is on account of its fondness for human society, including the accumulations of food and refuse which invariably accompany the latter. It still holds its own, although hard pressed by the automobile, farm tractor and garbage incinerator.

The immense swarms of migratory locusts which appear regularly in certain parts of the world have, from time immemorial, been among the most spectacular examples of insect abundance, and population counts or estimates of their extent have been frequently made.

In Anatolia, Syria and Palestine there was a great invasion of migratory locusts in 1915, involving several species. In western Anatolia, according to Bücher ('17), some 6,000 tons of eggs and 11,000 tons of locusts were collected during three months. From the known weights of other species these amounts would represent about 18,000 billion eggs, 88,000 million grasshoppers, numbers so vast that it is utterly hopeless to grasp their significance.

Workers in the Federated Malay States, in their campaign against destructive locusts, have recorded (South '14) the amounts of insects collected and when calculated in actual numbers, one day's catch included something over 100 million specimens. In another instance a nearly equal amount was obtained from 65 swarms, indicating well over one million individuals in a swarm. As about 10,000 swarms were destroyed in these Malay States during a single year soon afterwards, it is evident that even such wholesale destruction does not permanently reduce the locust population. Similar occurrences in North Africa (Mancheron '14) and other parts of the world subject to locust invasions are comparable to those just cited. The actual abundance of a non-migratory grasshopper, *Camnula pellucida*, present in a locality in one of our western states has been estimated by Ball ('15). He found in one heavily infested part of a breeding ground of this species that there were about one billion eggs per acre, or in more readily comprehensible terms, 25,000 to each square foot of soil surface. As these insects hatch they form slowly migrating swarms and one swarm was found to include some three or four billion young grasshoppers distributed over an area of four square miles with fully half of them concentrated over an area of half a square mile. It is further stated that this swarm was insignificant in extent to one previously observed at the same place.

Among other insects, which are general feeders like the locusts, may be mentioned the large common "June beetles" of the genus *Phyllophaga* (see Fig. 17) so widely distributed in eastern North America. The adult beetles feed mainly upon the foliage of various deciduous trees which they assail just at dusk on the warm evenings of early summer. As many as 400,000 beetles were collected during one summer in a small town in Wisconsin (Sanders and Fracker '16) and similar numbers might easily be obtained generally in many localities in the middle western states.

Excessive numbers of very small insects may readily be present without attracting attention, particularly if they live in concealed situations. Thus the minute gall midge known as the Hessian fly, *Phytophaga destructor*, is widely distributed over the wheat belt of the United States. Its abundance varies greatly, but accurate counts made by McConnell ('21) show that a very moderately infested field harbored over 200,000 larvæ per acre despite the fact that this number represents less than 5 per cent of an original population of nearly 5,000,000 per acre which was present previous to decimation by parasites.

The occurrence of insect drift along the beach line of large bodies of water serves to illustrate the abundance of many insects whose numbers are otherwise not noticed. This drift has been commented upon by Needham ('00, '04) and later by several others (Snow '02; Bueno '15 and Parshley '17), and years ago I had opportunity to examine it a number of times in company with Professor A. L. Melander on the western shore of Lake Michigan where Needham's observations were made. Usually a great variety of insects are present and when conditions are favorable they form windrows on the beach mingled with débris of various kinds. Sometimes one species greatly predominates, and in a sample minutely examined by Needham there were over 2,500 crickets (*Nemobius*), 680 grasshoppers (*Melanoplus*) and less than 900 other insects of the most diverse kinds. As such accumulations must be derived from very extensive areas they do not necessarily indicate an originally dense population of insects. Indications that such drift at sea may extend for very long distances have been presented by Felt ('25) in connection with observations by Elton ('25) on insect drift in the far north.

As insects of various kinds are commonly encountered over bodies of water at considerable distances from land these accumulations result from the drifting shoreward of the bodies of those that fall into the water. A description of several swarms of a large groundbeetle (*Harpalus calceatus*) in the Black Sea by Adams ('08) illustrates the origin of such drift. Swarms of this beetle were met with near the western shore of the Sea of Azof just north of the Black Sea, hundreds were swept from the decks of the steamer and for

several days immense numbers were seen floating on the surface of
the water. Two weeks later another swarm was in progress.

Accumulations of insect drift are conspicuous in midsummer along
the shore of Great Salt Lake in Utah, but these have a very different
origin from those just referred to. They consist almost entirely of
the puparia of a brine-fly, *Ephydra gracilis* (Pl. II, A). The larvae
of this species develop in the highly saline water of the lake and the
puparia collect at the surface whence they drift on shore to form a
strip often a foot or more in thickness. Aldrich ('12) has described
their abundance and estimated the numbers of adult flies that
emerged from the puparia and settled upon the beach and surface
of the water near the shore line. On the basis of at least 25 flies to
the square inch covering a strip 20 feet in width, the fly population
for every mile of beach would amount to 370 million. Fortunately
the flies do not bite, but the innumerable puparia that are cast up
along the shore of the lake to die and decay diffuse at times an in-
sufferable odor. Many of the smaller saline and alkaline lakes of
this region and Mexico support large numbers of other forms of
Ephydra, the preparatory stages of which form an article of diet for
the aborigines.

The preparatory stages of another species, *Ephydra hians*, occur
in enormous numbers in the waters of Mono Lake in eastern California. This strange body of alkaline water contains such a large
amount of borax and sodium bicarbonate and its specific gravity is
so high (1.0409) that its fauna and flora are limited to a very few
species. Besides this *Ephydra* the only other macroscopic animal
inhabitant seems to be a brine-shrimp, *Artemia gracilis*. This
crustacean is present in countless numbers, apparently quite uniformly distributed throughout the waters of the lake. Estimates
made by us during late August several years ago indicated a population of at least several hundred shrimp per cubic metre. These
appear to extend quite uniformly over the entire lake, although the
brine-flies are restricted to the shallow waters close to the shore.
The brine-flies of Great Salt Lake and Mono Lake (see Pl. II A)
and the brine-shrimp of the latter illustrate very clearly the great
abundance that may be attained by an organism when freed from
the struggle for existence that prevails among the faunal components of an environment less hostile to life. Once able to exist and
multiply where other forms are excluded, they maintain the maximum abundance compatible with the food supply, in the same way
that introduced insect pests multiply so extravagantly when unhampered by their natural parasites.

A most extraordinary example of the great abundance sometimes attained by aquatic insects is presented by certain Hemiptera of the families Corixidae and Notonectidae. In certain parts of Mexico, notably in Lake Texcoco and the smaller neighboring lakes which communicate with it, these water-bugs are so numerous that their eggs form a staple article of diet for the human population. In Lake Texcoco two species are particularly concerned, *Krizousacorixa azteca* and *Notonecta unifasciata*, the eggs of which are collected from the surface of water plants to which they are attached. Ancona ('33) states on the basis of observations by Penafield that the nymphs of these water-bugs may at times reach a concentration of 200,000 individuals per cubic metre of water or the enormous total of over $3\frac{1}{2}$ million millions in the whole Lake Texcoco. This abundance has persisted in spite of the use of their eggs in great quantities for food since time immemorial, as the earliest published reports of this practice date from the early seventeenth century (1625–1649). The eggs are deposited in such numbers on aquatic plants that they may be collected from bundles of sedge allowed to float in the water, dried, "threshed" out and ground into flour. The eggs and adults also make very satisfactory fish and bird food and have for this purpose been exported to England by the ton. Estimated on the basis of about 40 milligrams each, a ton will comprise some 25,000,-000 adult bugs.

Several British entomologists have observed remarkable aggregations of a small acalyptrate muscid fly, *Chloropisca circumdata* in buildings. Scot ('16), Imms ('22), Wainright ('22), Kearns ('29, '30), and earlier, Sharp in the *Cambridge Natural History* have mentioned numerous cases where this species has appeared in enormous numbers, carefully estimated in one case as 50,000 in a single room. These indoor swarms are probably in preparation for hibernation as they have been noted in the autumn. The flies appear generally to exhibit a preference for the upper floors of buildings, and the frequency with which the swarms have been reported indicates a well-fixed instinct. In common with most members of the family this species lives in the larval condition on grasses.

Various small Diptera with aquatic larvae are sometimes extremely abundant. Certain chironomid midges (Fig. 3) commonly become associated in swarms containing hundreds of individuals and these may assume gigantic proportions. The well-known dipterist, Williston ('08) observed them in meadows in the Rocky Mountains "rise at nightfall in the most incredible numbers, producing a noise like that of a distant waterfall, and audible for a considerable distance."

Lesser swarms of these midges are frequently met with in the most diverse places as many entomologists can testify. An interesting case of such swarms during early May was observed by Bill ('32) in eastern Massachusetts, where numerous streamers, thought at first to be smoke, were observed swaying in the breeze above the tree tops. These were found to be mixed swarms of several species of Chironomidae, containing myriads of midges. On several succeeding evenings, repetitions of the phenomenon occurred. Two other

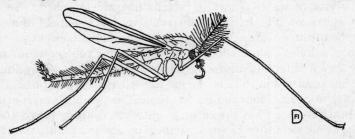

FIG. 3. A midge, *Chironomus colei*, representative of the family Chironomidae, minute flies which often appear in immense swarms. Redrawn from Cole.

cases of particularly large swarms of chironomids have come to my attention, one in England (Scott) and another in Germany (Thienemann). In both cases the swarms were associated with the spires of churches, above which the clouds of midges were mistaken for smoke as they swayed and wavered in the air.

The larvae of these midges (Fig. 4) are aquatic, living usually in soft mud under water, and the great abundance in such situations of one common North American species, *Chironomus plumosus*, has been described by Johnson and Munger ('30). In Lake Pepin, an expansion of the Mississippi River near the northern boundary of Wisconsin, they found larvae present to the extent of more than 7,000 in a single square yard, or at the rate of well over three million individuals per acre. As there are some fifty miles of shore line along Lake Pepin, a further calculation of its total chironomid population would exceed 3×10^9, a magnitude quite incomprehensible to the nonmathematical mind. Similar abundance of a related form in Japan has been observed by Miyadi who found often, in certain Japanese lakes, more than 1,500 larvae per square metre. Great variations occur among these midges with reference to the presence of sand, mud or clay on the bottoms of the ponds in which they develop.

Very clear-cut cases of migration among the larvae of small fungus-gnats of the allied family Mycetophilidae have frequently been observed in both Europe and America (Lintner; Felt). The swarming takes place among the full-grown larvæ which form long snake-like masses composed of thousands of grubs which crawl one over another at a slow rate. The term "Snakeworm" or "Heerwurm"

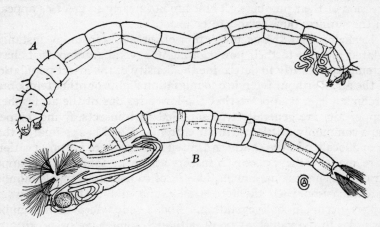

FIG. 4. One of the very abundant North American chironomid midges, *Chironomus tentans*. *A*, larva; *B*, pupa. Redrawn from Johannsen.

has been applied to these insects, which seem always to be species of Sciara. The crawling columns may be as long as 10–15 feet or rarely even 30 yards, several inches wide and half an inch in thickness. As the larvæ are about 6 mm. long and 0.7 mm. thick, one of these larger swarms about 10 feet in length would include over half a million larvae.

A number of blood-sucking midges are likewise often very abundant, particularly in marshy maritime areas, as some of them breed in brackish water. Known as sand-flies or "no-see-ums" on account of their minute size, they are a great menace to human comfort, and also to the public health, for some act as vectors of serious human diseases. One of these, *Culicoides dovei*, has been extensively investigated by Hull and others in the salt marshes of our southeastern states. Their larvae may occur in concentrations of about 5,700,000 per acre during the height of their seasonal abundance.

The populations of such insects and of mosquitoes are difficult to estimate with accuracy on the basis of the bloodthirsty, winged females as these seek us out, congregating from considerable distances to make life miserable for those of us who are forced to venture into the open when such pests are about. Thus, the seemingly innumerable mosquitoes, biting midges and black-flies, prevalent alike in the arctic tundra and tropical jungle, give a very false impression of their numbers. These are not nearly so great as appears on the surface; our surface, to be exact.

Realizing that the accumulation of insects is in many instances related directly to their power of flight, several entomologists have attempted recently to determine the density of the insect population of the air. Without reference to migration flights or other temporary circumstances, it appears that the lower regions of the atmosphere (upper air) are generally well supplied with insects, flying, or perhaps commonly drifting with air currents. Coad has estimated that in one locality in the southern United States where his observations were made, some 25,000,000 insects are to be found in the upper air over a single square mile. This is, of course, a negligible number in comparison with the terrestrial fauna and that of the lowest ranges occupied by vegetation. Above 3,000 feet their number dwindles to the vanishing point, although some have been captured at elevations up to nearly 15,000 feet. As the weaker flying species range more commonly into the upper reaches, it is obvious that drifting in air currents brings them there. Such appears to be at least a minor factor in the frequent accumulation of flying insects on high mountain slopes where they are often to be seen moving upwards on warm, sunny days, especially on southern slopes. Simple, tropistic interpretations with reference to gravity, light, etc. do not seem to apply.

The attraction of insects to brilliant lights at night is particularly well suited to furnish data concerning numbers normally present over the areas included in the sphere of influence of the lights. As every entomological collector knows, particularly those whose experience dates back to the first general use of electric arc lamps for community lighting, the numbers of insects which appear as if by magic are almost unbelievable. They include a great variety of types, among them a very noticeable predominance of certain beetles, aquatic Hemiptera, may-flies, midges, neuropteroids and to a lesser extent of moths. Thirty-five years ago it was possible to pick up under the street lights of Chicago such beetles as *Calosoma*, *Hydrophilus*, *Cybister* and *Phyllophaga* by the quart and the same was true of the large Hemiptera of the genera *Lethocerus* and *Benacus*. These latter are aquatic forms which are ordinarily rarely taken with the water net, although over practically their entire range they are so universally abundant at street lights that they are widely known as "electric-light bugs." Brilliant lights are so attractive to many of these insects that after a few years their numbers diminish enormously, due to the vast concourse thus drawn from their natural environment, never to return.

With injurious insects that are readily attracted to lights, it is a common practice to construct various types of traps fitted with lamps for the collection and destruction of such species. In Europe there are several species of moths which feed as caterpillars on the foliage of grapes and these may be so abundantly attracted to lights that their economic control by this method is possible, although expensive. In one vineyard on the Moselle over 18,000 moths were trapped on five nights, and again in another locality nearly 275,000 moths were thus collected from a vineyard of 13.5 acres during a period of one month (Dewitz '12), indicating a population of about 20,000 moths per acre.

The attractive effect of light of different wave length varies greatly on insects since they are particularly sensitive to the ultraviolet. Consequently there is great variation in the results obtained by different types of illumination. Many of the earlier experiments with traps failed to take this very important factor into consideration.

The most complete and accurate data on the density of insect populations are those gathered by observers who have investigated the fauna of the soil. This lends itself especially well to such work as the movements of individual animals are greatly lessened by the character of the environment.

The fauna of certain forest soils in Denmark, as studied in great detail by Bornebusch ('30) shows an astounding abundance and diversity. Thus, in areas of different forest types the population of insects varies in round numbers between 4,000,000 and 34,000,000 individuals per acre, with an average of nearly 15,000,000. Of these, the greater part are *Collembola* of only a few species, as, for example, in one locality the number of *Onychiurus armatus* alone amounted to over 16,000,000 individuals.

Similarly in England, Morris ('20, '22) has made comparative examinations of the hypogaeic insect and other invertebrate fauna of pasture and cultivated lands. The results of this census indicate populations of generally similar magnitude to those of the Danish forest soil. In the pasture soil 3,586,000 insects per acre were indicated by samples; in cultivated land, which had not been treated with manure for nearly a century, the number was less, or 2,474,000 per acre. However, on land constantly manured for 90 consecutive years, the insect population was treble that of the nontreated field or 7,727,000 per acre. Later studies by Morris ('27) of a near-by locality showed conspicuously smaller, but nevertheless by no means meagre populations of from 673,000 to 4,677,000 insects per acre in cultivated land which had been variously treated either with none or with one of several types of fertilizer. As before, the *Collembola* constituted a very considerable part of the insect life, ranging from 379,000 to 1,727,000. The latter figures are particularly interesting since another British entomologist, MacLagan ('32), working in the same region, found the population of a single species of collembolan, the "lucerne flea," *Sminthurus viridis* (Fig. 5), in

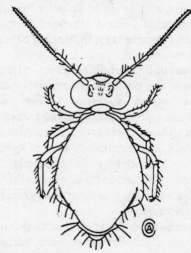

FIG. 5. A very abundant and widely distributed soil insect (*Sminthurus viridis*). Redrawn from Davidson.

permanent pasture to be about 600,000 per acre. These and other similar investigations by British entomologists have offered much of interest which cannot be reviewed here. References are included in the list of literature at the end of this chapter.

In the tropical jungle near Pará, Brazil, a similar census of life in the soil was made by Beebe ('16). The place selected was beneath a large tree, frequented by great numbers of birds, and four square feet of soil were carefully investigated. The resulting count indicated a population of 5,545,000 invertebrate animals to each acre of soil, of which 3,267,000 were insects.

Thus, from widely scattered spots on the earth's surface it appears that the soil supports an almost incredibly large and apparently rather uniform population of insects.

In North America, quite comparable figures for an entirely different type of soil have been given by Hendrickson ('31). In the soil of a wet, boggy area in Iowa supporting a growth of the grass, *Spartina Michauxiana*, he found a total animal population of 1,570,-000 per acre. Unfortunately, in this case, the proportion of insects is not stated.

Mention has already been made of the abundance of many pests of growing agricultural plants. The same opportunities for excessive multiplication are enjoyed by those that affect stored food products. The immense quantities of dried beans that go into storage warehouses each year have been investigated by Larson and Fisher. They estimate that about 170,000 pounds of bean weevils (see Fig. 14) are stored away annually with these beans. This represents only 0.17 per cent of the crop, but is enough to furnish a good mouthful for every patriotic American who tries to bolster a deficient protein diet with vegetable substitutes.

Ants and termites are among the most abundant insects and the association of large numbers of individuals in discrete colonies makes it possible to determine actual numbers with considerable accuracy. Yung ('99, '00) has estimated the populations of a number of nests of the European *Formica rufa* and finds that average nests contain from about 30,000 to over 100,000 individuals. The population of the nests of many other species is quite similar. The North American *Formica exsectoides*, one of our common mound-building ants, forms colonies of from 40,000 to 240,000 individuals according to estimates made by Cory and Haviland. On this basis one acre which was found to harbor over 70 mounds would have an ant population of this species of over 10,000,000 individuals. It must be remembered also that several abundant species of ants commonly live in the same areas where their feeding areas overlap.

Among forms like the leaf-cutter ants of the genus *Atta* (Pl. II B) and the carnivorous driver ants (*Eciton*) in the American tropics, the colonies are without question many times larger, especially those of *Atta*.

Among termites there are many species that build enormous nests in the tropics of both hemispheres and these form the most populous houses in the world (Pl. III). Andrews ('11; Andrews and Middleton '11) estimates that the average nest of the common *Nasutitermes* (*Eutermes*) *morio* of Jamaica, which constructs large arboreal termitaria, houses about half a million inhabitants. With such a nest the "traffic" in and out amounts to about 8,000 termites per hour at the time of greatest activity, which is shortly after midnight. It may be centuries before an American apartment house can boast of so extensive a human population with such highly coördinated nocturnal activities. Another South American species is said by Emerson to house approximately 3,000,000 termites in one of its large carton nests. The populations of certain African and Australian termitaria are undoubtedly much greater, but such accurate data are not generally available. In one Australian species of *Eutermes*, nests housing from 750,000 to 1,750,000 termites have been reported.

The population of colonies of the honey-bee is well known and amounts in the ordinary hive to about 30,000 or 40,000 at the beginning of the season. The capabilities of the queen bee for egg-laying are stupendous and observations by Nolan indicate that she may deposit 1,000 or more eggs each day during her more active period. In addition to the adult bees, a colony may contain another 30,000 or 40,000 bees in earlier stages of development (eggs, larvæ and pupæ), bringing the total population to some 75,000 individuals (Pl. IV). As the life of the worker bee is at most only a few weeks, and that of the male far shorter, rapid multiplication is necessary to keep up the size of the colony. It may be noted that the population of a beehive is approximately the same as that of the nests of temperate region ants referred to in a previous paragraph, although it appears very probable that the life of the individual ant workers is much longer than that of the honeybee.

Perhaps the most prolific insects in existence are the aphids which live by sucking the juices of leaves or other succulent parts of plants. They reproduce parthenogenetically during the warmer parts of the season, producing living young which at birth already contain growing embryos of the third generation. Many contemplative entomologists have from time to time calculated the brief period that might elapse before these insects could multiply to the point of suffocating the world beneath their weight. Fortunately they are harassed by parasites and limited by their food-supply; nevertheless they show periods of excessive fecundity. Most notable for their abundance among our aphids are the green-bug (*Toxoptera grami-num*) living on wheat and other grains, and the pea-aphid (*Illinoia pisi*) (Fig. 6).

It is seen from the foregoing that conspicuous swarms or aggregations of insects may be due to one of several quite distinct causes.

The populous colonies of social insects are an attribute of social life and their size appears to be limited mainly by the food supply, tempered, of course, by their natural enemies, for the reproductive

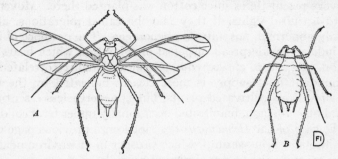

Fig. 6. The pea aphid (*Illinoia pisi*). *A*, winged female of viviparous generation; *B*, wingless female of viviparous generation. After Davis.

capacity of such colonies to produce eggs and to rear young is enormous. It is limited, however, by the length of the season in those social forms like our native vespid wasps where the life of each colony begins in the spring and is terminated abruptly by the onset of cold weather. Even in such cases the populations built up during the course of the summer are considerable. Some species of *Vespa* (hornets and yellow-jackets) in Europe and North America are known to produce at least several thousand adult wasps in a single season. On the basis of counts by several observers, it appears that the European species are consistently more prolific than our own, but we shall not draw any sociological conclusions on such slender evidence.

A very considerable proportion among the observed cases bear a more or less direct relation to mating, migration or dispersal, and this is true of the marriage flight of the sexual phases of ants and termites and probably also of many swarms of midges. Among insects preparing for or undergoing hibernation they are a clearly marked expression of the tendency for many non-social forms to become intensely gregarious at that time. Migration may commonly be induced by, or at least attendant upon, the necessity for a larger food supply than is available without migration. This is true of the migratory locusts. Migration to a new and more extensive food supply is evidenced by the spring movements of the cotton moth, but as this species resumes its migrations in the autumn and travels far beyond any food supply, the causal relation is very doubtful and seemingly only a coincidence, particularly as this species can have found food in the southern United States during only very recent times since cotton was planted there. Movements toward artificial lights, if they may be called migrations, are essentially abnormal, not natural phenomena. The presence of insect drift indicates widespread dispersal or migration which bears some relation, at least, to air currents. Great abundance correlated with an abundant food supply is most clearly indicated by the insect pests of various cultivated plants, but is nevertheless essentially a natural phenomenon, manifested occasionally under feral conditions like the case of the *Hymenorus* described on a previous page or the brine-flies and brine-shrimp, which prosper in an environment that rigidly excludes the vast majority of animals.

Very frequently the simultaneous emergence of the aërial forms of insects which have developed in the water (may-flies) or in the soil (periodical cicada) produce swarms which may later become dissipated through dispersal or remain during the brief remainder of their lives at practically the same spot.

The number of individual insects that become associated in these ways is often very large and sometimes assumes stupendous proportions. We know, however, really very little concerning the actual populations of insects as they exist about us, even those whose activities in destroying useful plants have been quite thoroughly investigated.

It is principally through great aggregations of one species or another that estimates of actual insect populations have been based, but there is every reason to believe that many less conspicuous kinds are represented by extremely large populations. Observers have, from time to time, attempted to take a census of the inhabitants of some circumscribed area, invariably demonstrating the presence of innumerable individuals.

How insects have been able to attain the great abundance and diversity which they now possess, and how they have been able to maintain these when once acquired, becomes clear only after we have inquired at some length into the methods which they have adopted. It appears to depend in great part upon the way in which they have utilized the abundant food supply furnished by nature, mainly in the form of vegetable material, and upon the many and devious ways which they have devised for living as predators and parasites at the expense of members of their own tribe.

Like all animals they possess powers of reproduction widely in excess of fulfillment at all times, and like them also insects are decimated by disease, parasites and predatory enemies. More thoroughly than other animals they have exploited their food supply and profited wonderfully thereby.

REFERENCES

Adams, L. E.
1908 Swarms of Insects in the Crimea. Zoologist (4), vol. 12, pp. 9–12.

Aldrich, J. M.
1912 The Biology of Some Western Species of the Dipterous Genus *Ephydra*. Journ. New York Entom. Soc., vol. 20, pp. 77–103.

Alexander, C. P.
1925 An entomological survey of the salt Fork of the Vermillion River in 1921, with a Bibliography of Aquatic Insects. Bull. Illinois Nat. Hist. Survey, vol. 15, pp. 439–535.

Allee, W. C.
1927 Animal Aggregations. Quart. Rev. Biol., vol. 2, pp. 367–398.

Ancona, L. H.
1933 El Ahuatle de Texcoco. An. Inst. Biol., Mexico, vol. 4, pp. 51–69.

Andrews, E. A.
1911 Observations on Termites in Jamaica. Journ. Anim. Behavior, vol. 1, pp. 193–228.
1929 Population of Ant Mounds. Quart. Rev. Biol., vol. 4, pp. 248–257.

Andrews, E. A., and A. R. Middleton
1911 Notes on Rhythmic Activity in Termite Communities. Johns Hopkins Univ. Circ., February, 1911, pp. 26–34. (Reprint pp. 1–9).

Baerg, W. J.
1921 Nest and Population of a Colony of *Vespa diabolica*. Journ. Econ. Entom., vol. 14, pp. 509–510.

Ball, E. D.
1915 Estimating the Number of Grasshoppers. Journ. Econ. Entom., vol. 8, pp. 525–527.

Banks, Nathan
1907 A Census of four square feet. Science, n. s., vol. 26, p. 637.

Barber, G. W.

1925 The Efficiency of Birds in Destroying Over-wintering Larvae of the European Corn Borer in New England. Psyche, vol. 32, pp. 30–46, 1 pl., 2 figs.

Barnes, H. F.

1933 Two Further Instances of Flies Swarming at the Rothamstead Experimental Station, with some References to this Phenomenon. Entom. Monthly Mag., vol. 69, pp. 230–232.

1934– Studies of Fluctuations in Insect Populations. Published in a number of 41 parts in The Journal of Animal Ecology. The following are especially pertinent. Pt. III, vol. 3, pp. 165–181; Pts. IV, V, VI, vol. 4, pp. 119–126, 244–263; Pt. VII, vol. 9, pp. 202–214; Pt. IX, vol. 10, pp. 94–120.

Barrett, R. E.

1933 A General Method of Measuring Insect Populations and its Application in Evaluating Results of Codling Moth Control. Journ. Econ. Entom., vol. 26, pp. 873–879.

Beall, G.

1935 Study of Arthropod Populations by Methods of Sweeping. Ecology, vol. 16, pp. 216–255.

Beebe, C. W.

1916 Fauna of Four Square Feet of Jungle Debris. Zoologica, New York, vol. 2, pp. 107–119.

Bernard, C.

1920 Verschillende Rupsen-Plagen in Theetuinen. Meded. Proefstation voor Thee, Buitenzorg, No. 68, pp. 11–27, 10 pls.

Betz, B. J.

1932 The Population of a Nest of the Hornet, *Vespa maculata*. Quart. Rev. Biol., vol. 7, pp. 197–209.

Bigelow, H. B.

1926 The Plankton of the Offshore Waters of the Gulf of Maine. Bull. U. S. Bur. Fisheries, vol. 40, pt. 2, 509 pp., 134 figs.

Bill, J. P.

1932 Swarming Chironomidae. Psyche, vol. 39, p. 68.

Bodenheimer, F. S.

1937 Population Problems of Social Insects. Biol. Rev., Cambridge Philos. Soc., vol. 12, pp. 303–430, 10 figs.

Bonnet, A.

1911 Recherches sur les causes des variation de la faune entomologique aërienne. C. R. Acad. Sci., Paris, vol. 152, pp. 336–339.

Bornebusch, C. H.

1932 Das Tierleben der Waldböden. Forstwiss. Centralbl., vol. 54, pp. 253–256.

Bridwell, J. C., and L. J. Bottimer

1933 The Hairy-vetch Bruchid. *Bruchus brachialis* in the United States. Journ. Agric. Res., vol. 46, pp. 739–751.

Brues, C. T.

1926 Remarkable Abundance of a Cistelid Beetle, with Observations on Other Aggregations of Insects. American Naturalist, vol. 60, pp. 526–545.

1933 Progressive Changes in the Insect Population of Forests Since the Early Tertiary. American Naturalist, vol. 67, pp. 385–406.

Brues, C. T., and A. L. Melander

1932 Classification of Insects. Bull. Mus. Comp. Zool. Harvard, vol. 73, 672 pp., 1121 figs.

Bryson, H. R.

1931 The Interchange of Soil and Subsoil by Burrowing Insects. Journ. Kansas Entom. Soc., vol. 4, pp. 17–21. Cf. also *ibid.*, vol. 6, pp. 81–90 (1933).

Bücher

1917 Zusammenfassender Bericht über die Heuschreckenbekämpfung in Anatolien, Syrien und Palästina im Jahr 1916. Tropenpflanzer, vol. 20, pp. 373–387.

Buckle, P. A.

1921 A Preliminary Survey of the Soil Fauna of Agricultural Land. Ann. Appl. Biol., vol. 8, pp. 135–145.

Bueno, J. R. de la Torre

1915 Heteroptera in Beach Drift. Entom. News., vol. 26, pp. 274–279.

Burrill, A. C.

1913 Notes on Lake Michigan Swarms of Chironomids; Quantitative Notes on Spring Insects. Bull. Wisconsin Nat. Hist. Soc., vol. 11, pp. 52–69.

Calvert, P. P.

1923 The Number of Living Insects. Entom. News, vol. 34, p. 122.

Cameron, A. E.

1925 A General Survey of the Insect Fauna of the Soil Within a Limited Area Near Manchester. Journ. Econ. Biol., vol. 8, pp. 159–204, 3 figs., 2 pls.

Carnes, E. K.

1912 Collecting Ladybirds (Coccinellidae) by the Ton. Monthly Bull. Commis. Hortic., California, vol. 1, pp. 71–81, 7 figs.

Chapman, R. N.

1933 The Causes of Fluctuations of Populations Among Insects. Proc. Hawaiian Entom. Soc., vol. 8, pp. 279–292, 5 pls.

1939 Insect Population Problems in Relation to Insect Outbreaks. Ecol. Monogr., vol. 9, pp. 261–269.

Clarke, G. L., and D. J. Zinn

1937 Seasonal Production of Zooplankton off Woods Hole, with Special Reference to *Calanus finmarchicus*. Biol. Bull., vol. 73, pp. 464–487.

Coad, B. R.

1931 Insects Captured by Airplane are Found at Surprising Heights. Yearbook U. S. Dept. Agric., 1931, pp. 320–323, 4 figs.

Cory, E. N., and E. E. Haviland

1938 Population Studies of *Formica exsectoides*. Ann. Entom. Soc. America, vol. 31, pp. 50–57.

Criddle, N.

1932 The Correlation of Sunspot Periodicity with Grasshopper Fluctuation in Manitoba. Canadian Field Nat., vol. 46, pp. 195–199, 1 fig.

Davies, Wm. M.

1939 Studies on Aphides Infesting the Potato Crop. VII. Report on a Survey of the Aphis Populations. Ann. Appl. Biol., vol. 26, pp. 116–134.

Ditman, L. P.

1932 An Observation on the Hibernation of the Corn Ear-worm in Maryland. Journ. Econ. Entom., vol. 25, p. 413.

Dobrzhansky, Th.

1925 Ueber das Massenauftreten einigen Coccinelliden im Gebirge Turkestans. Zeits wiss. Insektenbiol., vol. 20, pp. 249–256.

Edwards, E. E.

1929 A Survey of the Insects and Other Invertebrate Fauna of Permanent Pasture and Arable Land of Certain Soil Types. Ann. Appl. Biol., vol. 16, pp. 299–323.

Eidmann, H.

1931 Zur Kenntnis der Periodizität der Insektenepidemien. Zeits. angew. Entom., vol. 18, pp. 537–567, 5 figs.

Elton, C. S.

1924 Periodic Fluctuations in the Numbers of Animals; their Causes and Effects. British Journ. Exper. Biol., vol. 2, pp. 119–163, 8 figs.

1925 The Dispersal of Insects to Spitzbergen. Trans. Entom. Soc. London, 1925, pp. 289–299.

1942 Voles, Mice and Lemmings. 496 pp., University Press, Oxford.

Elton, C. S., and M. Nicholson

1942 The Ten-year Cycle in Numbers of the Lynx in Canada. Journ. Anim. Ecol., vol. 11, pp. 215–244.

Emerson, A. E., et al.

1939 Insect populations (a symposium). Ecol. Monogr., 9, No. 3, pp. 259–320, 2 figs.

1939 Populations of Social Insects. Ecol. Monogr., vol. 9, pp. 287–300.

Escherich, K.

1911 Termitenleben auf Ceylon. pp. xxxii + 262, G. Fischer, Jena.

Felt, E. P.

1901 Snakeworm. Bull. New York State Mus., vol. 7, No. 36, pp. 992–994, 2 figs.

1925 The Dissemination of Insects by Air Currents. Journ. Econ. Entom., vol. 18, pp. 152–156.

Fleming, W. E., and F. E. Baker

1936 A Method for Estimating Populations of Larvae of the Japanese Beetle in the Field. Journ. Agric. Res., vol. 53, pp. 319–331.

Forbes, W. T. M.

1938 Note on the Population of *Formica exsectoides*. Ann. Entom. Soc. America, vol. 31, pp. 356–357.

Ford, J.

1937   Research on Populations of *Tribolium confusum* and its Bearing on Ecological Theory: A Summary.   Journ. Anim. Ecol., vol. 6, pp. 1–14, 1 fig.

Fraenkel, G.

1932   Die Wanderungen der Insekten.   328 pp., 36 figs.   Ergebrisse der Biologie, Julius Springer, Berlin.

Gause, G. F.

1934   The Struggle for Existence.   Baltimore: Wms. & Wilkins Co., pp. ix + 163, 41 figs.

Gerould, J. H.

1915   The Cotton Moth in 1912.   Science, vol. 41, pp. 464–465.

Glasgow, J. P.

1939   A Population Study of Subterranean Soil Collembola.   Journ. Anim. Ecol., vol. 8, pp. 323–353.

Glick, P. A.

1939   The Distribution of Insects, Spiders, and Mites in the Air.   Tech. Bull. U. S. Dept. Agric., No. 673, 151 pp., 13 figs.

Graham, S. A.

1939   Forest Insect Populations.   Ecol. Monog., vol. 9, pp. 301–310.

Guyénot, E.

1913   Etudes biologiques sur une mouche, *Drosophila ampelophila*.   C. R. Soc. Biol., Paris, vol. 74, pp. 97, 178, 223, 270, 332, 389, 443.

Hartman, C. G.

1923   Swarming Insects Simulating Smoke.   Science, n.s., vol. 57, pp. 149–150.

Hartzell, Albert

1921   Further Notes on the Life History of the Potato Leafhopper.   Journ. Econ. Entom., vol. 14, pp. 62–68.

Hawkes, O. A. M.

1926   On the Massing of the Ladybird, *Hippodamia convergens*, in the Yosemite Valley.   Proc. Zool. Soc., London, 1926, pp. 693–705.

Headlee, T. J.

1910   Burning Chinch Bugs.   Circ. Kansas State Expt. Sta. No. 16, 7 pp., 6 figs.

Hendrickson, G. O.

1931   Subterranean Insects of Marsh Grass (*Spartina Michauxiana*).   Canadian Entom., vol. 63, pp. 109–110.

Herrick, G. W.

1926   The "Ponderable" substance of aphids (Homoptera).   Entom. News, vol. 37, pp. 207–210.

Hewatt, W. G.

1937   Ecological Studies on Selected Marine Intertidal Communities of Monterey Bay, California.   American Midl. Nat., vol. 18, pp. 161–206.

Holdaway, F. G., F. J. Gay, and T. Greaves

1935 The Termite Population of a Mound Colony of *Eutermes exitiosus.* Journ. Council Sci. & Indust. Res., Australia, vol. 8, pp. 42–46, 1 fig.

Howard, L. O.

1898 Swarming of Western Willow Flea Beetle. Bull. Div. Entom., U. S. Dept. Agric., No. 18, 100 pp.

1901 Migration of the Western Willow Flea Beetle. Bull. Div. Entom., U. S. Dept. Agric., No. 30, p. 97.

1911 The Housefly. xix + 312 pp., 40 figs. Philadelphia, Pennsylvania, F. A. Stokes Co.

Hudson, G. V.

1905 Notes on Insect Swarms on Mountain Tops in New Zealand. Trans. New Zealand Inst., vol. 38, pp. 334–336.

Hull, J. B., W. E. Dove, and F. M. Prince

1934 Seasonal Incidence and Concentration of Sand Fly Larvae in Salt Marshes. Journ. Parasitol., vol. 20, pp. 162–172, 7 figs.

Hutson, J. C.

1917 Insect Pests in British Guiana in 1916. Agric. News, Barbadoes, vol. 16, pp. 266–267.

Ihering, H. von

1896 Zur Biologie der socialen Wespen Brasiliens. Zool. Anz., vol. 19, pp. 449–453.

Imms, A. D.

1922 Note on Swarms of *Chloropisca circumdata.* Entom. Monthly Mag., vol. 56, p. 20.

Ingram, J. W., and E. K. Bynum

1932 Observations on the Sugar-cane Beetle in Louisiana. Journ. Econ. Entom., vol. 25, pp. 844–849.

Jacot, A. P.

1940 The Fauna of the Soil. Quart. Rev. Biol., vol. 15, pp. 28–58.

Janet, C.

1895 Sur *Vespa crabro.* Histoire d'un nid depuis son origine. Mém. Soc. Zool., France, vol. 8, pp. 1–139.

Johnson, M. S., and F. Munger

1930 Observations on Excessive Abundance of the Midge, *Chironomus plumosus* on Lake Pepin. Ecology, vol. 11, pp. 110–126.

Kearns, H. G. H.

1929 The Swarming of *Chloropisca circumdata.* Entom. Monthly Mag., vol. 65, pp. 205–206.

1930 Early Spring Swarming of *Chloropisca circumdata* with Notes on the Control of the Flies in Autumn. *Ibid.,* vol. 66, pp. 166–168.

King, K. M.

1939 Population Studies of Soil Insects. Ecol. Monogr., vol. 9, pp. 270–286.

Kofoid, C. A., S. F. Light, et al.
1934 Termites and Termite Control, xxv + 734 pp. Univ. of California Press, Berkeley, California.

Larson, A. O., and C. K. Fisher
Insects Screened from Bean Samples. Entom. News, vol. 41, pp. 74–76.

Lintner, J. A.
1896 Notes on *Sciara*. 10th Rept. New York State Entom., pp. 387–399.

Loughnane, J. B.
1940 A Survey of the Aphis Population of Potato Crops in Ireland. Journ. Dept. Agric., Eire, vol. 37, pp. 370–382.

McAtee, W. L.
1907 A Census of Four Square Feet. Science, vol. 26, pp. 447–449.

McClure, H. E.
1938 Insect aerial populations. Ann. Entom. Soc. America, vol. 31, pp. 504–511, 5 figs.

McConnell, W. R.
1921 Rate of Multiplication of the Hessian Fly. Bull. U. S. Dept. Agric., No. 1008, 8 pp.

MacLagan, D. S.
1932 An Ecological Study of the "Lucerne Flea" (*Smynthurus viridis*). II. Bull. Entom. Res., vol. 23, pp. 151–190, 10 figs.

MacLagan, D. S., and E. Dunn
1934–35 The experimental analysis of the growth of an insect population. Roy. Soc. Edinburgh. Proc., vol. 55, No. 2, pp. 126–139, 5 figs.
1940 Sunspots and Insect Outbreaks: An Epidemiological Study. Proc. Univ. Durban Philos. Soc., vol. 10, pp. 173–190, 3 figs.
1941 Recent Animal-Population Studies and their Significance in Relation to Socio-biological Philosophy. (1). Proc. Univ. Durham Philos. Soc., vol. 10, pp. 310–331.

Mancheron, P.
1914 La Lutte contre les criquets dans la commune mixte du Djebel Nador. Rev. Agric. Vitic. Afrique Nord, Algiers, vol. 3, pp. 460–461.

Marlatt, C. L.
1907 The Periodical Cicada. Bull. U. S. Dept. Agric. Bur. Entom., No. 71, 181 pp., 6 pls., 68 figs.
1908 The Imported Elm Leaf-beetle. Circ. Bur. Entom. U. S. Dept. Agric., No. 8, 6 pp.

Miyadi, D.
1933 Studies on the Bottom Fauna of Japanese Lakes. Japanese Journ. Zool., vol. 5, pp. 171–187. (Also previous papers in the same journal.)

Morris, H. M.
1920 Observations on the Insect Fauna of Permanent Pasture in Cheshire. Ann. Appl. Biol., vol. 7, pp. 141–155.

1922 The Insect and Other Invertebrate Fauna of Arable Land at Rothamstead. *Ibid.*, vol. 9, pp. 282–305, 7 figs.
1927 The Insect and Other Invertebrate Fauna of Arable Land at Rothamstead, Part II, *ibid.*, vol. 14, pp. 442–464, 3 figs.

Needham, J. G.

1900 Insect Drift on the Shore of Lake Michigan. Occas. Mem. Chicago Entom. Soc., vol. 1, pp. 19–26, 1 fig.
1904 Beetle Drift on Lake Michigan. Canadian Entom., vol. 36, pp. 294–296.

Nicholson, A. J.

1933 The Balance of Animal Populations. Journ. Anim. Ecol., vol. 2, pp. 132–178.

Nolan, W. J.

1925 The Brood Rearing Cycle of the Honeybee. Bull. U. S. Dept. Agric., No. 1349, 55 pp., 29 figs.

Park, T.

1933 Studies in Population Physiology, II. Factors Regulating Initial Growth of *Tribolium confusum* Populations. Journ. Expt. Zool., vol. 65, pp. 17–42.

Parshley, H. M.

1917 Insects in Ocean Drift. Canadian Entom., vol. 49, pp. 45–48.

Pickles, W.

1935– Populations and Territories of the Ants, *Formica fusca*, *Acanthomyops flavus*
36 and *Myrmica ruginodis*. Journ. Anim. Ecol., vol. 4, pp. 22–31, 1 pl., 2 figs.; vol. 5, pp. 262–270, 1 pl., 1 fig.
1940 Fluctuations in the Populations, Weights and Biomasses of Ants in Yorkshire. Trans. R. Entom. Soc., London, vol. 90, pp. 467–485.

Poulton, E. B.

1904 A Possible Explanation of Insect Swarms on Mountain Tops. Trans. Entom. Soc. London, proc., 1904, pp. xxiv–xxvi.

Pussard, R.

1929 Sur la déstruction du Coléoptère, *Niptus hololeucus* pullulant dans une villa aux environs de Lyon. Rev. Path. Végét. Entom. Agric., vol. 16, pp. 29–33.

Quaintance, A. L., and C. T. Brues

1904 The Cotton Bollworm. Bull. Bur. Entom., U. S. Dept. Agric., No. 50, 155 pp., 25 pls., 27 figs.

Rau, P.

1929 Nesting Habits of the Bald-faced Hornet, *Vespa maculata*. Ann. Entom. Soc. America, vol. 22, pp. 659–675.
1941 The Swarming of Polistes Wasps in Temperate Regions. *Ibid.*, vol. 34, pp. 580–584.

Ritchie, J.

1915 Observations and Deductions Concerning the Habits and Biology of the Common Wasp. Scottish Naturalist, vol. 46, pp. 318–331.

Sanders, J. G., and S. B. Fracker

1916 Lachnosterna Records in Wisconsin. Journ. Econ. Entom., vol. 9, pp. 253–261, 3 figs.

Schmidt, H.

1917   Beobachtungen an einen ausgehobener Reste von *Vespa germanica*. Zeits. wiss. Insektenbiol., vol. 13, pp. 153–160.

Scott, Hugh

1916   Note on the Swarming of Chloropid Flies, Psocida, etc. in Houses. Entom. Monthly Mag., vol. 52, pp. 18–21 and 43.

1926   Note on the Swarming of Gnats or Midges Round Lofty Towers. *Ibid.*, vol. 62, pp. 18–19.

Sherman, F.

1938   Massing of Convergent Ladybeetle at Summits of Mountains in Southeastern United States. Journ. Econ. Entom., vol. 31, pp. 320–322.

Smirnov, E., and W. Prolejaeff

1934   Density of Population and Sterility of the Females in the Coccid, *Lepidosaphes ulmi.* Journ. Anim. Ecology, vol. 3, pp. 29–40, 6 figs.

Smith, C. E., and N. Allen

1932   The Migratory Habit of the Spotted Cucumber Beetle. Journ. Econ. Entom., vol. 25, pp. 53–57.

Snow, L. M.

1902   The Microcosm of the Drift Line. American Natural., vol. 36, pp. 855–864.

South, F. W.

1914   Summary of Locust Work, March 12 to April 30, 1914. Agric. Bull. Fed. Malay States, Kuala Lumpur, vol. 2, pp. 294–297. *Cf.* also *ibid., t. c.*, pp. 227–230.

Spuler, A.

1925   Baiting Wireworms. Journ. Econ. Entom., vol. 18, pp. 703–707.

Stanley, J.

1932–   A Mathematical Theory of the Growth of Populations of the Flour Beetle,
34   *Tribolium confusum.* Canadian Journ. Res., vol. 6, pp. 632–671; vol. 7, pp. 426–433, 550; vol. 11, pp. 728–732.

Swinton, A. H.

1883   Data Obtained from Solar Physics and Earthquake Commotions Applied to Elucidate Locust Multiplication and Migration. Third Rept. U. S. Entom. Comm., pp. 65–85 (Chapter V).

Talbot, M.

1943   Population Studies of the Ant, *Prenolepis imparis.* Ecology, vol. 24, pp. 31–44.

Thienemann, August

1924   *Chironomus* und Feuerwehr. Zeits. Wiss. Insektenbiol., vol. 19, p. 192.

Thomas, W. A.

1932   Hibernation of 13-spotted Lady Beetle. Journ. Econ. Entom., vol. 25, p. 136.

Thompson, M.

1924   The Soil Population. An Investigation of the Biology of the Soil in Certain Districts of Aberystwythe. Ann. Appl. Biol., vol. 11, pp. 349–394, 4 figs.

1934 The Development of a Colony of *Aphelinus mali*. Parasitology, vol. 26, pp. 449–453.

Trägårdh, Ivar

1933 Methods of Automatic Collecting for Study in the Fauna of the Soil. Bull. Entom. Res., vol. 24, pp. 203–214.

Uichanco, L.

1936 Secular Trends of Locust Outbreaks in the Philippines and Their Apparent Relation with Sunspot Cycles. Philippine Agriculturist, vol. 25, pp. 321–356, 1 map, 12 charts.

Uvarov, B. P.

1928 Locusts and Grasshoppers. Imp. Bur. Entom., London, 352 pp., 118 figs.

Vallot, J.

1912 Sur une immense quantité de *Desoria glacialis* à la surface d'un glacier. C. R. Acad. Sci., Paris, vol. 155, pp. 184–185.

Wainright, C. J.

1922 *Chloropisca circumdata* Occurring in Houses. Entom. Monthly Mag., vol. 58, pp. 38–39.

Waksman, S. A., and R. L. Starkey

1931 The Soil and the Microbes. 260 pp., Wiley & Sons, New York.

Wene, George

1940 The Soil as an Ecological Factor in the Abundance of Aquatic Chironomid Larvae. Ohio Journ. Sci., vol. 40, pp. 193–199.

Werner, F.

1913 Massenansammlung von Coccinella. Zeitschr. wiss. Insektenbiol., vol. 9, p. 311.

Wilbur, D. A., and C. W. Sabrosky

1936 Chloropid Populations on Pasture Grasses in Kansas. Journ. Econ. Entom., vol. 29, No. 2, pp. 384–389, 6 figs.

Wilbur, D. A., and R. A. Fritz

1940 Grasshopper Populations. Journ. Kansas Entom. Soc., vol. 13, pp. 86–100.

Wildermuth, V. L., and E. G. Davis

1931 The Red Harvester Ant and How to Subdue it. Farmers' Bull., U. S. Dept. Agric., No. 1668, 21 pp.

Williams, C. B.

1923 Records and Problems of Insect Migration. Trans. Entom. Soc. London, 1922, pp. 207–233.

1925 Notes on Insect Migration in Egypt and the Near East. Trans. Entom. Soc. London, 1924, pp. 439–456.

1930 The Migration of Butterflies. xi + 473 pp. Edinburgh.

1940 The Effect of Weather Conditions on Insect Activity; and the Estimation and Forecasting of Changes in the Insect Population. Trans. Roy. Entom. Soc., London, vol. 90, pp. 227–306, 24 figs.

Williston, S. W.

1908 Manual of North America Diptera, 3rd Ed. 405 pp. J. T. Hathaway, New Haven, Conn.

Wolcott, G. N.

1927 An Animal Census of Two Pastures and a Meadow in Northern New York. Proc. Entom. Soc. Washington, vol. 29, pp. 62–65.

Yung, E.

1899 Dénombrement des nids de la fourmi fauve (*F. rufa*). Arch. Zool. Exper. (3) vol. 7, pp. xxxiii–xxxv.

1900 Combien y a-t-il des Fourmis dans une fourmilière? (*Formica rufa*.) Arch. Sci. Phys. Nat. Genève (4) vol. 10, pp. 46–56.

CHAPTER II

TYPES OF FOOD HABITS AMONG INSECTS AND THEIR RELATION TO STRUCTURE AND ENVIRONMENT

Rien n'est plus joli, plus merveilleux, à une infinité d'égards, que les différents appareils des fonctions de nutrition chez les Insectes.

EMILE BLANCHARD, in *Metamorphoses, Moeurs et Instincts des Insectes*, 1868

WE HAVE already mentioned briefly the diversity of insects and their great abundance both as species and as populations. Our next step will consist of an inquiry into the general food relations of this vast horde which has so successfully and continuously found abundant sustenance in competition with the great variety of other animals which live with them in this world of ours. At least we take it for granted that the world is ours to do with as we please, and one of the great natural obstacles to the complete spoilation of nature by man has been the persistent failure of insects to fit in with this plan. Particularly have they failed to accept our doctrine of unlimited multiplication, which they themselves have been denied in the struggle for existence among living organisms. We shall not concern ourselves directly with the conflict between the insects and man since this has been ably and frequently presented by others who are far better qualified to speak upon this subject than is the writer.

The search for food by insects has persisted since the earliest kinds appeared, but evidently became highly specialized before early Mesozoic times. We know that certain methods of feeding are correlated with the form of those bodily parts which function mainly in securing food. We may therefore reasonably assume on the basis of such structural modifications seen in fossil insects that most of the types of food-habits characteristic of the insects of the present day were developed during the Mesozoic or earliest Tertiary. Since then they appear only to have been perfected and more firmly fixed in the genetic constitution of the several series. Thus a great

PLATE III

Arboreal termite nest near Cienfuegos, Cuba, showing galleries extending down the tree-trunk to the soil beneath. Original photograph.

PLATE IV

A swarm of honeybees, containing probably | between ten and twenty thousand workers clustered about a single fertile female.   Original photograph.

constancy of food relations has become generally characteristic of many groups of insects, broken here and there by sporadic shifts to a different diet. These sudden changes or mutations only serve to emphasize the general trend to retain specific food habits over periods long enough for vast series of forms to become differentiated into independent species.

A complete survey of the materials which furnish food for insects would involve a compilation of stupendous proportions, certainly far beyond the strength of any ordinary individual and suited only for a Federal Aid or Economy Program. It would, moreover, serve no useful purpose except to show us a catalogue of practically every plant and animal alive or dead and the innumerable products that are derived from them through natural decay or mechanical and chemical processing.

There are a considerable number of insects that are quite indiscriminate in their choice of food, partaking of a mixed diet or passing from one kind to another as necessity demands or fancy dictates. Such polyphagous kinds are usually members of the more generalized orders. The majority restrict themselves greatly in the choice of food, depending either upon living plants, living animals or dead and decaying matter but not upon these in combination. We may thus recognize several categories of insects on a somewhat arbitrary basis of their food requirements. These divisions do not coincide in any close degree with the natural relationships of the various groups, but from our present viewpoint are valuable, as they enable us to consider in an orderly manner the devious ways in which food is related to structure and behavior.

Quite generally four types of food-habits among insects may be distinguished. Each of these is characteristic of a large number of species belonging to at least several of the natural orders, although very few of the larger groups exhibit essentially uniform food-habits throughout.

Living plants, especially the flowering plants, furnish the food materials for fully half of the living species. Every portion of the plant is utilized, the insects usually selecting some particular parts or materials such as the leaves, buds, flowers, stems, roots, or woody portion. Frequently, as we shall see in more detail later, these vegetarian types may restrict their diet not only to these particular parts of the plants, but may confine themselves to a greatly limited series of plant species, or even to the members of one genus or to a single species. The term phytophagous is applied to these vegetarian insects and is usually, although quite inaccurately, restricted to those which feed on flowering plants and ferns and more rarely on the lower green plants. Those which depend for food upon mushrooms and the smaller or microscopic fungi, yeasts, etc., form a quite distinct category, conveniently referred to as mycetophagous. Some of these forms which require yeasts commonly occur in fermenting vegetable material. Others that consume certain fungi actually propagate these in their burrows or nests and thus enter into clearly defined symbiotic relations with their food-plants. Similar relations exist also between certain ants and flowering plants known as myrmecophytes with which these ants are definitely associated.

Living animals, mainly other smaller or weaker insects, serve as food for a great variety of carnivorous species. Such forms are usually referred to as predatory, although this term is not always exact in its application if we compare them with the predatory birds or mammals. This will be more clearly evident when we come to discuss insect predatism in more detail. The carnivorous forms are not nearly so numerous as the vegetarian ones but they are diverse structurally and in general more highly adapted in morphology and behavior. We should naturally expect this since many of these forms capture living prey, behavior which requires aggressiveness, activity and astute instincts not demanded in the acquisition of vegetable food. Purely carnivorous habits are characteristic of numerous families and larger groups among both primitive and highly specialized insects. That this type of sustenance is of very ancient origin is shown by its general occurrence in the early fossil dragon-flies for these insects were as admirably built for predatory life a hundred million years ago as they are to-day. The frequent appearance of similar habits among the higher holometabolous insects attests their adaptability in comparison with other methods of securing food.

The saprophagous insects form a third category which is also widely scattered among the several orders of insects (Pl. V A). Typically such forms feed upon dead or decaying animal matter and ostensibly they appear to be readily classed as scavengers. In fact they have on many occasions received great praise as useful little agents in reducing the dead remains of organisms to their more simple constituents, thus aiding in the economy of nature for the benefit of man. In essence the matter is less simple than it appears on the surface, for many saprophagous insects depend at least to a great extent upon microörganisms present in the decomposing materials that they eat, and on the other hand, the distinction between saprophagy, phytophagy and predatism is not always easily drawn.

A fourth type of food-habits is parasitism. Here we encounter difficulties in making any general statements since there are two fundamentally different kinds of parasitism among insects. One of these is really a refined sort of predatism to which Wheeler has applied the term parasitoidism. This is characteristic of a considerable number and variety of groups belonging to several of the larger orders, particularly the Hymenoptera and Diptera. In the former, three large superfamilies, embracing the ichneumon-flies, chalcid-flies and others are almost exclusively parasitic. One very extensive family of Diptera (Tachinidae) and several others of less extent have similar habits. Such species, most commonly known as entomophagous parasites, feed during the larval stages within or attached to the bodies of other arthropods, almost exclusively other insects, which they consume while still in the living state and finally destroy on completing their growth. Insect parasites of another kind are those that live externally on the bodies of vertebrate animals, restricted to birds and mammals. These epizoic parasites usually suck blood for food, either in the adult condition, during growth, or during their entire life. Then there are a few peculiar types that live as internal parasites, mainly of terrestrial vertebrates, during their developmental period. It is thus evident that insect parasites are very diverse and that they deserve a much fuller consideration which we shall hope to give them in a later chapter. Some writers would even speak of the insects that feed on plants as parasites, but as we have already indicated they are here excluded from this category which is restricted to those forms dependent upon animal hosts.

We might make a more elaborate classification of food habits but this would add confusion rather than clarity, for whatever types or definitions we may select, there always remains a residuum of forms which do not fall decisively into any one group.

Occasionally commensalism, or "eating at the same table," may be noted, although such cases are really a type of parasitic behavior. A most remarkable example was described by Wheeler, involving an ant and a minute phorid fly (*Cataclinusa pachycondylae*). The ant larva is fed by the ant workers while lying on its back in the nest. Bits of food are placed on its belly, and as it reaches down to feed, the fly larva which lives coiled about its neck, reaches out likewise and partakes of the food so generously supplied by the ants (Fig. 7). Many other myrmecophiles or ant-guests which live in the nests of ants are true parasites, others are scavengers (Pl. V B) and some are predatory. Some insects pass from one type of food haibts to another during development; indeed, in the foregoing

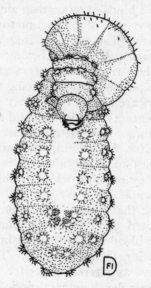

Fig. 7. Larva of a phorid fly, *Cataclinusa pachycondylae*, curled around the neck of the ant, *Pachycondyla harpax*. It is a commensal, sharing the food given the ant larva by the attendant ant workers. Redrawn from Wheeler.

statements no distinction between the feeding of larval and imaginal stages has been drawn, since these are so commonly dissimilar in insects that undergo complete metamorphosis.

In fact, the acquisition of indirect development by the higher insects opened new avenues for the initiation of profound, although transitory structural modifications in the larval stages, since the latter are so interpolated in development that they may acquire at this time entirely novel characteristics which are highly adaptive, but leave no impress upon the imago. These are generally correlated with peculiarities in food habits. Similarly, hypermetamorphosis may result in the differentiation of even the several larval stages into divergent types, quite independently of one another.

The utilization of most unusual materials for food is encountered among insects, some of which we shall mention in connection with saprophagous habits. Thus horn, beeswax, cork, pepper, cured tobacco and other equally unexpected delicacies are the exclusive choice of certain species. Occasionally substances entirely devoid of nutritive content are eaten incidentally. Insects as diverse as beetles of the families Bostrychidae, Cerambycidae and Dermestidae as well as carpenter bees are on occasion highly destructive to lead pipes or lead-covered electrical cables, whither they are attracted for various reasons, and termites have been known to vary their diet of cellulose by consuming the rubber insulation of electrical wiring. The burrowing into lead is so frequent that it constitutes an economic problem.

We shall consider the several types of food habits in the following chapters, but it is appropriate at the present time to inquire into some of the more general relations of food to growth, development and metamorphosis. Furthermore, we cannot disregard the dependence of food habits on anatomical structure, and the conditions presented by the great variety of natural environments in which insects live.

All animals must feed to produce energy and to acquire the materials necessary for growth. Such feeding is normally intermittent, dependent upon the opportunities of securing food at regular or irregular intervals and the satiation of appetite during digestion. In insects the cyclical character of ingestion and digestion is not always very apparent and is further modified by a second, slower and more irregular cycle. Insects that undergo incomplete metamorphosis commonly stop feeding for a time just preceding each molt or ecdysis, so that every individual during life experiences a series of short periods during which no food is taken. In the higher insects a longer additional fast occurs during the pupal stage as there is no possibility of ingesting food at this time due to the absence of functional mouthparts. The inhibition of feeding before molting may be induced directly by the impending ecdysis, but more probably the short period of starvation serves as a stimulus for the transformation since starvation is known in some cases to hasten molting and Kopec ('24) has demonstrated that repeated periods of starvation during the larval stages of gipsy moth caterpillars greatly shorten the pupal stage although larval life is prolonged. Similar results have been obtained by other workers with other Lepidoptera. In the imaginal stage certain insects may temporarily abstain from feeding in connection with certain activities. Dodd has observed an Australian pentatomid bug, *Tectoris lineola*, which broods over its eggs for several weeks, not partaking of any food until the eggs have hatched. Another related genus, *Elasmostethus*, has similar habits. Also, as is well known, the fertile young queens of many ants partake of no nourishment during the period that they are establishing their nest and rearing their first brood of young to maturity. At this same time they are, moreover, feeding their larvæ with regurgitated food that they necessarily elaborate from the tissues of their own bodies, and utilizing some of their freshly laid eggs for the same purpose. In spite of such a long fast some of these young queen ants survive till their brood of workers is matured and ready to bring sustenance to the nest where the queen may then remain to enjoy a long life, sometimes extending to a decade or more.

Finally, there are a number of insects belonging to several orders which do not feed at all after maturity, depending entirely upon materials which they have stored within the body during their developmental stages to carry them through a short imaginal life. This condition known as aphagia prevails among the mayflies, proverbial for their ephemeral period of sexual life, but is encountered generally among the caddis-flies, in scattered families and genera among Lepidoptera and in several families of Diptera, particularly the bot-flies of the families Oestridae and Gastrophilidae as well as a few Hymenoptera and Coleoptera. In some cases aphagia represents the spartan self-denial so commonly characteristic of the instinctive behavior of the social insects. There are probably many more such instances than have been observed but this phenomenon is usually associated with vestigial mouthparts whose non-functional condition effectually prevents lapses in dietary procedure. Nevertheless, some of these species may survive for some time without food or water. The large silkworm moths of the family Saturniidae, such as the *Cecropia* and *Polyphemus* moths, normally mate and produce their eggs under such conditions over a period of one to several weeks before death ensues. Rather similar periods without food have been noted for the adult life of fleas and other insects. There is usually a rapid loss of weight under such circumstances, amounting to nearly 50% in certain beetles (*Geotrupes*) where death ensues after a couple of weeks. These data refer, of course, to temperatures at which the species are active. As with other cold-blooded animals pronounced cooling greatly prolongs such fasting periods. The reduction of mouthparts in this great variety of insects is thus an interesting case of degeneration coincident with disuse, although it has apparently never been cited in connection with evolutionary theory, and really furnishes no concrete evidence bearing on Lamarckian principles. Finally, there is of course a cessation of feeding along with other activities during hibernation, or more rarely during estivation. This may occur either in the larval stages where feeding extends over more than one season or in adults. In the latter case it usually follows the autumnal transformation of individuals whose reproductive functions are delayed until the following year. Experiments of Wodsedalek ('17) on larvae of the little dermestid beetle, *Trogoderma*, have shown an almost unbelievable ability of these insects to remain alive without food or water for long periods of time. In one case a well-grown larva lived for more than five years. These specimens molted more or less regularly during this time and gradually grew smaller at each ecdysis until they reached about the size of a newly hatched larva. The same entomologist succeeded in reversing this process several times with certain specimens by alternate starvation and feeding. With other insects that have been examined an excess or moderate reduction of food does not usually effect any change in the number of larval molts.

Among insects that feed openly on plants or capture prey for food there is usually a pronounced diurnal rhythm. This is by no means uniform throughout the groups, but tends on the whole probably to daylight feeding. Most winged species that visit flowers for pollen or nectar are actively feeding during the sunlight hours, and remain quiescent at night; this is particularly true of Diptera, Hymenoptera and certain beetles. Some forms are crepuscular, like the sand-flies (*Culicoides*) and certain mosquitoes (*Anopheles*) or many anthophilous hawk-moths that appear toward nightfall at the hour when their preferred flowers emit their sweetest perfumes. May-beetles (*Phyllophaga*) frequently ascend in whirring swarms from their hiding places in grassy woodlands in the cool of the evening to feed voraciously on the leaves of trees. Many caterpillars eat at their food-plants quite continuously during daylight to rest at night, while some feed stealthily by dark, like the hated garden cutworms and many others whose activities are not so evident to the casual observer. The evening appearance of the insectivorous garden toad is striking evidence of the nocturnal activity of insects, most of which is directly associated with the process of food-getting. With species that are subterranean or live

within the tissues of plants or bodies of host animals a knowledge
of diurnal rhythm in feeding is, of course, far less complete. We
have reason to believe, however, that changes in temperature of the
soil or the plants in which they are enclosed serve to stimulate or
retard feeding activity, since optimum temperature ranges for the
growth of certain species have been demonstrated by eliminating
natural fluctuations and rearing them at constant temperatures.

Such experiments cannot answer this question directly as we
know that feeding does not depend entirely on temperature, that
there are diverse optima for different forms, and also that various
other factors enter into growth, which is really only a measure of
the final transformation of food into living tissues. They do, however, form one approach to the general question of food in relation
to growth and size. In insects as diverse as the clothes moth,
Tineola, the pomace-fly, *Drosophila*, and the bean weevil, *Bruchus*,
several workers (Titschak, Alpatov and Pearl, and Menusan) have
found that lower temperatures prolong the developmental period
and result in the production of larger individuals, perhaps as a
result of greater food intake before growth is completed. In the
specific case of *Drosophila*, flies reared at 18° C. (64° Fahr.) are
larger than those reared at 28° C. (82° Fahr.). Over its present
habitat both temperatures are well within the normal breeding
range of this fly, which since its naturalization into the northern
United States multiplies abundantly until late autumn. We speak
now, of course, of the great outdoors and not of the genetical
laboratories where the breeding of *Drosophila* in milk bottles has
provided a novel laboratory use for these handy containers. Other
conditions being equal, temperature affects the size of bean weevils
in the same way. Capable of developing over a range of from 17°–
41° C., these beetles grow fastest at about 30°, but adult size is
increased as the temperature affecting the growing stages is lowered
below this point. These and similar observations show that feeding
and growth are profoundly affected by temperature and moisture
since the influence of heat has been abundantly shown to be modified by humidity. There seems to be no definite correlation between cool and warm environments and the food requirements of
similar insects inhabiting them. Certainly there is no increase in
size among the insects of cold climates in comparison with their
close relatives in warm regions, although such has been noted in
certain types of mammals. There, however, the difference is probably related to the conservation or dissipation of bodily heat, an

important consideration for warm-blooded animals, but not for others.

Difference in definitive size resulting from different amounts of food eaten during growth has been investigated in a number of cases, some of which will be dealt with in later chapters. As a rule such variation is less than that of most animals, especially in holometabolous insects, due to the fixed number of ecdyses and the delay in developing the imaginal body until the completion of growth. Added to this is the fact that size (not weight) as we commonly observe it is that of the adult skeleton. Conspicuous variations in adult size are very often characteristic of forms in which the larval food supply is unusually inconstant, like the blow-flies which develop in the carcasses of vertebrates from eggs often laid greatly in excess of the available food-supply. As a consequence many individuals suffer partial or complete starvation before maturity. The larvae of the giant solitary wasps known as "tarantula hawks" are each fed during growth on a large spider. Although these spiders vary greatly in size the mother wasp provides a single one for each of her progeny. To this procedure we must attribute the notoriously variable size of these wasps. Wood wasps of the family Xiphydriidae whose larvae bore in timber likewise vary extremely in size in spite of a uniform food supply. As they depend upon symbiotic microörganisms to digest their food, the symbionts are undoubtedly concerned in such cases. Certain entomophagous parasites that will attack insect larvae of indefinite sizes occasionally show the same excessive variation in size. Such differences seem to be clearly correlated with an insufficiency of food and their failure to occur more generally may depend upon a rather definite nutritional requirement for successful final transformation. If this be true we must assume that excessive irregularities in size are usually weeded out before maturity by intrinsic factors and not that constancy of size is determined by a uniform food-supply. This assumption seems to be demanded also in consideration of the fact noted by Jensen-Haarup that among beetles, the members of several typically predatory families, with obviously more precarious food-supply are less variable in size than the members of several vegetarian families with which he has compared them. Aside from such influences the discontinuous growth of insects when fully fed is a most remarkably regular process. Many years ago, Dyar noted a very constant growth ratio in the stages of caterpillars and found this to be of quite general application to other lepidopterous larvae.

Later, the Austrian physiologist Przibram determined from careful measurements of the nymphal stages of a praying mantis that the weight of each stage is twice that of the preceding one and the length 1.26 ($= \sqrt[3]{2}$) of the preceding one. This ratio has been extended by others and found to prevail quite universally with only slight variations (ranging in mass from about 1.9 – 2.2) and has been accepted as a most fundamental principle of growth and molting. There are many exceptions, especially in the holometabolous insects, but these are in a sense only apparent. Frequently the number of steps in this geometrical progression is greater than the number of actual molts, but wherever growth during one or more stages is in excess of the postulated amount it appears to be proportional to one or more additional doublings and we can regard such stages as due to suppressed molts or latent stages. In some species like the silkworm with four actual molts, the latent divisions may increase the doublings to 12 or 13. In such cases, even slight variations from an exact doubling may result in great differences in absolute growth. In the silkworm this may vary between 3,165 and 8,417 times according to Bodenheimer, so that the principle although probably sound, must be applied with caution to any problems beyond the theoretical sphere. Others, notably Harries and Henderson ('38) doubt that any real significance attaches to the idea, and still others have shown that it cannot be related as once thought to a regular process of cell division in the developing insect. According to Trager ('35), the growth of the blow-fly larvae (*Lucilia*) occurs through increase in cell size, while in the silkworm some tissues grow in this way, but others grow by increase in cell number. Ludwig and Abercrombie ('40) found that the head capsule of the larva of the Japanese beetle (*Popillia japonica*) does not increase in length and width at an equal rate, nor does it lengthen at a regular rate. Thus, they have been obliged to introduce more complex equations which lead us nowhere, except to show that the matter is not so simple as was once believed.

The correlation of form and anatomical structure with food habits is usually quite pronounced, although as we shall see later it is not necessarily so, since in some groups we find fundamental differences in food without corresponding changes in the mouthparts or other associated structures. The most extreme differences in insect mouthparts depend upon their modification for either biting and chewing food, or for imbibing liquid nourishment. In the former type, known as mandibulate, the mandibles serve as cutting and grinding organs, hinged to the head at the sides of the mouth and actuated by muscles that serve to oppose or separate their tips. In the second type known as haustellate the mandibles, and also the first pair of maxillae, labrum and hypopharynx are modified into piercing or stiletto-like blades, incorporated into a beak for sucking sap, blood or other liquid food. Secondary simplification of this apparatus in the higher Diptera appears by the degeneration of the blades, leaving only a lapping organ as in the housefly. Finally in a still more recently derived type, this lapping organ has become elongated and needle-like as in the stable fly (*Stomoxys*), capable of puncturing the skin of animals and sucking blood. Except for the nymphs of Hemiptera and Homoptera, this method of feeding is restricted to adult insects. Haustellate mouthparts may consist only of the maxillae as in moths and butterflies, or they may be otherwise structurally modified although serving the same purpose. Many such peculiarities will be mentioned later. We can state with assurance that these haustellate mouth structures are polyphyletic in origin, having arisen independently in several groups of insects.

The requirements of the chase are generally reflected in the outward bodily form of predatory species. They are lithe and muscular, which means in the case of an armored animal like an insect, having elongated legs and a more or less pedunculate body, powerful wings in aerial forms and heavily sclerotized integument. These requirements are, however, very frequently modified if the prey is captured by stealth or by waiting in ambush or if grasping and holding organs are necessary. The latter are usually modifications of the front legs. The mouthparts are commonly enlarged, and in biting forms the mandibles are sharply edged, toothed or prolonged; in sucking forms the beak tends to become enlarged and heavily sclerotized.

In phytophagous insects the body usually assumes a stouter, more well-fed appearance, with less heavily sclerotized, often very delicate integument, especially in the larval or nymphal stages. Locomotor activity is greatly lessened, at least among vegetarian larvae and nymphs. In larvae this usually entails a caterpillar-like (eruciform) body with weakened or reduced legs, or with the legs completely atrophied and the body vermiform. The mouthparts are generally weaker, except for the retention of powerful chewing or grinding mandibles capable of comminuting large quantities of plant material, even the solid woody tissue which is eaten by many species.

Saprophagous habits usually make a less characteristic impress upon bodily structures, partly because of the rather nondescript and drab picture which they present behavioristically and undoubtedly also because they have arisen in many cases from widely different, more highly adapted types. Thus the change from predatism or parasitism to saprophagy may be readily accomplished without the necessity of pronounced bodily changes.

In every group of organisms the acquisition of parasitic habits induces profound changes in their organization which we rather roughly regard as degenerative. They include in most instances simplification in structure and loss of locomotory powers in connection with an adequate supply of easily digestible food. Parasitism in definite animal hosts frequently involves, however, many highly specific adaptations to bring the parasite to its host, or even alternately to more than one kind of host in forms with a complicated life cycle. As is well known several species may be concerned in such transfer under conditions which are closely correlated with the habits and distribution of each. In such life-histories the parasitic animal itself is commonly quite passive, but the location of their hosts by most insect parasites is an active process. This results from the peculiarities of insect metamorphosis, whereby the parasitic stages are confined to larval development while the imaginal stage retains the complex organization of non-parasitic forms. This permits the parasitic larvae to undergo great changes in structure, including not only the loss or degeneration of organs but also the acquisition of structures entirely peculiar to themselves and often most exquisitely adapted to particular types of parasitic life. It is undoubtedly due to this opportunity for independent specialization during the developmental stages that the insects have surpassed all other animals in the great variety of their parasitic habits. Those insects whose parasitism is confined to the larval stages commonly exhibit during imaginal life but slight changes in food habits from those of their non-parasitic relatives.

Without going to great length it would be impossible to give any satisfactory account of the alimentary system of insects and this seems unnecessary in the present connection. There are certain features however that are of especial interest in connection with the several types of food habits which we have just enumerated. The alimentary canal of all insects consists of three regions, distinguishable on the basis of their embryonic origin. The anterior and posterior parts arise as separate invaginations of the body wall, one growing back from the mouth and the other forward from the anus. These then become connected by the development of a middle portion, the midgut, to form a continuous canal. The two end portions are lined with chitin which imposes a rather impervious layer between the gut and the body cavity. The middle section or midgut is not thus lined and the digestive functions are restricted mainly to this region. In some insects like the entomophagous larvae of Hymenoptera and a few others the mid- and hind-gut do not coalesce until larval growth is completed, so that the mid-gut ends blindly behind, and all waste materials are retained in the alimentary canal until final metamorphosis. In the case of these parasites this seems to be an adaptation to avoid premature damage to the host through the injection of waste products into its body cavity. This condition does not prevail, however, in many other entomophagous parasites belonging to other groups, so that we cannot regard it as a necessity for this type of development.

In many insects the chitinous lining of the fore-gut is prolonged as a thin more or less complete tubular growth or peritrophic membrane, secreted continuously from the posterior end of the fore-gut and serving to enclose the food mass as it progresses backwards. This is generally absent in insects that feed entirely upon nectar, blood or sap, but present in many of those that ingest solid food. Some carnivorous and predatory forms feed by regurgitating a digestive fluid which liquefies their food before it is taken into the body and in many of these at least, such as the larvae of beetles of the families Dytiscidae and Carabidae, the peritrophic membrane is not developed.

In forms like vegetarian caterpillars, the mid-gut becomes very greatly enlarged and extended to accommodate the vast quantities of food necessary for growth, since the amount of nutriment that can be extracted from leaves and similar materials is slight in comparison with their volume.

Food which contains large quantities of water is correlated with a very peculiar modification of the gut in certain aphids (Fig. 8). Here there is a long loop in the fore-gut which passes backward to come into close contact with the rectum and it is believed that some of the more watery constituents of the ingested fluid are shunted past the central portion of the gut to be excreted as "honey-dew" which is characteristically produced by these insects. In forms that imbibe large quantities of liquid food at a single meal, for example mosquitoes and butterflies, there is commonly an enlargement or stalked reservoir developed from the fore-gut which serves for temporary storage awaiting the more gradual process of digestion. Blood and also some vegetable juices require the addition of a substance to prevent premature coagulation. The origin of this material is, however, not well understood; in the case of blood sucking species it may possibly be secreted by cells in the anterior part of the gut which contain symbiotic microörganisms, in others it is probably contained in the salivary secretion.

Although we regard them as creatures of the air, the insects are

Fig. 8. Alimentary canal of an aphid, *Longistigma*, showing the long posterior loop of the intestine. Redrawn from Knowlton.

able to enter this environment only in the last phase of their life cycle and many species that never develop wings are denied this privilege entirely. Their opportunities for dissemination are greatly enhanced by these recurrent periods of flight. This has enabled many insects to invade certain environments from which they might otherwise have been utterly excluded.

Thus we find a considerable number of larger or smaller groups that are aquatic during their developmental stages and a few that remain in the water during their entire life. All, or practically all, of these retain the power of flight and consequently enjoy much wider opportunities for dissemination than most other highly organized aquatic animals. The aquatic insect fauna is extensive, and due to the varied and abundant supply of food in fresh-water lakes, ponds and streams is usually a populous one. Several orders undergo their development in water, as the Odonata, Ephemerida and Trichoptera. Several families of Coleoptera, Hemiptera and Diptera are exclusively aquatic and numerous cases in other orders might be cited of families, genera or individual species that have become secondarily aquatic. We practically never find the adult stage aquatic where the developmental ones are not. However, certain parasitic Hymenoptera of the family Mymaridae that live as internal parasites in the eggs of aquatic insects actually enter the water, and using their wings as paddles swim about to deposit their own eggs in those of their hosts.

It is a very striking fact that almost no insects are actually marine. A few live about the tidal zone or in salt and brackish pools and one small family of Hemiptera, the Halobatidae, live during their entire life on the surface of the sea, often at great distances from land, but truly marine insects are almost unknown, except for certain representatives of the dipterous family Chironomidae whose larvae, and in one or two cases the adults also, live below the tidal zone. The failure of insects to invade the sea is very remarkable when we consider the way in which they have found shelter in practically every other habitat. It has never been quite satisfactorily explained, but a very interesting discussion of the question has been published by Buxton ('33). It is undoubtedly due in part to the continually troubled condition of the surface of the sea to which the aërial method of respirations cannot well be adapted in spite of the various makeshifts which have rendered the fresh-water of smaller and quieter bodies of water readily habitable. It is also undoubtedly associated with the higher osmotic pressure of sea water since many kinds of aquatic insects breathe by gills during the larval stages, although no adult insects possess functional gills, and some of these at least should have been able to establish themselves in the sea if there were not an osmotic bârrier. The conclusion is borne out by the restricted and very similar insect fauna that lives in smaller bodies of saline and alkaline water.

At the other extreme of environmental conditions, deserts and other arid regions support extensive insect-faunas. So far as plants extend into such habitats we find practically all types of insects following them, vegetarian and saprophagous species together with associated predators and parasites. Since the upper temperature tolerance of insects and the higher plants is essentially the same, the presence of plant food is the main factor in limiting their extension into such places. The ability to conserve water in the tissues is of

course an important requirement, not always directly associated
with food, which is manifest in deserticolous species. It is also
associated with certain types of extremely desiccated food like that
of the powder-post beetles, and carpet beetles. The same conditions
are imposed by an irregular and precarious food-supply like that of
the worm-lion and ant-lion which may survive in their dry dusty
pits for protracted periods awaiting the chance capture of a stray
ant. The prolonged survival of trogoderma larvae without food or
water, which we have already mentioned, is another case in point,
as this beetle is undoubtedly a semidomesticated member of a desert
fauna. The great longevity of certain snails that inhabit desert
regions where rains are infrequent is well known, and individuals
have been known to survive a desiccation of several years or longer
in museum collections. Very rarely the same appears to be true of
certain insects, like the "earth pearls," scale insects of the genus
Margarodes, which in the preimaginal encysted stage have been
known to remain alive for many months, in one case for seventeen
years after removal from the soil to an entomologist's collection.

Much of the water required by insects living in extremely dry
habitats is produced as a by-product of their own metabolic processes. This is known as metabolic water and its formation is due
to the oxidation or burning up of carbohydrates in the body whereby
these are broken down into water and carbon dioxide. The latter
is eliminated, but the water is retained and becomes an important
source of supply where there is a deficiency of water in the environment.

The ability to conserve water in the body is variable in different
species because of the way in which it is incorporated in the tissues.
There is always some free water which evaporates easily, but the
remainder is colloidally bound and separable with great difficulty
either by desiccation or freezing. The proportions of the two types
may vary in the same species at different seasons as the bound water
increases preparatory to hibernation. Insects that live in a very
dry environment contain less water, but a larger part is bound,
while those consuming moist food have more abundant free water.
Thus Robinson has shown that a granary weevil, *Calandra granaria*
contains 46% of water, half of which is in the bound form, while
certain caterpillars may contain as much as 90% with less than one-
tenth of this in the bound form.

Many insects are subterranean (Fig. 9), not ordinarily extending
to any great depths in the soil and likewise generally limited by

the extension of the food-supply which consists mainly of decaying plant materials and living roots. On this account such hypogaeic forms include particularly a wide variety of saprophagous species with their attendant predators and, to a lesser degree, their parasites. The scarcity of entomophagous parasites in this type of environment is unquestionably due to the better protection afforded

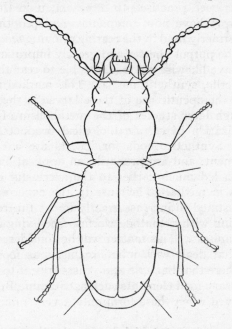

Fig. 9. A blind ground beetle, *Anillus*, one of an extensive fauna of carabid beetles that inhabit caves or deep soil. Original.

by the soil in concealing their hosts. The latter offers great obstacles to discovery and oviposition by parasites, since the adult females of these are almost always actively winged and not well adapted for entering and penetrating the soil.

Caves are also inhabited quite generally by an insect fauna of limited extent, but of much greater diversity than might be expected in such an uninviting environment. On account of the absence of light, the cavernicolous fauna includes no species dependent on green plants for food, but consists mainly of various scavengers and a few parasites of small mammals, especially bats. The free living forms commonly exhibit a loss of the wings, a degeneration of visual organs as is seen in most cave animals, and a multiplicity of species associated with isolated ranges, but not the extreme loss of pigmentation usually characteristic of most other cave animals.

Partly as a result of studies on digestion in insects and partly as an aid in such investigations the feeding of insects on artificial or synthetic foods has been practised to some extent in the laboratory. Foods of this sort have now completely replaced the fermenting banana-paste formerly used in the rearing of *Drosophila* for genetical work, serving the purpose better and greatly improving the atmosphere. They have likewise made it possible to rear the maggots of flesh-flies on sterile, synthetic media. This marked an important advance in the therapeutic use of these larvae in the treatment of wounds, although later studies of the metabolism of the maggots led to the utilization of certain biological products, rather than living insects. Synthetic foods for insects are a comparatively recent development, and are primarily an application of the same principles which led much earlier to the successful cultivation of microörganisms in media. Their use in the experimental field is undoubtedly destined to increase greatly in the future with insects as it is now doing with nematode worms, free-living Protozoa and other small organisms. This matter will be considered further in a later chapter that deals with microörganisms as food for insects. It may be said here that bacteria and yeasts appear to be the source of certain necessary food elements such as vitamine B_1 in the diet of some species, even when these form only a very minor part of the regular diet.

The coloration of insects which is often brilliant and elaborately patterned is due most generally to pigments deposited in the chitinous exoskeleton or in growths arising from the cuticle, such as scales or hairs. It may also be structural, depending upon the presence of thin laminae or minute striae in the integument sufficiently small to break up the incident light, which is reflected as the very brilliant, so-called metallic colors of certain butterflies, beetles and other insects. The relation of pigmental coloration to food has been demonstrated in a number of cases where more or less modified plant pigments are present in the blood or hemolymph of vegetarian insects and in others where such coloring matter accumulates in certain tissues or in the integument. It has been claimed by several workers that the green color of many caterpillars is due to chlorophyll and carotin derived from their food-plants and retained in a slightly modified condition. Others have strenuously combated this idea. There seems to be good evidence that chlorophyll is concerned in the greenish color of some caterpillars with thin, transparent integument and possibly also in the greenish or yellowish color of the silk, spun into cocoons, for example by the silkworm of commerce. Chemical examination shows that many green animal pigments are very different from chlorophyll and as some green predatory insects have no opportunity to secure chlorophyll directly from plants it is evident that animal "greens" are not all similar, even though some may be derived directly from chlorophyll. That the matter is not so simple is shown by Gerould's careful studies on a mutation in the butterfly, *Colias protodice*, where the hemolymph of the caterpillar is changed from the normal green color to blue-green. In this case the color modification behaves as a recessive character and Gerould believes the change due to a factor which destroys the normal carotinoid pigment, thus reducing the amount of yellow. Carotin, also derived from food plants, occurs abundantly in many insects, commonly in the hemolymph which is often highly colored as a result. Palmer and Knight have even traced its secondary transfer from the hemolymph of the vegetarian larvae of the Colorado potato beetle to a predatory bug, *Perillus*, where it is deposited in the integument and forms the basis for the red and yellow markings on the body of the bug.

Anthocyanin has been found in the bodies of a number of insects, where it is derived without question from the plants on which they feed. Among these are included both sucking and biting forms. Glaser has reported its presence in the aphid, *Pterocomma*; in the larvae of a weevil, *Cionus*, Hollande traced the origin of the anthocyanin to the host-plant, *Verbascum*, and found that it accumulates in the fat body to which it imparts a bluish violet color.

It seems quite probable that certain pigments derived from plants may have been concerned in the origin of protective coloration in insects and also that they may quite possibly enter into the phenomena of aggressive coloration and mimicry that have caused so much controversy among biologists. This possibility was broached many years ago by Poulton but the matter seems to have been since entirely neglected.

The amount of food consumed by insects, especially vegetarian ones, is in the aggregate enormous and is the reason for their great economic importance to agriculture. This is due entirely to their extreme abundance as there is no indication that they require any more food in comparison to their size than other kinds of equally active animals. An adult grasshopper feeding on grass leaves eats about one-third or one-fourth its own weight per day and during its development has consumed twenty times its adult weight. On this basis Rubtzov has estimated that an infestation of ten adult grasshoppers per square metre in a one acre field would destroy 607 lbs. of grass. On another basis Wolcott found that lepidopterous caterpillars feeding on grass consume twenty times their own dry weight in dry weight of food, while the consumption of certain other leafy plants amounts to only about seven times. He estimated that the growth of entomophagous parasites required five times their own final weight or less, of food. Such a difference is of course to be expected, as the food of parasites is very high in nutritive value, and in the case of entomophagous parasites the tissues of the host are essentially similar to those of the parasite and more completely assimilable. Such is particularly the case in certain secondary hymenopterous parasites which are very closely related to their hosts. From data given on a later page, the food consumption of certain predatory insects might be similarly calculated. Among solitary bees there is definite mass provisioning of each cell which is then sealed up with the bee's egg placed on its supply of food. The weight of this bee-bread is only from two to three times the weight of the adult bee, indicating the very efficient utilization of a nutritious type of food, which consists of a mixture of honey and pollen.

The purely physiological aspects of digestion are beyond the scope of the present work, but there are some peculiar adaptations to the utilization of certain unusual types of food materials which we shall refer to later in connection with specific insects. These are associated mainly with saprophagous and wood-eating forms.

Several types of relationships are known to exist between food and the development of the gonads and their reproductive functions. The production of an infertile worker caste in the social ants, bees and wasps is believed to depend upon differences in the quantity or quality of the food consumed during larval growth. These workers are imperfect female individuals which do not develop the ovaries to a functional condition, or at least very rarely do so, and associated with this condition there are changes in the external morphology. These may be very slight, but in most ants there are extensive and profound modifications in the thorax and other parts of the body. In the domesticated honeybee and other members of the genus *Apis* to which it belongs the queen or fertile female is reared by the workers in a special cell upon a diet of regurgitated material of glandular origin. The larvae which produce workers are fed with this secretion for several days, but during the remainder of their growth they receive a different and presumably less nutritious mixture of pollen and honey containing less protein. On a diet particularly low in proteins the production of venom required for the sting is reduced and the bees grow more docile. In social wasps the distinction between the worker and functional female is less clearly marked and probably depends only upon the quantity of food as determined by the trophic condition of the colony. In the highly social termites a worker caste is developed, including individuals of both sexes and often of several polymorphic types. Here the distinction between the workers and reproductive caste is thought to depend upon food but the polymorphism of the workers is hardly explicable on this basis alone and must involve differences in genetic constitution about which we are able only to speculate at the present time. In ants the worker caste, which may here also be highly polymorphic, is generally attributed to food and this belief is strengthened by the observed effect of certain nematode and insect parasites that produce changes in queen ants whereby they become structurally more similar to workers. This phenomenon, known as parasitic castration, is well known among other animals and is frequently the cause of profound changes in external form, especially of secondary sexual characters. In the ants it cannot be stated however, that the differentiation of the worker caste is determined entirely by food since its action may prove to be conditioned by genetic factors. As with the termites, this question requires further investigation.

The effect of food on fecundity has been studied in a number of diverse insects, with reference to both larval and imaginal feeding. As large food reserves are built up in the body and as many insects feed but little or not at all after maturity, we should expect the food of the imago to play a minor part. Such appears to be quite generally true and much of this effect may be due to the taking of water. Thus in a species of bean-weevil (*Bruchus*) (see Fig. 14), studied by Larson in which the adult beetle does not regularly eat or drink, access to water increases the production of eggs very considerably although not to the same extent as sugar-water. Similarly Norris found in moths of the genus *Ephestia* that life lengthens with the addition of sugar-water without any increase in fecundity. On the other hand several species of blow-flies (*Lucilia, Calliphora*) studied by Mackerras fail completely to produce eggs unless they receive some protein in their diet after emergence, although males produce normal spermatozoa on a sugar-water ration. The striking contrast between the beetles and flies is due to a difference in the time of maturing of the eggs. The eggs of *Bruchus* are mature on the emergence of the imago and oviposition begins on the first day of life with a large deposit of eggs, while on succeeding days the number falls gradually. The female blow-fly on the other hand does not mature any of her eggs for a week or more after emergence and the food materials that go into the eggs are derived from imaginal and not from larval food. Between these extremes are found cases where adult feeding modifies the fecundity to a greater or less extent, dependent upon the utilization of stored larval food for this purpose. Butterflies and moths never require nitrogenous food after transformation.

Some of the broader ecological aspects of insect abundance and the growth of populations are intimately related to the more specific matters just dealt with. A balance between fecundity and population density commonly obtains among all the lower animals, whereby multiplication is regulated or adjusted with reference to the density of populations. The basic causes for this balance have been partially determined by experimental analysis of populations under controlled conditions. They are not entirely nutritional, even where the food-supply is limited. Studies of certain beetles that infest stored grains and granary products, first made by Chapman ('28) and later by others, including McLagan ('36) have shown that as a population grows its increasing density induces changes in the frequency of copulation, number and fertility of eggs laid and mortality rate during development. Thus a combination of these factors determines the rate of increase in population-density which is finally brought to a standstill. This limitation under such conditions is automatic and the instincts and physiological processes are so constant that the course of events can be predicted by a mathematical formula. This is remarkable, since instinctive behavior enters largely into the picture both with reference to food and to other relations of the beetles to one another and to the environment. Also the conditions of the experiments are widely different from any in nature and the stimuli which the experimental insects receive are radically different from those to which the species have been subjected during their racial history. More or less similar conditions may be produced under natural conditions by the periodic invasions or epidemics of certain insects, but it would be hazardous to apply the data mentioned to such occurrences, as their growth and decline are in many instances obviously more dependent upon food and other environmental factors, including diseases. Fundamentally, food and the instinctive behavior, including reproduction, that has developed with reference to it, both by the insects and the living organisms that prey upon them appear to be the prime determinants of population growth and density. Whether self-limiting factors or internal resistance within populations are the primary forces that regulate population-density of certain insects mentioned in the preceding chapter, like the brine-flies (*Ephydra*) which regularly overrun the shallow waters of Great Salt Lake, or the myriads of water boatmen (*Corixa*) in Mexican lakes cannot be stated. This does not appear probable, although it offers a promising field for investigation.

There are numerous instances particularly among social insects where the development of internal resistance in relation to changes in the environment does not follow the same course. This is due to the acquisition of a more stable food supply and a limiting of the reproductive functions to one or only a few individuals in each colony which thus becomes a sort of colonial organism. The storage of food for future use and the growing of food in the nest, or even in the alimentary system among termites, renders the development of large populations possible and assures their continued prosperity. Such developments simulate in many outward respects the accomplishments of human society in tempering the disastrous effects of a fluctuating food supply. They differ, of course, in the continuity of their operation.

REFERENCES

FOOD, GENERAL

Barfurth, D.

1887   Der Hunger als förderndes Prinzip in der Natur. Arch. mikr. Anat., vol. 29, pp. 28–34.

Bischoff, H.

1927   Biologie der Hymenopteren. vii + 598 pp., 224 figs. Berlin, Julius Springer.

Bowers, R. E., and C. M. McCay

1940   Insect Life Without Vitamin A. Science, vol. 92, p. 291.

Brues, C. T.

1919   The Classification of Insects on the Characters of the Larva and Pupa. Biol. Bull., vol. 37, pp. 1–21.

1921   Insects and Human Welfare. Cambridge, Mass. Harvard Univ. Press, xii + 104 pp., 42 figs.

1923   Choice of Food and Numerical Abundance Among Insects. Journ. Econ. Entom., vol. 16, pp. 46–51.

1930   The Food of Insects Viewed from the Biological and Human Standpoint. Psyche, vol. 37, pp. 1–14.

1939   Food, Drink and Evolution. Science, vol. 90, pp. 145–149.

Chapman, R. N.

1931   Animal Ecology with Special Reference to Insects. 464 pp., 153 figs. New York, McGraw-Hill.

Dyar, H. G., and F. Knab

1906   The Larvae of Culicidae Classified as Independent Organisms. Journ. New York Entom. Soc., vol. 14, pp. 169–230, 78 figs.

Elton, C.

1933   The Ecology of Animals. 97 pp. London, Methuen & Co.

Forbes, S. A.

1880   On Some Interactions of Organisms. Bull. Illinois State Lab. Nat. Hist., vol. 1, No. 3, Second Edition, 1903, pp. 3–18.

Handlirsch, A.

1926   Biologie (Ökologie-Ethologie). Die Nahrung. Handbuch der Entomologie, Ch. Schroder, vol. 2, pp. 35–52.

Hewitt, C. G.

1917   Insect Behaviour as a Factor in Applied Entomology. Journ. Econ. Entom., vol. 10, pp. 81–94.

Hoskin, W. M.

1935   Recent Progress in Insect Physiology. Physiol. Rev., vol. 15, pp. 525–586.

Imms, A. D.

1931   Recent Advances in Entomology. viii+374 pp., 84 figs. London, J. & A. Churchill.

1937   Recent Advances in Entomology. x + 431 pp. Philadelphia, Blaikiston Co. (2nd Edition).

McAtee, W. L.

1928 Biological Species from the Standpoint of the Insect Taxonomist. Proc. Entom. Soc. Washington, vol. 30, p. 32.

Shelford, V. E.

1929 Laboratory and Field Ecology. The Responses of Animals as Indicators of Correct Working Methods. xii + 608 pp., 219 figs.

Snodgrass, R. C.

1925 Anatomy and Physiology of the Honeybee. xvi + 327 pp. McGraw, N. Y.

Uvarov, B. P.

1929 Insect Nutrition and Metabolism. A Summary of the. Literature. Trans. Entom. Soc. London, vol. 76, pp. 255–343.

Wardle, R. A.

1929a The Principles of Applied Zoology. xii + 427 pp., 56 figs. London, Longmans Green & Co.

1929b The Problems of Applied Entomology. xii + 587 pp., 29 figs. Manchester, University Press.

Wardle, R. A., and P. Buckle

1923 The Principles of Insect Control. xvi + 295 pp., 33 figs. London, Longmans Green & Co.

Weber, H.

Biologie der Hemipteren. vii + 543 pp., 329 figs. Springer, Berlin.

Weiss, H. B.

1922 A Summary of the Food Habits of North American Coleoptera. American Naturalist, vol. 56, pp. 159–165, 1 fig.

1925 Notes on the Ratios of Insect Food Habits. Proc. Biol. Soc. Washington, vol. 38, pp. 1–4.

FOOD AND GROWTH, DEVELOPMENT
AND METAMORPHOSIS

Abderhalden, E.

1923 Studien über das Wachstum der Raupen. Zeitschr. physiol. Chemie, vol. 127, pp. 93–98.

Alpatov, W. W.

1929 Growth and Variation of Drosophila Larvae. Journ. Expt. Zool., vol. 52, pp. 407–437.

Ass, M., and Funtikow, G.

1932 Ueber die Biologie und technische Bedeutung der Holzwespen. Zeits. angew. Entom., vol. 19, pp. 557–578. 19 figs.

Baumberger, J. P.

1914 Studies in the Longevity of Insects. Ann. Entom. Soc. America, vol. 7, pp. 323–353, 1 pl.

1917 Hibernation, a Periodical Phenomenon. Ann. Entom. Soc. America, vol. 10, pp. 179–186.

Bodenheimer, F. S.

1933 The Progression Factor in Insect Growth. Quart. Rev. Biol., vol. 8, pp. 92–95, 7 tables.

Bodine, J. H.

Physiological Changes during Hibernation in Certain Orthoptera. Journ. Expt. Zool., vol. 37, pp. 457–476.

Börner, C.

1927 Ueber den Einfluss der Nahrung auf die Entwicklungsdauer von Pflanzenparasiten nach Untersuchungen an der Reblaus. Zeits. angew. Entom., vol. 13, pp. 108–128.

Bottger, G. T.

1940 Preliminary Studies of the Nutritive Requirements of the European Corn Borer. Jour. Agric. Res., vol. 60, pp. 249–257.

Buddington, A. R.

1941 The Nutrition of Mosquito Larvae. Journ. Econ. Entom., vol. 34, pp. 275–281.

Calvert, P. P.

1929 Different Rates of Growth among Animals with Special Reference to the Odonata. Proc. American Philos. Soc., vol. 68, pp. 227–274.

Campbell, W. G.

1941 The Relationship between Nitrogen Metabolism and the Duration of the Larval Stage of the Death-watch Beetle, *Xestobium*. Biochem. Journ., vol. 35, pp. 1200–1208.

Chapman, R. N.

1920 The Life Cycle of the Coleoptera. Ann. Entom. Soc. America, vol. 13, pp. 174–180.

1926 Inhibiting the Process of Metamorphosis in the Confused Flour-Beetle (*Tribolium confusum*). Journ. Expt. Zool., vol. 45, p. 293.

Chiu, S. F., and C. M. McCay

1939 Nutritional Studies of the Confused Flour Beetle (*Tribolium confusum*) and the Bean Weevil (*Acanthoscelides obtectus*). Ann. Entom. Soc. America, vol. 32, pp. 164–170.

Davidson, J.

1927 The Biological and Ecological Aspect of Migration in Aphids. Science Progress, vol. 21, pp. 641–658, 2 diagr.

Dyar, H. G.

1890 The Number of Moults of Lepidopterous Larvae. Psyche, vol. 5, pp. 420–422.

Eidmann, H.

1924 Untersuchungen über Wachstum und Häutung der Insekten. Zeits. Morph. Oekol. d. Tiere, vol. 2, pp. 567–610.

Evans, A. C.

1932 Some Aspects of the Chemical Changes during Insect Metamorphosis. Jour. Exp. Biol., vol. 9, pp. 314–321.

Farkas, G., and H. Tangl

1926 Die Wirkung von Adrenalin und Cholin auf die Entwicklungszeit der Seidenraupen. Biochem. Zeitschr., vol. 172, pp. 350–354.

Ferris, G. F.

1919 A Remarkable Case of Longevity in Insects. Entom. News, vol. 30, pp. 27–28.

Fox, H. M., and G. P. Smith

1933 Growth Stimulation of Blowfly Larvae Fed on Fatigued Frog Muscle. Journ. Expt. Biol., vol. 10, pp. 196–200.

Fracker, S. B.

1920 The Life Cycle of the Lepidoptera. Ann. Entom. Soc. America, vol. 13, pp. 167–173.

Gahan, J. C.

1924 A Long-Lived Longicorn Larva. Trans. Entom. Soc. London, pp. iii–iv.

Gaines, J. C., and F. L. Campbell

1935 Dyar's Rule as Related to the Number of Instars of the Corn Ear Worm, *Heliothis obsoleta*, Collected in the Field. Ann. Entom. Soc. America, vol. 28, pp. 445–461.

Girault, A. A.

1910 Preliminary Studies on the Biology of the Bedbug, *Cimex lectularius*. I. The Effect of Quantitatively Controlled Food Supply on Development. Journ. Econ. Biol., vol. 5, pp. 88–91.

Goe, M. T.

1925 Eight Months' Study of Earwigs (Dermaptera). Entom. News, vol. 36, pp. 234–238.

Gregory, L. H. (Miss)

1917 The Effect of Starvation on the Wing Development of *Microsiphum destructor*. Biol. Bull., vol. 33, pp. 296–303.

Harries, F. H., and C. F. Henderson

1938 Growth of Insects with Reference to Progression Factors for Successive Growth Stages. Ann. Entom. Soc. America, vol. 31, pp. 557–572.

Hinman, E. H.

1933 The Role of Bacteria in the Nutrition of Mosquito Larvae — the Growth Stimulating Factor. American Journ. Hyg., vol. 18, pp. 224–236.

Hobson, R. P.

1935 Growth of Blow-fly Larvae on Blood and Serum, in Association with Bacteria. Biochem. Journ., vol. 29, pp. 1286–1291.

Hodge, C.

1933 Growth and Nutrition of *Melanoplus differentialis*. I. Growth on a Satisfactory Mixed Diet and on Diets of Single Food Plants. Physiol. Zool., vol. 6, pp. 306–328.

Hofman, C.

1933 Der Einfluss von Hunger und engem Lebensraum auf das Wachstum und die Fortpflanzung der Lepidopteren. Zeits. angew. Entom., vol. 20, pp. 51–84.

Jensen-Haarup, A. C.

1908 Ueber die Ursache der Grössenverschiedenheit bei den Coleopteren. Zeits. wiss. Insektenbiol., vol. 4, pp. 100–102.

Kellogg, V. L., and R. G. Bell

1904 Variations Induced in Larval, Pupal, and Imaginal Stages of *Bombyx mori* by Controlled Varying Food-supply. Science, vol. 18, pp. 741–748.

Kojima, T.

1932 Beiträge zur Kenntnis von *Lyctus linearis*. Zeits. angew. Entom., vol. 19, pp. 325–356, 17 figs.

Krizenecky, J.

1914 Ueber die beschleunigende Einwirkung des Hungers auf die Metamorphose. Biol. Centralbl., vol. 34, pp. 46–59.

Lackey, J. B.

1940 The Microscopic Flora of Tree Holes. Ohio Journ. Sci., vol. 40, pp. 186–199.

Lauter, W. M., and V. L. Vrla

1939 Factors Influencing the Formation of the Venom of the Honeybee. Journ. Econ. Entom., vol. 32, pp. 806–807.

Lebedev, A. G., and A. N. Savenkov

1932 Die Nahrungsnormen des Kieferspinners (*Dendrolimus pini*). Zeits. angew. Entom., vol. 19, pp. 85–103.

Lincoln, E.

1940 Growth in *Aeschna tuberculifera*. Proc. American Philos. Soc., vol. 83, pp. 589–605, 5 figs.

Loeb, J., and J. H. Northrop

1917 On the Influence of Food and Temperature Upon the Duration of Life. Journ. Biol. Chem., vol. 32, pp. 103–121.

Ludwig, D.

1934 The Progression Factor in the Growth of the Japanese Beetle (*Popillia japonica*) (Coleop.: Scarabaeidae). Entom. News, vol. 45, pp. 141–153.

Ludwig, D., and W. F. Abercrombie

1940 The Growth of the Head Capsule of the Japanese Beetle Larva. Ann. Entom. Soc. America, vol. 33, pp. 385–390.

Menusan, H., Jr.

1936 The Influence of Constant Temperatures and Humidities on the Rate of Growth and Relative Size of the Bean Weevil, *Bruchus obtectus*. Ann. Entom. Soc. America, vol. 29, pp. 279–288.

Metcalfe, M. E.

1932 Method for Determining the Number of Larval Instars in *Sitodrepa panicea*. Ann. Appl. Biol., vol. 19, pp. 413–419.

Meyer, Eduard

1927 Die Ernährung der Mutterameise und ihrer Brut während der solitären Koloniegrüngung. Biol. Zentralbl., vol. 47, pp. 264–307, 4 figs.

Miles, H. W.

1931 Growth in the Larvae of Tenthredinidae. Journ. Expt. Biol., vol. 8, pp. 355–364.

1933 Observations on Growth in the Larvae of *Plodia interpunctella*. Ann. Appl. Biol., vol. 20, pp. 297–307, 2 figs.

Musconi, Lusia

1925 L'accrescimento delle larve di mosca (*Calliphora eritrocephala*) in rapporto con la temperatura e l'alimentazione. Boll. Lab. Zool. Gen. Agrar., Portici, vol. 18, pp. 95–115, 9 figs.

Nelson, J. A., A. P. Sturtevant, and B. Lineburg

1924 Growth and Feeding of Honeybee Larvae. Bull. U. S. Dept. Agric., No. 1222, 34 pp., 13 figs.

Nicoll, W.

1912 The Length of Life of the Rat Flea Apart from its Host. British Med. Journ., No. 2702, pp. 926–928.

Northrop, J. H.

1917 The Effect of Prolongation of the Period of Growth on the Total Duration of Life. Journ. Biol. Chem., vol. 32, pp. 123–126.

1926 Duration of Life of an Aseptic Drosophila Culture Inbred in the Dark for 230 Generations. Journ. Gener. Physiol., vol. 9, pp. 763–765.

Osborn, T. B., and L. B. Mendel

1914 The Suppression of Growth and the Capacity to Grow. Journ. Biol. Chem., vol. 18, pp. 95–106, 4 figs.

Park, T.

1936 The Oxygen Consumption of the Flour Beetle (*Tribolium confusum*). Journ. Cell. and Compar. Physiol., vol. 7, pp. 313–323.

Passerini, N.

1925 Influenza della qualita degli alimenti sull' accrescimento delle larve e sul metabolismo del *Tenebrio molitor*. Rend. Accad. nazion. Lincei, vol. 1, pp. 58–59.

Pospelov, V.

1911 Die postembryonale Entwicklung und die imaginale Diapause bei den Lepidopteren. Trav. Soc. Natural., Kieff, vol. 21, pp. 163–418.

Przibram, A., and F. Megusar

1912 Wachstumsmessungen an *Sphodromantis bioculata*. Arch. Entwick.-Mech. d. Organismen, vol. 34, pp. 680–741.

Richards, A. G., and A. Miller

1937 Insect Development Analyzed by Experimental Methods; A Review, parts I and II. Journ. New York Entom. Soc., vol. 45, pp. 1–60; 149–210.

Richardson, C. H.

1926 A Physiological Study of the Growth of the Mediterranean Flour Moth (*Ephestia kühniella* Zeller) in Wheat Flour. Journ. Agric. Res., vol. 32, pp. 895–929.

Roques, X.

1912　Recherches biometriques sur l'influence du régime alimentaire chez un insect, *Limnophilus flavicornis*. Assoc. Franç. Avanç. Sci., vol. 40, pp. 566–578.

Rozeboom, L. E.

1934　The Effect of Bacteria on the Hatching of Mosquito Eggs. American Journ. Hyg., vol. 20, pp. 496–501.

1935　The Relation of Bacteria and Bacterial Filtrates to the Development of Mosquito Larvae. American Journ. Hyg., vol. 21, pp. 167–179.

Rubtzov, I. A.

1932　On the Amount of Food Consumed by Locusts. (In Russian). Plant. Prot. 1932, No. 2, pp. 31–40. Abstract in Review Appl. Entom., A, vol. 21, p. 52, 1933.

Ruzicka, V.

1917　Beschleunigung der Häutung durch Hunger. Arch. Entwickl.-Mech., vol. 42, pp. 671–704.

Schneider, B. A.

1943　The Effect of Vitamin B Complex on Metamorphosis, Growth, and Adult Vitality of *Tribolium confusum*. American Journ. Hyg., vol. 37, pp. 179–192, 3 figs.

Severin, H. H. P., and H. C. Severin

1911　The Life History of the Walking Stick, *Diapheromera femorata*. Journ. Econ. Entom., vol. 4, pp. 307–320, 3 figs.

Shull, A. F.

1929　Determination of Types of Individuals in Aphids, Rotifers and Cladocera. Biol. Rev. & Biol. Proc. Cambridge Philos. Soc., vol. 4, pp. 218–248.

Skoblo, J. S.

1935　The Effect of Intermittent Starvation upon the Development of Larvae of *Loxostege sticticalis*. Bull. Entom. Res., vol. 26, pp. 345–354.

Smith, G. P.

1936　The Growth Stimulation of Blow-fly Larvae fed on Fatigued Frog Muscle. Journ. Expt. Biol., vol. 13, pp. 249–252.

Takagi, G.

1933　On the Volume of Leaves Eaten by the Larva of the Pine Moth. (In Japanese). Bull. For. Expt. Sta. Korea, No. 15, pp. 84–96. Abstr. in Rev. Appl. Entom. A., vol. 21, p. 548.

Thomas, H. D.

1943　The Hatching of the Eggs of *Aëdes aegypti* under Sterile Conditions. Journ. Parasit., vol. 29, pp. 324–328.

Thomsen, M., and O. Hammer

1936　The Breeding Media of Some Common Flies. Bull. Entom. Res., vol. 27, pp. 559–587.

Titschak, E.

1924　Untersuchungen über das Wachstum, den Nahrungsverbrauch und die Eierzeugung. I. *Carausius morosus*. Zeitschr. wiss. Zool., vol. 123, pp. 431–487. II. *Tineola biselliella, ibid.*, vol. 128, pp. 509–569.

Trager, W.

1935 The Relation of Cell Size to Growth in Insect Larvae. Journ. Expt. Zool., vol. 71, pp. 489–508.

1935 The Culture of Mosquito Larvae Free from Living Microörganisms. American Journ. Hyg., vol. 22, pp. 18–25.

1936 The Utilization of Solutes by Mosquito Larvae. Biol. Bull., vol. 71, pp. 343–352.

Wheeler, W. M.

1908 Honey Ants, with a Revision of the American Myrmecocysti. Bull. American Mus. Nat. Hist., vol. 24, pp. 345–397, 28 figs.

Wilson, S. E.

1933. Changes in the Cell Contents of Wood (Xylem Parenchyma) and their Relationships to the Respiration of Wood and its Resistance to Lyctus Attack and to Fungal Invasion. Ann. Appl. Biol., vol. 20, pp. 661–690, 11 figs.

Wodsedalek, J. E.

1912 Life History and Habits of *Trogoderma tarsale*. Ann. Entom. Soc. America, vol. 5, pp. 367–381.

1917 Five Years of Starvation of Larvae. Science, vol. 46, pp. 366–367.

Wolcott, G. N.

1925 On the Amount of Food Eaten by Insects. Journ. Dept. Agric. Porto Rico, vol, 9, pp. 47–58.

Woodruff, L. C.

1938 The Normal Growth Rate of *Blatella germanica*. Journ. Expt. Zool., vol. 79, pp. 145–165.

1939 Linear Growth Ratios for *Blatella germanica*. *Ibid.*, vol. 81, pp. 287–298.

MOUTHPARTS AND OTHER STRUCTURES
ASSOCIATED WITH FOOD

NOTE: Other references may be found at the end of Chapter VI.

Adler, S., and O. Theodor

1926 The Mouth-parts, Alimentary Tract, and Salivary Apparatus of the Female *Phlebotomus papatasii*. Ann. Trop. Med. Parasitol., vol. 20, pp. 109–142, pls. 8–14, 3 figs.

Awati, P. R.

1914 The Mechanism of Suction in the Potato Capsid Bug, *Lygus pabulinus*. Proc. Zool. Soc. London, 1914, pp. 687–733, 29 figs.

Bertin, Léon

1923 L'adaptation des pièces buccales aux régimes alimentaires chez les Coléoptères Lamellicornes. Ann. Soc. Linn. Lyon, n.s., vol. 69, pp. 145–159, 8 figs.

Bugnion, E.

1924 Les Organes buccaux de la Scolie. Mitt. Schweiz. Entom. Gesellsch., vol. 13, pp. 368–396.

Burt, E. T.

1941 A Filter-feeding Mechanism in a Larva of the Chironomidae. Proc. R. Entom. Soc. London, vol. A 15, pp. 113–121.

Chatin, J.

1884 Morphologie comparée des pièces maxillaires, mandibulaires et labiales chez les insectes broyeurs. Paris.

Corset, J.

1931 Les Coaptions Chez les Insectes. Bull. Biol. France et Belgique, 1931, Suppl. 13, pp. 1–337, 2 pls., 182 figs.

Cummings, B. F.

1913 On Some Points in the Anatomy of the Mouthparts of Mallophaga. Proc. Zool. Soc. London, 1913, pp. 128–141, 9 figs.

Das, G. M.

1937 The Musculature of Mouth Parts of Insect Larvae. Quart. Journ. Micros. Sci., vol. 80, pp. 39–80, 12 pls.

Davidson, J.

1914 On the Mouth-parts and Mechanism of Suction in *Schizoneura lanigera*. Journ. Linn. Soc. London, Zool., vol. 32, pp. 307–330, 2 pls., 2 figs.

Demoll, R.

1909 Die Mundteile der Wespen, Tenthrediniden und Uroceriden. Zeitschr. wiss. Zool., vol. 92, pp. 187–209.

Dimmock, Geo.

1880 The Trophi and their Chitinous Supports in Gracilaria. Psyche, vol. 3, pp. 99–103.

Gennerich, J.

1922 Morphologische und biologische Untersuchungen der Putzapparate der Hymenopteren. Arch. f. Naturgesch., Jahrg., 88, A, Heft 12, pp. 1–64, 65 figs.

Genthe, K. W.

1897 Die Mundwerkzeuge der Mikrolepidopteren. Zool. Jahrb. Abth. f. Syst., vol. 10, pp. 373–471, 3 pls.

Golden, H. M.

1926 Die kauenden Insektenmundteile und ihre Beziehung zur Nahrung. Arch. Naturg. 91, A, Heft 7, pp. 1–47, 7 pls.

Graham-Smith, G. S.

1930 Further Observations on the Anatomy and Function of the Proboscis of the Blow-fly. Parasitology, vol. 22, pp. 47–115.

Hoke, G.

1924 The Anatomy of the Head and Mouth parts of Plecoptera. Journ. Morph., vol. 38, pp. 347–385, 6 pls.

Keilin, D.

1912 Structure du Pharynx en fonction du régime chez les larves de Diptères cyclorrhaphes. C. R. Acad. Sci. Paris, 1912, p. 1550.

1916 Sur la viviparité des Diptères et les larves des Diptères vivipares. Arch. Zool. Expér., vol. 55, pp. 393–415, 8 figs.

Jobling, B.

1926 A Comparative Study of the Structure of The Head and Mouth Parts in the Hippoboscidae. Parasitology, vol. 18, pp. 319–349, 6 pls., 4 figs.

PLATE V

A. A tropical American cockroach, typical of a large fauna of these insects that are abundant in the rain-forests of both hemispheres. Original photograph.

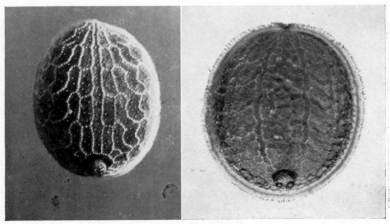

B. Larvae of two species of syrphid flies (*Microdon*). These feed as scavengers in the nests of various ants. Original photographs.

PLATE VI

A. "Alder blight," an aphid (*Prociphilus tessellatus*) that feeds on the bark of the black alder. Original photograph.

B. Group of scale insects (*Lepidosaphes ulmi*) attached to a twig of aspen. Original photograph.

1928 The Structure and Mouth Parts in *Culicoides pulicaria*. Bull. Entom. Res., vol. 18, pp. 211–236, 4 pls., 10 figs.

Krafchick, B.

1942 The Mouth parts of Blackflies with special reference to *Eusimulium lascivum*. Ann. Entom. Soc. America, vol. 35, pp. 426–434, 2 pls.

Lindner, Hugo

1918 Ueber die Mundwerkzeuge einiger Dipteren und ihre Beziehungen zur Ernährungsweise. Zool. Anz., vol. 50, pp. 19–27, 1 pl., 14 figs.

Lucas, R.

1893 Beiträge zur Kenntniss der Mundwerkzeuge der Trichoptera. Arch. Naturg., vol. 59, pp. 285–330, 2 pls.

Meek, W. J.

1903 On the Mouthparts of the Hemiptera. Bull. Kansas Univ., Sci., vol. ii, pp. 257–277, 5 pls.

Miller, F. W.

1932 Mechanical Factors Affecting the Feeding Habits of Two Species of Aphids, *Macrosiphum ambrosiae* and *Macrosiphum granarium*. Journ. Econ. Entom., vol. 25, pp. 1203–1206.

Mitter, J. L.

1918 Note on the Method of Feeding of *Corizoneura* (*Pangonia*) *longirostris* with a Description of the Mouthparts. Indian Journ. Med. Res. Calcutta, vol. 5, pp. 523–528, 1 pl.

Muir, F., and J. C. Kershaw

1911 On the Homologies and Mechanism of the Mouthparts of Hemiptera. Psyche, vol. 18, pp. 1–12, 5 pls.

Otanes, F. Q.

1922 Head and Mouth-parts of Mecoptera. Ann. Entom. Soc. America, vol. 15, pp. 310–327, 4 pls.

Parfent'ev, I. A.

1925 Aphagia in Insects. (In Russian). Rev. Zool. Russe, Moscow, vol. 4, pp. 234–261, 17 figs.

Peacock, A. D.

1918 The Structure of the Mouth-parts and Mechanism of Feeding in *Pediculus humanus*. Parasitology, vol. 11, pp. 98–117.

Peterson, A.

1915 Morphological Studies on the Head and Mouthparts of the Thysanoptera. Ann. Entom. Soc. America, vol. 8, pp. 20–59.

Plateau, F.

1885 Expériences sur le rôle des palpes chez les Arthropodes maxillés. I. Palpes des Insectes broyeurs. Bull. Soc. Zool. France, vol. 10, pp. 67–90.

Robinson, G. G.

1939 The Mouth parts and their Function in the Female Mosquito, *Anopheles maculipennis*. Parasitology, vol. 31, pp. 212–242, 9 figs.

Robinson, V. E.

1930 The Mouthparts of the Larval and Adult Stages of *Dermestes vulpinus*. Ann. Entom. Soc. America, vol. 23, pp. 399–416, 2 pls.

Silvestri, F.

1904 Contribuzione alla conoscenza della metamorfosi e dei costumi della *Lebia scapularis*. Redia, vol. 2, pp. 68–84, 4 pls.

Snodgrass, R. E.

1905 A Revision of the Mouth-parts of the Corrodentia and Mallophaga. Trans. American Entom. Soc., vol. 31, pp. 297–306, 1 pl.

1910 The Anatomy of the Honey Bee. Bull. U. S. Dept. Agric., Tech. Ser., No. 18, pp. 1–162, 57 figs., 2nd Ed. 1925, McGraw-Hill, New York.

1921 The Mouthparts of the Cicada. Proc. Entom. Soc. Washington, vol. 23, pp. 1–15, 2 pls.

1932 Morphology of the Insect Head and the Organs of Feeding. Smithsonian Inst. Rept. for 1931, pp. 443–489, 25 figs.

1943 The Feeding Apparatus of Biting and Disease-carrying Flies. Smithsonian Misc. Coll., vol. 104, No. 1, 51 pp., 18 figs.

Ulrich, W.

1924 Die Mundwerkzeuge der Sphecinen. Zeits. Morph. Oekol. Tiere, vol. 1, pp. 539–636, 113 figs.

Wheeler, W. M.

1928 A Study of Some Ant Larvae, with a Consideration of the Origin and Meaning of the Social Habit among Insects. Proc. American Philos. Soc., vol. 57, pp. 293–343.

Whitfield, F. G. S.

1925 The Relation between the Feeding-habits and the Structure of the Mouthparts in the Asilidae. (Diptera). Proc. Zool. Soc. London, 1925, pp. 599–638, 2 pls., 15 figs.

ALIMENTARY TRACT

Becton, E. M., Jr.

1930 The Alimentary Tract of *Phanaeus vindex*. Ohio Journ. Sci., vol. 30, pp. 315–324.

Bodenheimer, Fritz

1923 Beiträge zur Kenntnis der Kohlschnake (*Tipula oleracea*). Zur Anatomie und Oekologie der Imago. Zeits. wiss. Zool., vol. 121, pp. 393–441, 25 figs.

Bordas, L.

1911 L'appareil digestif et les tubes de Malpighi des larves de Lépidoptères. Ann. Sci. Nat. (Zool.), vol. 14, pp. 103–273, 3 pls., 32 figs.

1917 Nouvelles observations sur l'appareil digestif des Cetoninae. Bull. Soc. Zool. France, vol. 42, pp. 7–12.

Cameron, A. E.

1912 Structure of the Alimentary Canal of the Stick-insect *Bacillus rossii*; with a Note on the Parthenogenesis of This Species. Proc. Zool. Soc. London, 1912, pp. 172–178, 3 pls.

Cox, J. A.

1938   Morphology of the Digestive Tract of the Black-fly, *Simulium nigroparvum.*
Journ. Agric. Res., vol. 57, pp. 443–448, 3 pls.

Dean, R. W.

1932   The Alimentary Canal of the Apple Maggot (*Rhagoletis pomonella*).   Ann.
Entom. Soc. America, vol. 25, pp. 210–233, 4 pls.

1933   Morphology of the Digestive Tract of the Apple Maggot Fly (*Rhagoletis
pomonella* Walsh).   Tech. Bull. New York State Agric. Expt. Sta., No. 215, 17
pp., 17 figs.

Fletcher, F. W.

1930   The Alimentary Canal of *Phyllophaga gracilis*.   Ohio Journ. Sci., vol. 30, pp.
109–119.

Frenzel, J.

1882   Ueber Bau und Tätigkeit des Verdauungskanals der Larve des *Tenebrio molitor* mit Bercüksichtigung anderer Arthropoden.   Berlin. Entom. Zeits., vol.
26, pp. 257–316.

Fritze, A.

1888   Ueber den Darmcanal der Ephemeriden.   Ber. Ges. Freiburg, vol. 4, pp. 5–24.

Guyénot, E.

1907   L'appareil digestive et la digestion de quelques larves de mouches.   Bull. Sci.
France et Belgique, vol. 41, pp. 353–370.

Haseman, L.

1930   The Hessian Fly Larva and its Method of Taking Food.   Journ. Econ.
Entom., vol. 23, pp. 316–321, 2 pls., 1 fig.

James, H. C.

1926   The Anatomy of a British Phytophagous Chalcidoid of the Genus *Harmolita*
(*Isosoma*).   Proc. Zool. Soc. London, 1926, pp. 75–182, 68 figs.

Keilin, D.

1919   On the Alimentary Canal and its Appendages in the Larvae of Scatopsidae and
Bibionidae, with some Remarks on the Parasites of these Larvae.   Entom.
Monthly Mag., (3) vol. 5, pp. 92–96, 2 figs.

Keilin, D., and C. Picado

1920   Biologie et Morphologie larvaires d'*Anastrepha striata*, mouche des fruits de
l'Amérique centrale.   Bull. Sci. France et Belgique, vol. 48, pp. 423–441, 6
figs.

Knowlton, G. F.

1925   The Digestive Tract of *Longistigma caryae*.   Ohio Journ. Sci., vol. 25, pp. 244–
249, 1 pl., 2 figs.

Könnemann, R.

1924   Ueber den Darm einiger Limnobiidenlarven.   Zool. Jahrb. Abth. Anat., vol.
46, pp. 343–388.

Kumm, H. W.

1935   The Digestive Mechanism of one of the West Indian "Eye Gnats," *Hippelates
pallipes*.   Ann. Trop. Med. and Parasitol., vol. 29, pp. 283–302, 2 pls., 3 figs.

Lewis, H. C.

1926   The Alimentary Canal of *Passalus*. Ohio Journ. Sci., vol. 26, pp. 11–29.

Metcalfe, M. E.

1933   The Morphology and Anatomy of the Larva of *Dasyneura leguminicola*. Proc. Zool. Soc. London, 1933, pp. 119–130, 6 pls.

Miall, L. C., and A. Denny

1886   The Structure and Life-History of the Cockroach.   vi + 224 pp., 125 figs. London.

Misra, A. B.

1931   On the Internal Anatomy of the Female Lac Insect.   Proc. Zool. Soc. London, 1931, pp. 297–323, 10 pls.   Anatomy of male, *ibid.*, 1931, pp. 1359–1381, 11 figs.

Neiswander, R. B.

1935   The Alimentary Canal of the Oriental Fruit Moth Larva.   Ohio Journ. Sci., vol. 35, pp. 434–437, 2 pls.

Pelton, J. Z.

1938   The Alimentary Canal of the Aphid, *Prociphilus tesselatus*. Ohio Journ. Sci., vol. 38, pp. 164–169, 10 figs.

Peterson, A.

1912   Anatomy of the Tomato-Worm Larva, *Protoparce carolina*.   Ann. Entom. Soc. America, vol. 5, pp. 246–269, 3 pls.

Snodgrass, R. E.

1933   The History of an Insect's Stomach.   Smithsonian Inst., Ann. Rept. 1932, pp. 363–387, 15 figs.

Steel, A.

1931   On the Structure of the Immature Stages of the Frit Fly (*Oscinis frit*).   Ann. Appl. Biol., vol. 18, pp. 352–369.

Sternfeld, A. F.

1907   Die Verkümmerung der Mundteile und der Funktionswechsel des Darms bei den Ephemeriden.   Zool. Jahrb., Abth. f. Anat., vol. 24, pp. 415–430.

Woods, W. C.

1918   The Alimentary Canal of the Larva of *Altica bimarginata*.   Ann. Entom. Soc. America, vol. 11, pp. 283–313, 4 pls.

DIGESTION

Abbott, R. L.

1926   Contributions to the Physiology of Digestion in the Australian Roach, *Periplaneta australasiae*.   Journ. Expt. Zool., vol. 44, pp. 219–253, 2 pls.

Babers, F. H., and P. A. Woke

1937   Digestive Enzymes in the Southern Armyworm.   Journ. Agric. Res., vol. 54, pp. 547–550.

Bertholf, L. M.

1927 The Utilization of Carbohydrates as Food by Honeybee Larvae. Journ. Agric. Res., vol. 35, pp. 429–452.

Biedermann, W.

1911 Die Aufnahme, Verarbeitung und Assimilation der Nahrung. Neunter Teil. Die Ernahrung der Insekten. Winterstein, Handb. vergl. Physiol., vol. 2 (1), pp. 726–902.

1919 Beiträge zur vergleichenden Physiologie der Verdauung. VIII. Die Verdauung pflanzlichen Zellinhalts im Darm einiger Insekten. Arch. ges. Physiol., vol. 174, pp. 392–425.

Campbell, F. L.

1927 Notes on Silkworm Nutrition. Journ. Econ. Entom., vol. 20, pp. 88–90.

Cook, S. F., and K. G. Scott

1933 The Nutritional Requirement of *Zootermopsis angusticollis*. Journ. Cell. Comp. Physiol., vol. 3, pp. 223–246.

Craig, R., and W. M. Hoskin

1940 Insect Biochemistry. Ann. Rev. Biochem., vol. 9, pp. 617–640.

Crowell, H. H.

1943 Feeding Habits of the Southern Army-worm and Rate of Passage of Food through the Gut. Ann. Entom. Soc. America, vol. 36, pp. 243–249.

Dewitz, J.

1915 The Poisons of Plant-lice. Ann. Entom. Soc. America, vol. 8, pp. 343–346.

Esterly, C. O.

1916 The Feeding Habits and Food of Pelagic Copepods and the Question of Nutrition by Organic Substances in Solution in the Water. Univ. California Public. Zool., vol. 16, pp. 171–184, 2 figs.

Falck, R.

1931 Scheindestruktion des Holzes durch die Larven von Anobium. Celluloschemie, vol. 2, pp. 128–129.

Fink, D. E.

1932 The Digestive Enzymes of the Colorado Potato Beetle and the Influence of Arsenicals on their Activity. Journ. Agric. Res., vol. 45, pp. 471–482.

Hargitt, C. W.

1923 The Digestive System of the Periodical Cicada. II. Physiology of the Adult Insect. Biol. Bull., vol. 45, pp. 200–212.

Haydak, M. H.

1936 Is Wax a Necessary Constituent of the Diet of Wax Moth Larvae? Ann. Entom. Soc. America, vol. 29, pp. 581–588.

1937 The Influence of a Pure Carbohydrate Diet on Newly Emerged Honeybees. Ann. Entom. Soc. America, vol. 30, pp. 258–262.

1941 Requirement of Wax Moth Larvae for Vitamin B1. Proc. Minnesota Acad. Sci., vol. 9, pp. 27–29.

Hinman, E. H.

1932a The Utilization of Water Colloids and Material in Solution by Aquatic Animals, with Special Reference to Mosquito Larvae. Quart. Rev. Biol., vol. 6, pp. 210–217.

1932b The Role of Solutes and Colloids in the Nutrition of Anopheline Larvae. American Journ. Trop. Med., vol. 12, pp. 263–271.

Hitchcock, F. A., and J. G. Haub

1941 The Interconversion of Foodstuffs in the Blow-fly (*Phormia regina*) during Metamorphosis. Ann. Entom. Soc. America, II, vol. 34, pp. 17–25; 32–37.

Hobson, R. P.

1932 Studies on the Nutrition of Blow-fly Larvae. I. Structure and Function of the Alimentary Tract, pp. 109–123. II. Role of the Intestinal Flora in Digestion, pp. 128–138. III. The Liquefaction of Muscle, pp. 359–365. IV. The Normal Role of Microörganisms in Larval Growth, pp. 366–377. Journ. Expt. Biol., vols. 8 and 9.

1935 On a Fat-soluble Growth Factor Required by Blow-fly Larvae. Biochem. Journ., vol. 29, pp. 1292–1296.

Kawase, S., and K. Saito

1921 The Digestive Enzymes of the Silkworm. Journ. Chem. Soc. Japan, vol. 42, pp. 103–117.

Lengerken, H. v.

1924 Extraintestinale Verdauung. Biol. Zentralbl., vol. 44, pp. 273–293, 8 figs.

Mackerras, M. J., and M. R. Freney

1933 Observations on the Nutrition of Maggots of Australian Blow-flies. Journ. Expt. Biol., vol. 10, pp. 237–246.

Martin, H. E., and L. Hare

1942 Nutritive Requirements of *Tenebrio molitor* Larvae. Biol. Bull., vol. 83, pp. 428–437.

Melvin, R., and R. C. Bushland

1940 The Nutritional Requirements of Screwworm Larvae. Journ. Econ. Entom., vol. 33, pp. 850–852.

Painter, R. H.

1928 Notes on the Injury to Plant Cells by Chinch Bug Feeding. Ann. Entom. Soc. America, vol. 21, pp. 232–242, 1 pl.

Parkin, E. A.

1940 The Digestive Enzymes of Some Wood-boring Beetles. Journ. Expt. Biol., vol. 17, pp. 364–377.

Phillips, E. F.

1927 The Utilisation of Carbohydrates by Honeybees. Journ. Agric. Res., vol. 35, pp. 385–428.

Picado, C.

1935 Sur le princip bactéricide des larves des mouches. Bull. Biol. France et Belgique, vol. 69, pp. 409–438.

Plateau, M. F.

1874 Recherches sur les phénomènes de la digestion chez les insects. Mém. Acad. Roy. Sci. Belgique, vol. 41.

1876 Note sur les phénomènes de la digestion chez la Blatte américaine (*Periplaneta americana*). Bull. Acad. Sci. Belgique, vol. 41, pp. 1206–1233.

Portier, P.

1910 Recherches physiologiques sur les Insectes aquatiques. II. Digestion des larves de Dytiques, d'Hystrobius et d'Hydrophile. C. R. Soc. Biol. Paris, vol. 66, pp. 379–382.

Roy, D. N.

1937 The Physiology of Digestion in Larvae of *Gastrophilus equi*. Parasitology, vol. 29, pp. 150–162.

Sanford, E. W.

1918 Experiments on the Physiology of Digestion in the Blattidae. Journ. Expt. Zool., vol. 25, pp. 355–411, 21 figs.

Sarin, E.

1923 Ueber Fermente der Verdauungsorgane der Honigbiene. Biochem. Zeitschr., vol. 135, pp. 59–74.

Snipes, B. T., and O. E. Tauber

1937 Time Required for Food Passage through the Alimentary Tract of the Cockroach, *Periplaneta americana*. Ann. Entom. Soc. America, vol. 30, pp. 277–284.

Swingle, H. S.

1925 Digestive Enzymes of an Insect. Ohio Journ. Sci., vol. 25, pp. 209–218.

1928 Digestive Enzymes of the Oriental Fruit Moth. Ann. Entom. Soc. America, vol. 21, pp. 469–475.

1930 Anatomy and Physiology of the Digestive Tract of the Japanese Beetle. Journ. Agric. Res., vol. 41, pp. 181–196, 4 figs.

Trager, W.

1935 On the Nutritive Requirements of Mosquito Larvae (*Aëdes aegypti*). American Journ. Hyg., vol. 22, pp. 475–493.

Waksman, S. A.

1940 The Microbiology of Cellulose Decomposition and Some Economic Problems Involved. Bot. Rev., vol. 6, pp. 637–665.

Wigglesworth, V. B.

1927 Digestion in the Cockroach. (H-ion Concentration in the Alimentary Canal). Biochem. Journ., vol. 21, pp. 791–796, 797–811; vol. 22, pp. 150–161.

Yeager, J. F.

1931 Observations on Crop and Gizzard Movements in the Cockroach. Ann. Entom. Soc. America, vol. 24, pp. 739–750.

Young, C. M.

1938 Recent Work on the Digestion of Cellulose and Chitin by Invertebrates. Sci. Prog., vol. 32.

WATER CONSERVATION

Babcock, S. M.

  1912 Metabolic Water: Its Production and Role in Vital Phenomena. Ann. Rept. Wisconsin Agric. Expt. Sta., vol. 29, pp. 87–181.

Ewart, A. J.

  1907 A Contribution to the Physiology of the Museum Beetle *Anthrenus musaeorum*. Journ. Linn. Soc. Zool., vol. 30, pp. 1–5.

Henkel, J. S., and A. W. Bayer

  1932 The Wattle Bagworm, *Acanthopsyche junodi*. An Ecological Study. South African Journ. Sci., vol. 29, pp. 355–365.

Hickin, N. E.

        The Food and Water Requirement of *Ptinus tectus*. Proc. R. Entom. Soc. London, vol. 17A, pp. 99–108.

Robinson, W.

  1938a Response and Adaptation of Insects to External Stimuli. Ann. Entom. Soc. America, vol. 21, pp. 407–417.

  1938b Water Conservation in Insects. Journ. Econ. Entom., vol. 21, pp. 897–902.

Schulz, F. N.

  1930 Zur Biologie des Mehlworms: der Wasseraushalt. Biochem. Zeits., vol. 227, pp. 340–357.

Speicher, B. R.

  1931 The Effects of Desiccation Upon the Growth and Development of the Mediterranean Flour-moth. Proc. Pennsylvania Acad. Sci., vol. 5, pp. 79–82.

Tower, W. L.

  1917 Inheritable Modification of the Water Relation in Hibernation of *Leptinotarsa decemlineata*. Biol. Bull., vol. 33, pp. 229–257.

AQUATIC ADAPTATIONS

NOTE: Any adequate list of references to this subject would be too long to include here and has necessarily been omitted.

INSECTS IN THE SEA AND OTHER
SALINE ENVIRONMENTS

Aldrich, J. M.

  1912 The Biology of Some Western Species of the Dipterous Genus *Ephydra*. Journ. New York Entom. Soc., vol. 20, pp. 77–103, 3 pls.

  1918 The Kelp-flies of North America. Proc. California Acad. Sci., vol. 8, pp. 157–179, 10 figs.

Baudrimont, A.

  1921 Coléoptères et chasse à la marée. Proc. verb. Soc. linn. Bordeaux, vol. 73, pp. 36–38.

Brues, C. T.

1928 Studies on the Fauna of Hot Springs in the Western United States and the Biology of Thermophilous Animals. Proc. American Acad. Arts Sci., vol. 63, pp. 139–228.

1932 Further Studies on the Fauna of North American Hot Springs. *Ibid.*, vol. 67, pp. 185–303, 8 figs.

Buxton, W. A.

1926 The Colonisation of the Sea by Insects. Proc. Zool. Soc. London, vol. 51, pp. 807–813.

Chevrel, R.

1903 *Scopelodromus isemerinus*, genre nouveau et espèce nouvelle de Diptère marin. Arch. Zool. Expér., vol. 1, pp. 1–29, 1 pl.

Coutière, H., and J. Martin

1901 Sur une nouvelle sous-famille d'Hémiptères marins, les Hermatobatinae. C. R. Acad. Sci. Paris, vol. 132, pp. 1066–1068.

Edwards, F. W.

1926 On Marine Chironomidae. Proc. Zool. Soc. London, vol. 51, pp. 779–806.

Griffiths, T. H. D.

1921 Anopheles and Sea Water, with Observations on the Development of American Species. U. S. Pub. Health Repts., vol. 36, pp. 990–1000.

Henkel, J. S.

1931 Attacking the Bagworm with Salt. The Natal Witness, December 24, 1931.

King, W. V., and F. del Rosario

1935 The Breeding of Anopheles in Salt Water Ponds. Philippine Journ. Sci., vol. 57, pp. 329–349, 7 pls., 2 figs.

Moniez, R.

1890 Acariens et Insectes marins des côtes de Boulonnais. Rev. Biol. Nord France, vol. 2, pp. 321 ff.

Packard, A. S.

1873 Insects Inhabiting Great Salt Lake and Other Saline or Alkaline Lakes in the West. 6th Ann. Rept. U. S. Geol. Surv., pp. 743–746.

Plateau, F.

1890 Les Myriapodes marins. Journ. Anat. et Physiol., vol. 26, pp. 236–239.

Satterthwait, A. F.

1933 Life History and Distribution of the Low Tide Billbug, *Calendra setiger*. Journ. Econ. Entom., vol. 26, pp. 210–217.

Saunders, L. G.

1928 Some Marine Insects of the Pacific Coast of Canada. Ann. Entom. Soc. America, vol. 21, pp. 521–545.

Terry, F. W.

1912 On a New Genus of Hawaiian Chironomidae. Proc. Hawaiian Entom. Soc., vol. 2, pp. 88–89.

Thienemann, A.

1915   Zur Kenntnis der Salzwasserchironomiden.   Arch. Hydrobiol., vol. 2, suppl., pp. 443–471.

Vayssière, P.

1933   Une cochenille halophile en Tunisie.   Bull. Soc. Entom. France, vol. 38, pp. 57–59, 2 figs.

White, F. B.

1883   Report on the Pelagic Hemiptera.   Rept. Sci. Res. Voy. H.M.S. Challenger 1873–1876, Zool., vol. 7, 82 pp., 3 pls.

Wigglesworth, V. B.

1933   The Adaptation of Mosquito Larvae to Salt Water.   Journ. Exper. Biol., vol. 10, pp. 27–37, 2 figs.

Woodhill, A. R.

1943   Oviposition of Mosquitoes in Relation to the Salinity and Temperature of the Water.   Proc. Linn. Soc. New South Wales, vol. 66, pp. 287–292; 396–400.

SYNTHETIC FOODS

NOTE: Other references will be found in the list of literature appended to Chapter V.

Bottger, G. T.

1942   Synthetic Food Media for use in Nutrition Studies of the European Corn Borer.   Journ. Agric. Res., vol. 65, pp. 493–500.

Brown, J. H.

1934   Pure Cultures of Paramecium.   Science, vol. 80, pp. 409–410.

Delcourt, A., and E. Guyénot

1910   De la possibilité d'étudier certains Diptères en milieu défini.   C. R. Acad. Sci. Paris, vol. 151, pp. 255–257.

Eyer, J. R.

1921   Rearing Anthomyid Root Maggots on Artificial Media.   Entom. News, vol. 32, pp. 123–133.

Frost, F. M., W. B. Herms, and W. M. Hoskins

1936   Nutritional Requirements of the Larva of the Mosquito, *Theobaldia incidens*.   Journ. Expt. Zool., vol. 73, pp. 461–479.

Fulton, R. A., and J. C. Chamberlin

1934   An Improved Technique for the Artificial Feeding of the Beet Leafhopper, with Notes on its Ability to Synthesize Glycerides.   Science, vol. 79, pp. 346–348.

Harris, R. G.

1923   Sur la culture des larves de Cécidomyies paedogénétiques (*Miastor*) en milieu artificiel.   C. R. Soc. Biol., Paris, vol. 88, pp. 256–258.

Haydak, M. H.

1937   Further Contribution to the Study of Pollen Substitutes.   Journ. Econ. Entom., vol. 30, pp. 637–642.

1939 Comparative Value of Pollen and Pollen Substitutes. Journ. Econ. Entom., vol. 32, pp. 663–665.

Hobson, R. P.

1933– Growth of Blowfly Larvae on Blood and Serum. Biochem. Journ., vol. 27, pp.
1935 1899–1909 (1933); vol. 29, pp. 1286–1291 (1935).

Holloway, J. K.

1939 An Agar Preparation for Feeding Adult Parasitic Insects. Journ. Econ. Entom., vol. 32, p. 154.

Ju-Ch'i Li

1931 A New Food for Laboratory Cultures of *Drosophila*. Peking Nat. Hist. Bull., vol. 5, pp. 29–31.

Lennox, F. G.

1939 A Synthetic Medium for Aseptic Cultivation of Larvae of *Lucilia cuprina*. Res. Pamphlet, Australian Counc. Sci. Ind., No. 90, 24 pp., 1 pl., 8 figs.

Loeb, J.

1915 The Simplest Constituents Required for Growth and the Completion of the Life Cycle in an Insect (*Drosophila*). Science, vol. 41, pp. 169–170.

Loeb, J., and J. H. Northrop

1916 Nutrition and Evolution. Second note. Journ. Biol. Chem., vol. 27, pp. 309–312.

Ludwig, D., and H. Fox

1938 Growth and Survival of Japanese Beetle Larvae Reared in Different Media. Ann. Entom. Soc. America, vol. 21, pp. 445–456.

Michelbacher, A. E., W. M. Hoskins, and W. B. Herms

1932 The Nutrition of Flesh Fly Larvae, *Lucilia sericata*. 1. The Adequacy of Sterile, Synthetic Diets. Journ. Expt. Zool., vol. 64, pp. 109–128.

Pearl, R.

1926 A Synthetic Food Medium for the Cultivation of *Drosophila*. Preliminary note. Journ. Gen. Physiol., vol. 9, pp. 513–519.

Richardson, C. H.

1933 An Efficient Medium for Rearing Houseflies Throughout the Year. Science, vol. 76, pp. 350–351.

Ripley, L. B., G. A. Hepburn, and J. Dick

1939 Mass Breeding of False Codling Moth, *Argyroploce leucotreta* in Artificial Media. Sci. Bull., Dept. Agric. South Africa, No. 207, 8 pp.

Robinson, W.

1934 Improved Methods in the Culture of Sterile Maggots for Surgical Use. Journ. Lab. and Clin. Med., vol. 20, pp. 77–85.

Zabinski, J.

1928 Élévages des Blattides soumis à une alimentation artificielle. C. R. Soc. Biol. Paris, vol. 98, pp. 73–80.

FOOD AND COLORATION

Abeloos, M., and K. Toumanoff

1926 Sur la présence de "carotin-albumines" chez *Carausius morosus*. Bull. Soc. Zool. France, vol. 51, pp. 281–285.

Biedermann, W.

1914 Farbe und Zeichnung der Insekten. Winterstein, Handb. vergl. Physiol., vol. 3 (1), pp. 1657–1922.

Brecher, L.

1917 Die Puppenfärbungen des Kohlweisslings, *Pieris brassicae*. Arch. f. Entwicklungsmech., vol. 43, pp. 88–221, 5 pls.

Carlier, E. W., and C. L. Evans

1911 Note on the Chemical Composition of the Red-coloured Secretion of *Timarcha tenebricosa*. I$^{er.}$ Congr. Intern. Entom., Bruxelles, vol. 1, pp. 137–142.

Comas, M.

1927 Sur l'origine des pigments des larves de *Chironomus*. C. R. Soc. Biol., Paris, vol. 96, pp. 866–868.

Dubois, R.

1890 Sur les propriétés des principes colorants naturels de la soie jaune et sur leur analogie de la carotine végétale. C. R. Acad. Sci. Paris, vol. 111, pp. 482–483.

Garrett, F. C., and J. W. Heslop-Harrison

1923 Melanism in the Lepidoptera and its Possible Induction. Nature, vol. 112, pp. 240–241.

Gerould, J. H.

1921 Blue-green Caterpillars: The Origin and Ecology of a Mutation in Hemolymph Color in *Colias philodice*. Journ. Exper. Zool., vol. 34, pp. 385–415, 1 pl., 1 fig.

1927 Studies in the General Physiology and Genetics of Butterflies. Quart. Rev. Biol., vol. 2, pp. 58–78, 3 figs.

Glaser, R. W.

1917 Anthocyanin in *Pterocomma smithiae*. Psyche, vol. 24, p. 30.

Hollande, A. C.

1913 Coloration vitale du corps adipeux d'un insecte phytophage par une anthocyane absorbée avec la nourriture. Arch. Zool. Expér. Génér., vol. 51, Notes et Revue, No. 2, pp. 53–58.

Hopkins, F. G.

1896 The Pigments of the Pieridae: A Contribution to the Study of Excretory Substances which Function in Ornament. Philos. Trans. R. Soc., London, vol. 186, pp. 661–662.

Husain, M. A., and C. B. Mathur

1936 Influence of Carbon Dioxide on Development of Black Pigmentation in *Schistocerca gregaria*. Indian Journ. Agric. Sci., vol. 6, pp. 1005–1030, 3 pls.

Knight, H. H.

1924 On the Nature of the Color Patterns of Heteroptera. Ann. Entom. Soc. America, vol. 17, pp. 257–272.

Levrat, D., and A. Conte

1902 Sur l'origine de la coloration naturelle des soies de Lepidoptères. C. R. Soc. Biol., Paris, vol. 135, pp. 700–702.

Palmer, L. S., and H. H. Knight

1924a Carotin — the Principal Cause of the Red and Yellow Colors in *Perillus bioculatus* and its Biological Origin from the Lymph of *Leptinotarsa decemlineata*. Journ. Biol. Chem., vol. 59, pp. 443–449.

1924b Anthocyanin and Flavone-like Pigments in the Hemipterous Families Aphididae, Coreidae, Lygaeidae, Miridae and Reduviidae. *Ibid.*, vol. 59, pp. 451–455.

Perkin, A. G., and A. E. Everest

1918 The Natural Organic Colouring Matters. 655 pp., London.

Poulton, E. B.

1885 The Essential Nature of the Colouring of Phytophagous Larvae and Their Pupae: with an Account of Some Experiments upon the Relation Between Such Larvae and their Food-plant. Proc. Roy. Soc. London, vol. 38, pp. 269–315.

1886 A Further Enquiry into a Special Color Relation Between the Larva of *Smerinthus ocellatus* and its Food-plants. Proc. Roy. Soc. London, vol. 40, pp. 135–173.

1893 The Experimental Proof that the Colours of Certain Lepidopterous Larvae are Largely Due to Modified Plant Pigments Derived from Food. Proc. Roy. Soc. London, vol. 54, pp. 417–430, 2 pls.

Prochnow, O.

1926 Die Färbung der Insekten. Schröder, Handbuch der Entom., vol. 2, pp. 430–572.

Przibram, H.

1913 Grüne tierische Farbstoffe. Arch. ges. Physiol., vol. 153, pp. 385–400.

Schöpfe, C., and H. Wieland

1926 Ueber das Leukopterin, das weisse Flügelpigment der Kohlweisslinge (*Pieris brassicae* und *P. napi*). Ber. Deutsch. Chem. Ges., vol. 59, pp. 2067–2072.

Schulze, P.

1913 Studien über die tierischen Körper der Carotingruppe. I. Insecta. SB. Ges. Naturforsch. Freunde Berlin, 1913, pp. 1–22.

1914 Studien über die Tierischen Körper der Carotingruppe. II. Das Carotingewebe der Chrysomeliden. *Ibid.*, 1914, pp. 398–406.

Sitowski, M. L.

1910 Experimentelle Untersuchungen über vitale Färbung der Mikrolepidopterenraupen. Bull. Acad. Sci. Cracovie, 1910 B, pp. 775–790.

Süffert, F.

1924 Morphologie und Optik der Schmetterlingsschuppen, insbesondere die Schillerfarben der Schmetterlinge. Zeits. f. Morph. u. Ökol., vol. 1, pp. 171–308. 6 pls.

Thomson, D. L.

1926a The Pigments of Butterflies' Wings. I. *Melanargia galatea*. Biochem. Journ., vol. 20, pp. 73–74.

1926b The Pigments of Butterflies' Wings. II. Occurrence of the Pigment of *Melanargia galatea* in *Dactylis glomerata*. Biochem. Journ., vol. 20, pp. 1026–1027.

Toumanoff, K.

1928 La rapport entre la pigmentation et l'alimentation chez *Dixippus morosus*. C. R. Soc. Biol., Paris, vol. 98, pp. 198–200.

Vaney, C., and J. Pelosse

1922 Origine de la coloration naturelle de la soie chez le *Bombyx mori*. C. R. Acad. Sci., Paris, vol. 174, pp. 1566–1568.

Verne, J.

1926 Les pigments dans l'organisme animal. Encyclop. Scientif., Paris, 608 pp.

Wieland, H., and C. Schöpfe

1926 Ueber den gelben Flügelfarbstoff des Citronenfalters (*Gonepteryx rhamni*). Ber. Deutsch. Chem. Ges., vol. 58, pp. 2178–2183.

FOOD AND REPRODUCTIVE ACTIVITY

Bugnion, E.

1912 Observations sur les Termites. C. R. Soc. Biol., Paris, vol. 72, pp. 1091–1094.

Doncaster, L.

1916 Gametogenesis and Sex-determination in the Gall-fly, *Neuroterus leuticularis*. Proc. R. Soc. London, vol. 89B, pp. 183–200, 2 pls., 1 fig.

Doten, S. B.

1911 Concerning the Relation of Food to Reproductive Activity and Longevity in Certain Hymenopterous Parasites. Bull. Nevada Agric. Expt. Sta., No. 78, 30 pp., 10 pls.

Escherich, K.

1909 Die Termiten. 198 pp., Leipzig.

Evans, H. M., G. A. Emerson, and J. E. Eckert

1937 Alleged Vitamin E Content in Royal Jelly. Journ. Econ. Entom., vol. 30, pp. 642–646.

Finlayson, L. R., and T. Green

1940 Note on the Effect of Certain Foods upon Fecundity and Longevity in *Microcryptus basizonus*. Canadian Entom., vol. 72, pp. 236–238.

Glaser, R. W.

1923 The Effect of Food on Longevity and Reproduction in Flies. Journ. Exper. Zool., vol. 38, pp. 383–412.

Gregg, R.

1942 Origin of Castes in Ants, with Special Reference to *Pheidole morrisi*. Ecology, vol. 23, pp. 295–308.

Herms, W. B.

1928 The Effect of Different Quantities of Food During the Larval Period on the Sex Ratio and Size of *Lucilia sericata* and *Theobaldia incidens*. Journ. Econ. Entom., vol. 21, pp. 720–729.

Holdaway, F. C., and H. F. Smith

1933 Alteration of Sex Ratio in the "Flour Beetle" *Tribolium confusum* following Starvation of Newly Hatched Larvae. Australian Journ. Exp. Biol. Med. Sci., vol. 11, pp. 35–43.

Imms, A. D.

1919 On the Structure and Biology of Archotermopsis. Phil. Trans. Roy. Soc., London, vol. 209, pp. 75–180, 8 pls.

Isely, D.

1928 Oviposition of the Boll Weevil in Relation to Food. Journ. Econ. Entom., vol. 21, pp. 152–155.

Kopéc, S.

1924 On the Heterogeneous Influence of Starvation of Male and of Female Insects on their Offspring. Biol. Bull., vol. 46, pp. 22–34.

Larson, O. A., and C. K. Fisher

1924 Longevity and Fecundity of *Bruchus quadrimaculatus* as Influenced by Different Foods. Journ. Agric. Res., vol. 29, pp. 297–305.

Light, S. F.

1942– Determination of Castes of Social Insects. Quart. Rev. Biol., vol. 17, pp.
43 312–326; vol. 18, pp. 46–63.

Mackerras, M. J.

1933 Observations on the Life-histories, Nutritional Requirements and Fecundity in Blow-flies. Bull. Entom. Res., vol. 24, pp. 353–362.

Mathis, M.

1938 Influence de la nutrition larvaire sur la fécondité du Stegomyia. Bull. Soc. Path. Exot., vol. 31, pp. 640–646.

Morland, D. T. M.

1930 On the Causes of Swarming in the Honeybee; an Examination of the Brood Food Theory. Ann. Appl. Biol., vol. 17, pp. 137–149, 3 figs.

Norris, M. J.

1934 Contributions Towards the Study of Insect Fertility. III. Adult Nutrition, Fecundity and Longevity in the Genus *Ephestia*. Proc. Zool. Soc. London, 1934, pt. 2, pp. 333–360.

Park, T.

1936 The Effect of Differentially Conditioned Flour upon the Fecundity and Fertility of *Tribolium confusum*. Journ. Expt. Zool., vol. 73, pp. 393–404, 2 figs.

Planta, A. v.

1888 Ueber den Futtersaft der Bienen. Zeitschr. physiol. Chemie, vol. 12, pp. 327–354.

Rau, P., and N. Rau

1914 Longevity in Saturniid Moths and its Relation to the Function of Reproduction. Trans. Acad. Sci., St. Louis, vol. 23, pp. 1–78, 5 pls.

Smith, Falconer

1942 Effect of Reduced Food Supply upon the Stature of Camponotus Ants. Entom. News, vol. 53, pp. 133–135.

Stanley, J.

1938 The Egg-producing Capacity of Populations of *Tribolium confusum* as Affected by Intensive Cannibalistic Egg-consumption. Canadian Journ. Res., vol. 16, pp. 300–306, 5 figs.

Vaney, C., and A. Maignon

1907 Influence de la sexualité sur la nutrition du *Bombyx mori*. C. R. Assoc. Franç. Avanç. Sci., vol. 35. Notes et mém. pp. 465–469.

Wesson, L. G., Jr.

1940 An Experimental Study of Caste Determination in Ants. Psyche, vol. 47, pp. 105–111.

Wheeler, W. M.

1907 The Polymorphism of Ants, with an Account of Some Singular Abnormalities due to Parasitism. Bull. American Mus. Nat. Hist., vol. 23, pp. 1–93.

1923 Social Life Among the Insects. 375 pp. New York, Harcourt, Brace & Co.

1928 Mermis Parasitism and Intercastes among Ants. Journ. Exper. Zool., vol. 50, pp. 165–237, 17 figs.

Whiting, P. W.

1938 Anomalies and Caste Determination in Ants. Journ. Heredity, vol. 29, pp. 189–193.

Woke, P. A.

1937 Effects of Various Blood Fractions on Egg Production of *Aëdes aegypti*. American Journ. Hyg., vol. 25, pp. 372–380.

GROWTH AND ABUNDANCE; POPULATIONS

NOTE: *Cf.* Literature in Chapter I.

Allee, W. C.

1931 Animal Aggregations. ix + 431 pp., Chicago, University of Chicago Press.

Barnes, H. F.

1932 Studies of Fluctuations in Insect Populations. I. The Infestation of Broadbalk Wheat by the Wheat Blossom Midges (Cecidomyidae). Journ. Anim. Ecology, vol. 1, pp. 12–31.

1933 Studies of Fluctuations in Insect Populations. II. The Infestation of Meadow Foxtail Grass (*Alopecurus pratensis*) by the Gall Midge (*Dasyneura alopecuri*). *Ibid.*, vol. 2, pp. 98–108.

Blunck, H.

1930 Der Massenwechsel der Insekten und seine Ursachen. 4. Wanderversamml. Deutschen Entom., Kiel, June 1930, pp. 19–41.

Bodenheimer, F. S.

1930 Ueber die Grundlagen einer allgemeinen Epidemiologie der Insektenkalamitäten. Zeits. angew. Entom., vol. 16, pp. 433–450.

1931 Der Massenwechsel in der Tierwelt. Grundniss einer allgemeinen tierischen Bevölkerungslehre. Arch. zool. ital., vol. 16, pp. 98–111.

Brues, C. T.

1923 Choice of Food and Numerical Abundance Among Insects. Journ. Econ. Entom., vol. 16, pp. 46–51.

Buxton, P. A.

1933 The Effect of Climatic Condition upon Populations of Insects. Trans. R. Soc. Trop. Med. and Hygiene, vol. 26, pp. 325–356.

Chapman, R. N.

1928 The Quantitative Analysis of Environmental Factors. Ecology, vol. 9, pp. 111–122.

1929 Biotic Potential, Environmental Resistance and Insect Abundance. Congr. internat. Zool. 10, pt. 2, pp. 1209–1218.

Knight, H. H.

1932 Some Notes on Scale Resistance and Population Density. Journ. Entom. Zool., Pomona College, vol. 24, p. 1.

MacLagan, D. S.

1932 The Effect of Population Density upon Rate of Reproduction with Special Reference to Insects. Proc. Roy. Soc. (B), vol. 3, pp. 437–454, 4 figs.

MacLagan, D. S., and E. Dunn

1936 The Experimental Analysis of the Growth of an Insect Population. Proc. Roy. Soc. Edinburgh, vol. 55, pt. 2, pp. 126–139.

Nicholson, A. J.

1933 The Balance of Animal Populations. Journ. Anim. Ecol., vol. 2, pp. 132–178.

Park, T., and N. Woollcott

1937 The Relation of Environmental Conditions to the Decline of *Tribolium confusum* Populations. Physiol. Zool., vol. 10, pp. 197–211, 3 figs.

Schneider, B. A.

1941 The Nutritional Requirements of *Tribolium confusum*. Survival of Adult Beetles on Patent Flour and Complete Starvation Diets. Biol. Bull., vol. 80, pp. 208–227, 7 figs.

Schwerdtfeger, F.

1932 Betrachtungen zur Epidemiologie des Kieferspanners. Zeits. angew. Entom., vol. 19, pp. 104–129.

Smirnov, E., and W. Polejaeff

1934 Density of Population and Sterility of the Females of a Coccid, *Lepidosaphes ulmi*. Journ. Anim. Ecol., vol. 3, pp. 29–40.

Smirnov, E., and N. Wiolovitsch

1933 Ueber den Zusammenhang zwischen der Populationsdichte und Eierproduktion der Weibchen bei der Schildlaus *Chionaspis salicis*. Zeits. angew. Entom., vol. 20, pp. 415–424.

Zwölfer, W.

    1932 Zur Lehre vor den Bevölkerungsbewegungen der Insekten. Zeits. angew. Entom., vol. 19, pp. 1–21.

ATTRACTANT SUBSTANCES AND ABNORMAL FOOD

NOTE: Additional references may be found in Chapters III and V.

Back, E. A., and R. T. Cotton

    1930 Insect Pests of Upholstered Furniture. Journ. Econ. Entom., vol. 23, pp. 833–837.

Bauer, O., and O. Vollenbruck

    1930 Ueber den Angriff von Metallen durch Insekten. Zeits. f. Metallk., vol. 22, pp. 230–234, 14 figs.

Burke, H. E., R. D. Hartman, and T. E. Snyder

    1922 The Lead-cable Borer or "Short Circuit" Beetle in California. Bull. U. S. Dept. Agric., No. 1107, 56 pp., 10 pls., 15 figs.

Chamberlin, W. J.

    1924 Another Lead Boring Beetle. Journ. Econ. Entom., vol. 17, pp. 660–661, 1 fig.

Cole, A. C., Jr.

    1932 The Olfactory Responses of the Cockroach (*Blatta orientalis*) to the More Important Essential Oils and a Control Measure Formulated from the Results. *Ibid.*, vol. 25, pp. 902–905.

Crumb, S. E.

    1922 A Mosquito Attractant. Science, vol. 55, pp. 446–447.

    1924 Odors Attractive to Ovipositing Mosquitoes. Entom. News, vol. 35, pp. 242–243.

Ditman, L. P., and E. N. Cory

    1933 The Response of Corn Earworm Moths to Various Sugar Solutions. Journ. Econ. Entom., vol. 26, pp. 109–115.

Eyer, J. R., and H. Rhodes

    1931 Preliminary Notes on the Chemistry of Codling Moth Baits. *Ibid.*, vol. 24, pp. 702–711.

Griswold, G. H., and M. Greenwald

    1941 Biology of Four Common Carpet Beetles. Mem. Cornell Agric. Expt. Sta., No. 240, 75 pp., 41 figs.

Hesse, R.

    1925 Zerstörung von Bleiröhren durch Tiere. Biol. Zentralbl., vol. 45, pp. 19–20, 2 figs.

Horn, W.

    1933 Ueber Insekten, die Bleimäntel von Luftkabeln durchboren. Arch. Port. Telegr., 1933, pp. 165–190, 60 figs.

Ladell, W. R. S.

    1931 Carpenter Bees Eating Lead Cable-covers. Journ. Siam Soc. Nat. Hist. Suppl., vol. 8, pp. 213–214.

Lehman, R. S.

1932 Experiments to Determine the Attractiveness of Various Aromatic Compounds to Adult Wireworms. Journ. Econ. Entom., vol. 25, pp. 949–958.

McPhail, M.

1939 Protein Lures for Fruitflies. *Ibid.*, vol. 32, pp. 758–761.

Metzger, F. W.

1932 Trapping the Japanese Beetle. Misc. Pub. U. S. Dept. Agric., No. 147, 8 pp., 4 figs.

Moreira, C.

1930 Insectos que corroem o chumbo. Bol. Inst. Defensa agric., Brazil, No. 8, 8 pp., 4 pls.

Pickel, B.

1929 Sobre um Coleoptero perfurador de cabos telephonicos observado em Pernambuco. Bol. Mus. Nac. Rio de Janeiro, vol. 5, pp. 35–38.

Rendell, E. J. P.

1930 Depredations to Lead-covered Aerial Cables in Brazil. Proc. Entom. Soc. Washington, vol. 32, pp. 104–113, 1 pl., 1 fig.

Ripley, L. B., and G. A. Hepburn

1931 Further Studies on the Olfactory Reactions of the Natal Fruit-fly, *Pterandrus rosa*. Entom. Mem. Dept. Agric. South Africa, No. 7, pp. 24–81.

Roubaud, E., and R. Veillon

1922 Recherches sur l'attraction des mouches communes par les substances de fermentation et de putréfaction. Ann. Inst. Pasteur, vol. 36, pp. 752–765.

Severin, H. H. P., and H. C. Severin

1914 Relative Attractiveness of Vegetable, Animal and Petroleum Oils for the Mediterranean Fruit Fly. Journ. New York Entom. Soc., vol. 22, pp. 240–248.

Thorpe, W. H.

1930 The Biology of the Petroleum Fly (*Psilopa petrolei*). Trans. Entom. Soc. London, vol. 78, pp. 331–343, 4 figs.

Withycombe, R.

1928 [Termites Attacking Rubber Cable Insulation]. Ann. Rept. Electr. Railway & Wireless Dept., Zanzibar Prot., 1927, p. 1.

Wolcott, G. N.

1933 Otiorrhynchids Oviposit Between Paper. Journ. Econ. Entom., vol. 26, pp. 1172–1173.

CHAPTER III

HERBIVOROUS INSECTS

> Perhaps not a single plant exists which does not afford a delicious food to some insect, not excluding even those most nauseous and poisonous to other animals — the acrid euphorbias, and the lurid henbane and nightshade. Nor is it a presumptuous supposition that a considerable proportion of these vegetables were created expressly for their entertainment and support.
>
> KIRBY AND SPENCE, *Natural History of Insects*, 1818

As THE insects to be considered in this category include nearly half of all the species so far discovered and described, a general consideration of their food habits cannot easily be condensed into a brief account. It has already been mentioned that these numerous vegetarian members of the class form the mainstay of the insects as a whole, since they make use of plants for food, and in turn serve as prey for the major portion of all predatory and parasitic insects.

They thus enjoy a most bountiful food supply, whose extent is further enhanced by the small size of the insects in comparison with most of the flowering plants upon which they depend. Nevertheless, insects frequently multiply at such an excessive rate that they may destroy immense quantities of their food plants. This happens in spite of the large number of prolific animals and microörganisms that prey in turn directly upon the insects and serve to check their multiplication. Under natural conditions insects are a prime factor in regulating the abundance of all plants, particularly the flowering plants as the latter are especially prone to insect attack. Thus, specific plants may on occasion be threatened with local extinction. This fact is most readily apparent in connection with periodic outbreaks or epidemics of particular species, or under artificial conditions, where agricultural crops suffer more and more severely as a result of their continuous propagation by man. We shall not consider these economically important insects as such since they represent no more than one phase of the relationship which prevails between plants and all herbivorous animals. This competition is really a swinging or fluctuating relationship which developed over the course of countless years, long before our ancestors conceived the Utopian dream of making the world over purely for human consumption.

As a whole the vegetarian insects are a drab lot, but that is to be expected. Any animal, even our own precious species, becomes equally uninteresting when confronted with a superabundance of

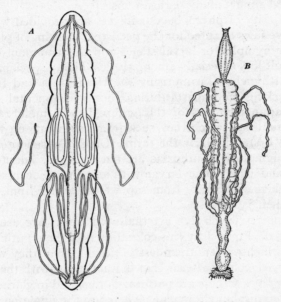

FIG. 10. Dissection of the alimentary canal and its appendages in the caterpillar of the oriental fruit moth (*Graptolitha molesta*). Arising anteriorly are the salivary and silk-spinning glands; beyond the middle are the excretory Malpighian tubules which branch and terminate in the walls of the rectum; *B*, alimentary canal of *Cnethocampa processionea*, salivary and silk gland omitted. *A* redrawn from Neiswander, *B* from Bordas.

food. Moreover, raw plants are bulky with water and indigestible materials, requiring an expansive masticatory and digestive mechanism on the part of animals that must grow rapidly in order to lay by the extensive stores of food required for their later reproductive activities (Fig. 10). As a result of such digestive excesses, vegetarian insects are as a rule sluggish and save for a few notable instincts relating to the maintenance of contact with the commissary, they are sadly lacking in the dash, go, or stealth characteristic of their predatory and parasitic relatives. It must be understood that we refer generally to the food habits of the developmental stages. These form the growing, feeding periods of the life span, and in many groups which undergo a complete metamorphosis the food-habits frequently change abruptly at the time of final transformation when the acquisition of food is not of such paramount importance in their lives. We shall see later that this change is still more pronounced among many parasitic insects.

The majority of phytophagous insects are provided with powerful, massive jaws adapted for the persistent chewing of plant tissues during the nymphal or larval stages and this commonly continues during adult life, especially among the less highly specialized orders. Only the Rhynchota, comprising the Homoptera and Hemiptera, possess sucking mouthparts during the developmental stages and the vegetarian members of this group depend mainly on sap or liquefied materials which they suck from the plants on which they feed (Pl. VI A). Similarly the thrips (Order Thysanoptera) imbibe sap, although their mouthparts are not so highly adapted to that purpose, and even some larvae with chewing mouthparts derive much of their sustenance from sap which exudes from tissues injured by their jaws.

Those insects which live externally on plants are usually strikingly colored. Frequently this coloration is of the cryptic or protective type which renders them less noticeable than they would be if furnished with color patterns that did not blend with the plants on which they live. Others are patterned or colored in glaring contrast with their natural environment. These are the numerous examples of aposematic or warning coloration generally conceived by biologists to represent a warning to insectivorous birds and other animals that they are either distasteful, poisonous or otherwise inedible. Others whose coloration resembles that of such inedible species are further quite generally believed to "mimic" the latter and to have developed these resemblances through the action of natural selection which has furthered the survival of variations that approach the patterns of the inedible, protected species. The reality of protective coloration as a biological principle rests upon a solid basis, but the general validity of warning coloration and mimicry is by no means completely acceptable to all those who have examined them in detail. This matter is also relevant to a discussion of predatory insects and we shall consider it further in a later chapter, and still again with reference to the diet of birds. There are undoubtedly several reasons for the gaudy coloration of such insects, but how far they may be by-products of physiological activity, natural selection or natural cussedness (*i.e.*, determinate evolution or orthogenesis) is not yet clear. We do not actually know what the insects look like to each other and even to some of their insectivorous enemies as the visual range of insects does not coincide with our own. They see best by ultraviolet light and poorly by the longer wave lengths that enable us to appreciate the beauties of a tropical sunset or to halt at a traffic light. I believe I speak for most biologists in stating that the principles of mimicry have been alternately embraced and discarded in more rapid succession by more people than any other biological theory ever propounded.

The members of many orders of insects are almost exclusively phytophagous. Conspicuous among these are the true Orthoptera (grasshoppers, locusts, crickets), Homoptera (cicadas, tree hoppers, aphids, scale insects), Lepidoptera (moths, butterflies), and some smaller orders like the Thysanoptera (thrips), Phasmatodea (walking sticks, leaf insects), and Isoptera (termites). A very great part of the vast horde of phytophagous insects are members of certain groups of the large orders Coleoptera, Hymenoptera and Diptera, all of which include also numerous series with predatory or parasitic habits. In these orders we may say in a general way that the vegetarian Hymenoptera, represented by the saw-flies and wood-wasps, are the most primitive members of their order, coincident, in all probability, with the extensive development of the woody plants at the time this order began its differentiation. On the other hand the phytophagous members of the Coleoptera (longicorns, leaf-beetles, weevils) represent the more highly specialized beetles, indicating that the appearance of phytophagy in this order is a secondary development, derived from the predatory or saprophagous food habits which are generally characteristic of the more primitive beetles.

Cases of phytophagy of obviously secondary origin are encountered in diverse groups of insects. Thus, among the parasitic chalcid wasps of the family Eurytomidae, species of the genus *Harmolita* (Fig. 11) develop in gall-like swellings on the culms of grains and other grasses although the group to which they belong is almost without exception parasitic in other insects. Here the vegetarian habit is obviously derived from parasitism, as only sporadic examples of vegetarian chalcids are known among the many thousands of species that form this enormous group. The development of phytophagy from predatism is well illustrated by one small group of ladybird beetles (Coccinellidae). This extensive family is characteristically predatory, feeding as both larvae and adults almost entirely on aphids and scale insects. One small subfamily, typified by the genus *Epilachna*, has turned vegetarian even more drastically than the human adherents of this cult. We have in America two

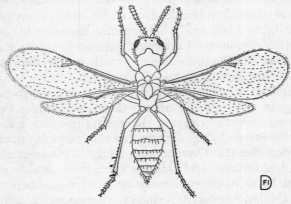

FIG. 11. The chalcid-wasp, *Harmolita grandis*, member of a large group of parasitic Hymenoptera. The species of this genus have become vegetarian and their larvae develop in the stems of grasses, causing gall-like swellings. Redrawn from Phillips.

species, one the noxious Mexican bean-beetle, and another the squash lady-bird, feeding respectively on the foliage of leguminous and cucurbitaceous plants. They are, indeed, just as highly specific in their choice of food plants as the most fastidious vegetarian insects who can proudly trace their distaste for flesh as far back as Mesozoic times. As a matter of fact in this instance it probably does extend into the Tertiary, perhaps some threescore millions of years ago, for *Epilachna* is now almost cosmopolitan, with species in Europe, Asia, Japan, Malaya, Australia, Africa and South America. All are vegetarian, but as we pass around the world the exotic species bring other plants into the picture, Solanaceae in the Orient and members of the cotton family (Malvaceae) in Africa. This interesting group undoubtedly underwent a shift or mutation in food habits at some time in the dim past. Since then it has encompassed the globe, undergoing minor changes, but clinging persistently to the changed diet. Such behavior shows the near immutability of the instinctive processes in insects, broken only at interminable intervals by some cataclysmic mutation. Whether this shift in food habits is exactly of the same nature as a mutation that involves bodily structure cannot be stated, but it gives every indication of being essentially similar. Certainly such changes in habits are not induced by any structural modifications that precede them. Space will not permit a further consideration of these interesting shifts in food-habits and the reader must be referred to a more lengthy account (Brues '36) in which numerous similar cases are cited.

The diet of many aquatic insects consists entirely or in considerable part of green algae, including unicellular plants as well as the more highly organized types. It must be remembered, however, that these plants are to a very large extent marine and consequently far removed from the sphere occupied by insects. Prominent among such aquatic vegetarian forms are the nymphs of the mayflies (Plectoptera) and stoneflies (Plecoptera), the larvae of many caddis flies (Trichoptera), of many midges and mosquitoes, and of a few other aquatic dipterous larvae. They are for the most part very prolific and dominant types, due no doubt to the abundance of their food plants. They appear to show little specificity in selecting particular plant species for food, although this apparent promiscuity may be due in some measure to a lack of accurate information. Certain of the higher aquatic plants like our common water-lilies for example are beset by a definitely circumscribed series of insect enemies.

Ferns are eaten to some extent by insects, but much less extensively than the flowering plants. Even in humid tropical regions where ferns form a very considerable part of the vegetation they are comparatively neglected by insects. This seems surprising as they produce an abundance of succulent foliage which appears fully as attractive as the quite similar leaves of the flowering plants. It is not easy to get reliable data on the comparative abundance of ferns and flowering plants aside from the actual number of species occurring within definite areas. In the United States the ratio is about 1.2 per cent ferns (1.0–2.3) and it does not appear to rise much above 2–3 per cent in most tropical areas. There are, however, everywhere certain dominant species and it would appear that the ferns do not receive the attention that they should were they as attractive as flowering plants. Certain of the earlier insects must have been dependent on the ferns as many vegetarian types antedate the modern development of the phanerogams. Thus the ferns after their long presence in the past and persistence today in considerable abundance have not become food plants for many of the innumerable insects to which they have been exposed. Swezey, who has had wide experience in Hawaii, believes that the immunity of ferns is apparent rather than real and lists 44 species known to attack them in these islands. Considering the small extent of the Hawaiian insect fauna this is really a goodly number.

In Europe, de Meijere has bred four species of insects from the common lady fern, *Asplenium filix-femina* and a total of six species from the extremely abundant and widespread *Pteridium aquilinum*. The number of species feeding on equally abundant flowering plants is certainly much higher. Although one or two of these are primitive members of the groups to which they belong, others are very highly specialized Diptera, not archaic types that might have fed on ferns for long periods. In this connection it is interesting to note that Franssen has reported a common aphid pest of beech developing a race that attacks a fern, *Asplenium*.

The terrestrial flowering plants are the *sine qua non* of the insect tribe for it is among the insects that feed upon these that phytophagy reaches its highest development. Such species present a series of complex relationships which are more easily understood if we first consider separately several of their peculiarities.

In the first place there is a wide variation in the diversity of food plants that are included in the diet of individual species. We ourselves are so accustomed to partake of a great variety of vegetable food according to the passing whims of our appetites, or with the turn of the seasons that brings numerous vegetable foods in or out of our menu. The same is true very generally of our vegetarian farm animals whose grazing is not closely limited to particular forage plants except that the more succulent kinds are commonly most attractive and some conspicuously malodorous or otherwise unpleasant species may be generally avoided. We thus think of a mixed vegetable diet as quite natural for herbivorous animals and are startled to learn that at least one primitive marsupial mammal, the Australian Koala (*Phascolarctos cinereus*) lives continuously from generation to generation on the leaves of a particular species of eucalyptus, trees notable for their pungent, oily flavor. Among insects this kind of restriction to a single species of food plant or to a series of related species is by no means rare and insects which do not show some such preference are the exception rather than the rule.

Although there is a completely graded series between species that will eat only a single kind of plant and those that regularly consume many very diverse ones, it is customary and convenient to speak of polyphagous insects as those which exercise little choice and monophagous ones where only a single food-plant is acceptable. Between these two extremes we encounter the intermediate condition including species having appetites more like our own. These are known as oligophagous. Only a very few insects appear to be normally pantophagous or omnivorous although under stress of necessity certain Orthoptera like the migratory locusts will eat every green plant that they encounter as they swarm over vegetation in countless numbers during their migratory flights. It was commonly assumed until recent years that most of our grasshoppers were very general feeders, but it now appears that this is by no means true.

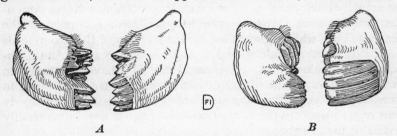

A   *B*

FIG. 12. Right and left mandibles of two species of grasshoppers having different food-habits. *A*, the lubber grasshopper (*Brachystola magna*), which feeds on foliage; *B, Mermaria maculipennis* which feeds on seeds. Redrawn from Isely.-

In 1928 Hubbell described a species of *Schistocerca* which feeds only on a single food-plant (*Ceratiola ericoides*) and a few years later Rubtzov ('32) showed that various Siberian grasshoppers utilize only a small part of the flora growing in the locale where they live, with a general preference for grasses. We may therefore quite legitimately call these ubiquitous creatures grasshoppers although some of our more pedantic entomological friends insist that they should be known as locusts. Further extensive studies by Isley, Painter and others support these observations. Thus even in the primitive Orthoptera the instincts which tend to restrict the list of acceptable food-plants are already highly developed (Fig. 12). Some related Orthoptera like the common European house cricket, *Gryllus domesticus*, now naturalized in America, are normally vegetarian, but on occasion become a nuisance in households, devouring clothing, upholstery, carpets, cotton cloth and even rubber goods. Presumably such idiosyncrasies are only a highly abnormal extension of polyphagy to the point of absurdity. It is even said by Gavin ('27) that rubber is preferred to digestible vegetable materials. This is comparable to the inordinate attraction of various unfamiliar substances to many insects which will be considered on a later page. Some other primitive insects like the cockroaches are very general feeders, but in this case their food does not consist of living plants but rather of dead vegetable substances with some animal matter. They are saprophagous, not vegetarian in the sense that we have defined the word. Other insects related to the Orthoptera and less closely so to the cockroaches form the vegetarian order Phasmatodea, including the walking-sticks and leaf insects. These are at least, in some cases, very specific in their food, and some appear to be actually monophagous. For example our common North American walking stick, *Diapheromera femorata*, feeds generally on the leaves of oaks while a giant East Indian member of this group feeds on the leaves of a single species of *Eugenia* in Sumatra. As this insect would make a fine experimental animal I have attempted to rear it on other species of this extensive genus of trees but failed to persuade the young walking sticks to do more than taste gingerly and reject promptly the unaccustomed foliage of any forms readily available in our region.

The Homoptera include a great variety of insects destructive to agricultural crops and other cultivated plants, consequently the specific food habits of a large number of species are well known. Some of the more generalized members of the order like the cicadas (see Fig. 10) feed during their developmental stages on the roots of plants and it is difficult to determine whether they may restrict themselves to particular host plants, although it is probable that they do so to some extent. It is true of the tree hoppers (Membracidae) and also of the leaf-hoppers (Jassoidea) that some species are monophagous or nearly so while a great many affect only a highly restricted series of plants, commonly related ones. Among the more highly specialized Homoptera like the plant-lice or aphids (Aphididae) a restriction to one or several food-plants is of very general occurrence although many species are less circumspect and may affect a considerable number of host plants. Among aphids the selection of food plants has been further specialized by the acquisition of definite alternate food-plants which involves an alternation of hosts by successive generations. Briefly, the succession is as follows. During the warm season several generations follow one another on the summer food-plants. These consist of wingless, rapidly maturing parthenogenetic individuals that bear living young in quick succession. In autumn a winged parthenogenetic generation appears, these individuals then migrate to the winter or alternate food-plant, and there produce aphids of both sexes which finally lay fertilized winter eggs that remain unhatched till the following spring. These eggs give rise to the broods of the following season that soon migrate by winged individuals back to the summer food-plants where they remain till the appropriate time to repeat the succession. Usually the food plants of each species of aphid are widely different plants, and frequently the selection of each by the aphids is highly specific. Occasionally the alternation involves only a migration between roots and aërial parts of the same host plant and it is not so clearly seasonal. In the tropics the life cycle is generally simplified and the sexual generation is frequently suppressed. The most highly specialized, and incidentally most degenerate group of Homoptera, are the scale insects (Coccidae, in the wide sense) (Pl. VI B). They include some monophagous, many oligophagous and a few rather general feeders with a wider range of food plants.

The food habits of the larvae of Lepidoptera, both moths and butterflies are far better known than those of any other order of insects. This knowledge of their food preferences and restrictions is largely due to the assiduous rearing of caterpillars by thousands of amateur entomologists. These accumulated data, together with the results of more pretentious field and laboratory experimentation give a clear indication of some of the factors which guide insects in choosing their food-plants with such constancy and unerring accuracy. They are sufficiently complete to show also the prevalence of monophagy and oligophagy throughout a considerable part of the order Lepidoptera, particularly the butterflies. There is great variation as in the several other orders already mentioned and we shall reserve further details for specific topics to be discussed on a later page.

Coleoptera form the largest order of insects and many families are phytophagous, particularly the leaf-beetles (Chrysomelidae, in the wide sense), longicorns (Cerambycidae) and the enormous suborder Rhyncophora, comprising the weevils or snout-beetles, which are as yet very incompletely known. They are unquestionably as specific in their selection of food-plants as the Lepidoptera, but unfortunately only a relatively few examples drawn mainly from economic pests of useful plants must serve for generalizations. Several of these like the Colorado potato-beetle, Mexican boll-weevil, and the Mexican bean-beetle already referred to on an earlier page, are monophagous or practically so, while others, like the Japanese beetle, balk at none of the fresh fruits and vegetables which radio barkers remind us so assiduously are "rich in all the essential vitamins and minerals." All in all, the Coleoptera are very highly specialized with reference to food-habits, not only those species that

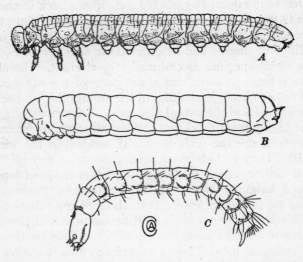

FIG. 13. Larvae of three insects showing the progressive degeneration of appendages that is associated with their feeding habits. *A*, a primitive sawfly (*Pamphilius*) that feeds externally on leaves; *B*, a wood-wasp that burrows in wood; *C*, a flea larva that feeds in organic detritus. *A* and *B* redrawn from Yuasa, *C* from Howard and Marlatt.

feed on foliage, but the numerous forms that live internally in wood, stems and roots.

The vegetarian Hymenoptera are likewise very generally closely restricted in their diet. The caterpillar-like larvae of sawflies (Fig. 13) behave like the true lepidopterous caterpillars in this respect, including some practically monophagous forms like the well known and destructive larch sawfly. The gall-making Hymenoptera of the family Cynipidae are likewise extensively monophagous. These will be dealt with in the succeeding chapter.

There is no period during the developmental cycle of the seed-plants that is free from attack by insects. The ripening seeds are supplied with moisture and contain the well balanced ration of essential materials necessary for the growth of the embryonic plant. Such substances are similarly active in promoting the growth of insects and other animals that feed on them. Large seeds are naturally the most attractive although those massed in capsules or in the heads of composite flowers also form particularly suitable food. Thus the seeds of our cultivated beans and peas are abundantly infested with the larvae of bean or pea weevils, members of a family Lariidae (Bruchidae) that develop almost exclusively in the seeds of leguminous plants (Fig. 14). The seeds of Indian corn or maize

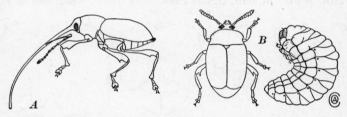

FIG. 14. Two vegetarian beetles. *A*, *Balaninus proboscideus* (Family Curculionidae), a weevil whose larvae develop in chestnuts. The minute jaws are at the tip of the greatly prolonged head. *B*, the pea weevil, *Mylabris pisorum* and its larva which feeds in the seeds of garden peas. *A* redrawn from Chittenden, *B* from Back and Duckett.

are likewise attacked by several common insects, in this case usually the larvae of small moths which develop within the kernels after they have ripened and lost the greater part of their water content. Many seed eating insects like the meal-worm, rice weevil, flour moths and beetles have become serious pests of dry cereals as they are able to develop in materials containing extremely small amounts of water. In a cruder way the large caterpillars of the corn earworm devour the kernels in the milky stage before ripening. Acorns and nuts likewise serve as food for a number of insects (Fig. 14). The members of one group of small Hymenoptera, *Megastigmus* and their allies, develop in smaller seeds like those of the apple, hawthorn, rose, etc. Strangely enough these little wasplike forms have become secondarily phytophagous as they are members of a large superfamily (Chalcidoidea) that is almost exclusively parasitic on other insects. Many other insects feeding within the fruits of plants depend on the contained seeds to a greater or less degree. Thus the small caterpillars of the codling moth, a notorious pest of apples, make their way to the center of the fruit where the seeds are ripening. The caterpillar of the yucca moth bores its way lengthwise through the succulent yucca pod following closely along one of the long rows of flat seeds which are left as rings or crescents after the larva moves on. In this case it appears that the rich content of fat and protein in the seeds is the portion sought out by the larva. Many other insects exhibit the same type of behavior.

Again, certain foraging insects like the seed-storing harvester ants put by stocks of seeds, usually those of grasses, to carry their colonies over the unproductive season (Fig. 15). Such ants are characteristic of arid or semi-desert areas. As usual, whenever anything

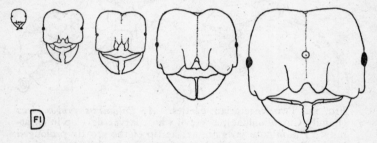

FIG. 15. Heads of various castes among the workers of an East Indian seed-eating ant, *Pheidologeton*. The gigantic heads of the large "soldier" ants serve as powerful nut-crackers for comminuting the food. Redrawn from Wheeler.

is put by for a rainy day, the seeds are attractive to other than their original harvester and the larvae of some of the common pinacate beetles, *Eleodes*, are known to feed on such seeds stored by the abundant mound-building ant, *Pogonomyrmex*, in our southwestern states.

Young seedling plants do not offer any special attraction to insects, probably on account of their small size and because insects restricted to special food plants usually have their life-cycle timed so that the feeding period corresponds to the time that food is abundant. Notorious exceptions are the garden cut-worms. These are the larvae of noctuid moths that begin their breeding during the late summer and go into hibernation before completing their growth. They are already of goodly size when they awaken in the spring to complete their feeding at the expense of tender developing plants or seedlings. In the case of trees, however, saplings often have

PLATE VII

Types of insect feeding on leaves. *A*, coarse chewing of gipsy moth caterpillars on oak leaves: *B*, more decorous feeding by rose-chafer beetles (*Macrodactylus*); *C*, feeding confined to one surface of the leaf by larvae of the elm-leaf beetle (*Galerucella*). The hole at the lower right was gnawed out by an adult beetle. Photographs by the author.

PLATE VIII

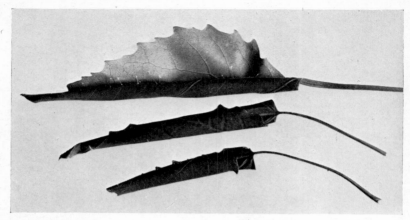

A. Successive stages in the formation of a "nest" by a leaf-rolling cater-pillar of the family Tortricidae. Photograph by the author.

B. Two types of "leaf-mines" made by insect larvae. Photographs by the author.

particular insect enemies. Thus certain longicorn beetles of the genus *Saperda* feed only in the wood of very small trees. Two of our common species injure poplar and apple, respectively, boring in the small trunk close to the ground.

The foliage of plants is far and away the most attractive to insects and supplies the food for a great variety of forms, commonly known as defoliators (Pl. VII). Most caterpillars of the Lepidoptera and vegetarian Hymenoptera eat the leaf tissues although they show many adaptations for doing this efficiently, rapidly and safely. Some consume the leaves entirely, trimming off pieces about the edges and using considerable care not to sever any piece to which they are attached. This is particularly noticeable in the case of gregarious sawfly caterpillars that will range themselves about the edge of a leaf with their mouths along the periphery. Or, in the case of those feeding on pine, they always begin at the tip of a needle so as not to cut off their own foothold or that of one of their companions.

Others confine their feeding to one side of the leaf, usually the lower surface, consuming the parenchyma entirely, but leaving the upper epidermis intact, together with the network of veins that holds it together as a filmy covering (Pl. VII). This serves probably as a protection against the heat of the sun and perhaps makes them less easily noticed by insectivorous birds. Such species are the bane of the agriculturist or gardener as they cannot be attacked by poisons sprayed or dusted on the plants from above. This method of feeding is especially characteristic of the larvae of leaf-beetles (Chrysómelidae) although the adult beetles which commonly eat the same foliage are less finicky and usually gnaw holes completely through the leaves. Another peculiarity in this respect has been noted in the larvae of the Mexican bean-beetle, a vegetarian ladybird of the genus *Epilachna* previously mentioned in the present chapter. This larva simply scrapes the underside of the leaf with its jaws, removing the parenchyma which it squeezes between the jaws to remove the sap which is swallowed. The dry residue is then left as streaks or windrows, plastered to the untouched epidermis (Howard '41). These larvae thus depend upon sap, rejecting the solids previous to ingestion.

Sometimes a single caterpillar folds a nest for itself in a leaf, drawing the surface together by silken threads. Again, certain leaf-rollers, especially of the family Tortricidae curl the leaf into a tight roll within which the larva is hidden while it undergoes its growth.

There is a well marked tendency among defoliating insects to become gregarious, feeding in groups often containing a great many individuals. This may come about through the association of all the young hatched from a single egg-mass as happens with the American tent-caterpillar (Pl. I). These larvae hatch in the early spring and start a communal web at a fork in the branch of a cherry, apple or similar tree. From this as a base they travel out to feed on the opening leaf-buds. As they grow, the silken web is enlarged to accommodate its inmates until they finally wander forth to pupate near-by. Occasionally two "nests" may combine if there has been a considerable mortality in adjoining webs. Many other small caterpillars web the leaves of trees together sometimes to form communal nests, spinning the silken threads necessary for this purpose (Pl. VIII A). Again, other gregarious larvae like those of the mourning cloak butterfly (*Euvanessa antiopa*) group themselves regularly on a single branch of an elm tree. Sawflies of the genus *Diprion* will concentrate in a small area on one side of a pine tree, and certain leaf-beetles (*Haltica*) will do the same on their food-plant. Many leaf-feeders however show no such communal tendencies, even though their eggs are commonly laid in masses of several hundred and association of the young might readily follow as a matter of course. In other gregarious species like the common squash-bug (*Anasa tristis*) it appears that such an association may offer some protection since these bugs produce a vile, malodorous secretion that surrounds them with an aura not conducive to molestation by insectivorous animals. Nevertheless, we must not predict the reactions of other animals to odors on the basis of our own noses.

In spite of the tremendously flattened form assumed by the leaves of most plants there are many insects that excavate the leaf parenchyma and allow the epidermis covering its two surfaces to remain intact. The larvae of some sawflies burrow into the leaf where first hatched as minute grubs and later feed by chewing away at the leaf as they grow larger. Many insect larvae, however, continue to burrow through the leaf, excavating galleries or blister-like cavities that may assume a great variety of forms, although their contour is usually quite constant for each species (Pl. VIII B). Such forms which are called leaf-miners include members of several orders of insects, most commonly moths or flies, all of comparatively small size as such habits are not suited to large insects. Occasionally the larvae may pass from one leaf to another before completing their feeding and growth, after which they usually drop to the ground for pupation and final metamorphosis. The leaf-mining larvae are often greatly modified in form with flattened body and mouth-parts suitable for chewing or sucking sap from the thin layer of leaf parenchyma. The caterpillars of some minute moths (*Coleophora*) spin a tiny case to contain the body, and feed by reaching out and tunneling into leaves. One such species hollows out the needles of larch trees one by one, taking time out to hibernate on the larch twigs as a half-grown caterpillar housed in its little silken case so small as to escape notice except on close scrutiny. Hundreds of leaf mining insects are known and extensive accounts of them may be found in the writings of Hering, Frost and others.

Many Homoptera and Hemiptera feed on the leaves particularly the smaller forms like the leaf-bugs (Miridae), plant lice (Aphididae) and white-flies (Aleurodidae). They insert the very slender piercing mouthparts into the tissues, between the cells or sometimes actually dissolve the cell walls by means of a salivary secretion, depending for nutriment on the cell sap. Much plasmolysis and injury to the tissues accompanies their feeding, especially to the vascular bundles which furnish most of the food. Such insects do not, of course, confine their feeding to leaves but they generally avoid tough, woody tissues. Sometimes a definite sheath develops about the mouth-parts, presumably induced as a reaction to the insect's feeding.

With such a diversity of insects infesting the leaves of herbs, shrubs and trees it is no wonder that they become sadly worn and ragged before the summer season comes to an end.

Many insects feed also on flowers, consuming the petals or other soft parts, but these are far less numerous than those affecting the leaves, due probably in great measure to the much shorter season during which such structures are available. Many winged adult insects frequent flowers which they visit most frequently in search of pollen and nectar or, like certain more voracious beetles such as the Japanese beetle and rose chafer, to consume the petals and other floral parts as well. Pollen forms a considerable part of the food of numerous insects notably of certain adult beetles, especially the longicorns and Meloidae.

In this connection we should not fail to remember that bees and other anthophilous insects play a most important rôle in the economy of a great series of flowering plants whose blossoms are cross-pollinated through their visits. We owe to the great naturalist, Charles Darwin, an extended account of these phenomena and his several contributions concerning them will always find their place among biological classics. It would be out of place here to consider these matters except in so far as they are directly related to the food of the insects concerned. We must shamefacedly admit that the part played by insects is in the nature of thievery which has been turned to advantage by the plants through many adaptations that insure cross pollination by the not too unwilling insects.

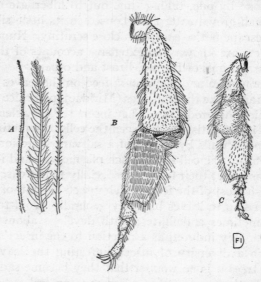

FIG. 16. Structural modifications of bees for collecting pollen. A, body hairs of several species showing their branched or fimbriate form; B, inner surface of hind leg of a worker honeybee, showing the enlarged first tarsal joint furnished with comb-like rows of bristles; C, outer surface of leg with a ball of pollen being carried in the "pollen-basket" at the tip of the tibia. A, Redrawn from Smith, B and C from Casteel.

Modifications on the part of the insects for harvesting the crop of pollen and nectar include pollen brushes on the legs and bodies of bees (Fig. 16), lengthened and otherwise modified sucking mouthparts in bees, butterflies and moths, coupled with persistent appetites and boundless energy in pursuit of food. Behavioristic modifications on the part of the visiting insects are numerous and are sometimes highly correlated with the structure of particular flowers. Among bees, for example, there are many kinds that visit a great variety of flowers more or less indiscriminately, but evince a well marked rhythm in passing very generally from one plant to another of the same species, for limited periods at least, whereby cross pollination is more readily effected. Whether we may consider this as anything more than "getting into a rut" may be seriously questioned, but it nevertheless conduces to the benefit of the plants by increasing their chances of receiving the needed pollen from other flowers of their own species. Some more specialized bees and other insects restrict their visits to a much more circumscribed assortment of plant species. With these oligotropic kinds the advantages to the plant are obviously greater than would accrue from random movements of their insect visitors. Numerous monotropic flower visitors are known where certain species of bees visit only a single species of plant. In such cases the season of flight must be timed exactly with the period that the host plant matures its blossoms. The amount of pollen appropriated by bees for the feeding of their brood is enormous. Recently, Eckert has determined the annual pollen consumption by hives of honeybees in a Californian apiary. Well over 100 pounds was collected each season by a normal colony which includes something like 40,000 individuals. At this rate, the amount of pollen contributed to pollen-eating insects in the aggregate must reach astronomical proportions. Sufferers from hay-fever may now await the appearance of a race of monotropic bees with appetites restricted to ragweed pollen. In a few bees, moths, etc. it appears that promiscuous flower visiting may be restricted by mechanical barriers, such as the size of the flower, its form or the extreme length of the floral tube that automatically exclude the majority of anthophilous insects. In perhaps only one known case does it appear that an insect deliberately performs an act necessary for the pollination of the plant and one which is entirely independent of its own search for honey and pollen. The yucca-moth (*Tegeticula*) actually collects a ball of pollen from the stamens of the yucca flower and thrusts it into the hollow tip of the pistil to insure the presence of pollen on the stigma which would otherwise be unable to effect the fertilization necessary for the production of its seeds. The larvae of this moth feeding in the developing seeds of the yucca fruit are equally dependent upon this behavior to provide the food required by their offspring. Furthermore, the female yucca moth has a special organ developed on its mouthparts that serves as a tool for this most remarkable pollinating performance so essential to both the insect and the plant.

A great variety of insects feed within the stems of plants. Herbs and succulent vines are affected commonly by the larvae of moths and beetles belonging to rather specialized families and more rarely by those of some Diptera and a very few Hymenoptera. Similarly the stems of shrubs and the woody portions of trees harbor the larvae of numerous insects, but the distribution of these in the woody portions is closely limited by the food substances available in specific tissues. The larvae of many Coleoptera, like bark-beetles of the family Scolytidae, develop in the inner bark where they excavate their food-burrows through the cambium layer, grooving the inner surface of the bark and adjacent sapwood. In this case the large supply of nutrient materials in this layer is obviously the reason that the larvae follow it so closely. The larvae of some of these bark-beetles, derive a major part of their food from the superficial sapwood. Their burrows are deep and usually extend longitudinally, while those that feed mainly on the bark excavate their food burrows transversely through this layer.

A few highly aberrant insects feed on corks in wine cellars, as *Tinea cloacella,* a species related to one of our common clothes-moths. Commercial cork consists of cortical tissue from the European cork oak in which this layer is enormously hypertrophied.

Even the larvae of certain much larger longicorn beetles, for example the pine sawyer (*Monochamus*) and the locust-borer (*Cyllene*), pass their first larval stage just beneath the bark and later dig deeply into the wood when their more powerful jaws are able to cope with the tough woody fibers. The larvae of the related family Buprestidae (flat-head borers) (Pl. IX A) feed similarly beneath bark and in the underlying sapwood. The cambium serves also as food for some very remarkable larvae that mine through it after the fashion of leaf-miners. These are minute acalyptrate Diptera of the genus *Agromyza* that lay their eggs in the thin bark of certain trees like birch or red maple high above the ground (Pl. IX B). The tiny larvae work their way downwards, often for a distance of many feet, emerging near the ground after this unusual journey. On the trunks of our New England birches the paths they have followed are well marked and may commonly be seen as conspicuous, blackened, serpentine streaks. Many related small flies, as well as certain beetles and other insects are often attracted to injured trees or cut stumps where they feed on the exuding sap at times when this material is flowing.

The economy of many wood-eating insects is complicated by the fact that they are unable to digest the cellulose and must depend upon sap, starch or other materials present in only small quantities in the sapwood and practically absent in the older heartwood. Some of these depend in part on fungi which are breaking down the wood. Others have established symbiotic relationships in their own bodies with microscopic fungi or bacteria that are capable of reducing cellulose to materials that the insects are able to utilize. These remarkable associations will be dealt with in a later chapter. It must be mentioned also that various insects are fond of books, literally eat them up, especially when musty from storage in a damp climate. The paper in books is, of course, mainly cellulose derived from wood or vegetable fibres together with glue, paste and a smear of printer's ink, not to mention ideas. Even when weighted with clay or the like, paper books are regularly devoured by a select coterie of insects that fed originally in wood. The sapwood of trees, particularly during the winter, contains limited quantities of starch. Logs from such trees felled at this season, commonly harbor certain "powder-post" beetles of the family Lyctidae. Even in wood which has been kiln-dried and subsequently fashioned into axe handles, furniture and the like, the grubs of these beetles continue to breed, reducing the wood to a dry powdery residue. This fills the burrows and finally the wooden articles may collapse into a dusty mass of finely comminuted wood from which the beetle larvae have extracted the starch during its passage through their bodies. Recently, Linsley has made the interesting discovery that many household pests feeding on stored food-products also occur in the nests of wild bees where they consume the bee-bread stored for the use of their brood. Quite possibly the original habitat of many such household pests may have been just such places.

The roots of plants of all kinds, herbs, shrubs and trees alike support many sorts of insects. Some of these live in the soil, eating off the smaller rootlets and often traveling from one plant to another under the ground (Fig. 17). These include particularly the larvae of certain beetles and moths. Others bore directly in the roots while certain Homoptera like the cicadas and a few aphids feed by piercing the roots with their slender beaks and withdrawing the sap (Fig. 17). Damage by root-infesting insects is insidious and usually noticed only by its weakening effects on the aërial portions of the plants.

If we look at the vegetarian insects from the standpoint of their food-plants we naturally find that every species of plant falls prey to a number of different insects. This would be true even if all insects were monophagous, confining their feeding to a single food-plant, since the number of existing insects so greatly outnumbers that of the flowering plants. This means that every species of these plants has a series of insects that depend upon it, either wholly or in part, for the food necessary to their continued existence. There is as yet very meagre information as to how many insects are included in such associations, biocoenoses, or "plant faunae" as they have been called. The insect pests of many plants of economic importance have been long and extensively studied by entomologists who have accumulated sufficient information to enable us to hazard

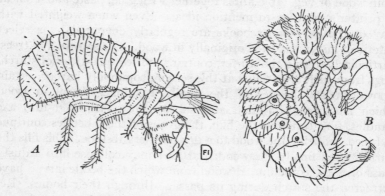

FIG. 17. Modifications for burrowing through the soil in two diverse insects. *A*, nymph of the periodical cicada ("17-year locust") with the front legs greatly enlarged; *B*, white grub, the larva of a may beetle (*Phyllophaga*). Both forms feed on roots, the former with sucking mouthparts and the latter with chewing jaws. *A* redrawn from Marlatt; *B* from Davis.

a rather vague answer to this question. To summarize very briefly we may say that wherever any economically important plant like maize, cotton, rice, wheat, walnut, citrus, apple, tea and the like is grown, from one to several hundred species of insects may be found feeding on each of these crops in a given area. Naturally the number of species is greatest where such cultivated plants are endemic, which should more nearly represent the natural balance before our pristine earth was turned topsy-turvy by the march of civilization. It must be remembered also that cultivated and other plants as well have received many insect enemies from alien areas to be added to those already present, and the crisscross shipments of everything movable are now adding a regular increment to all of these plant faunae even in remote regions. A number of references to these plant faunae that are listed at the end of this chapter may be briefly summarized to give a general idea of their extent. Some random samples follow: Mulberry, 115 species (Formosa), 97, 149 (Japan); Hawthorn, 382 (U. S.); Tea, 85 (Formosa); Rice, 157 (Japan); Cacti, 324 (U. S.); Poplar, 114 (Japan); Chestnut, 472 (U. S.); Wheat, 83 (France); Hemp, 100 (Italy); Oaks, 1,400 (France); Willows, 450 (Europe); Citrus fruits, 200 (Tropical Asia); Maize, 352 (U. S.), 123 (Russia); Walnuts, 388; Elm, 634 (U. S.); Spruce foliage, 101 (Canada). In the case of species feeding on cultivated plants that have been spread widely by man, many insects have undoubtedly been added during the course of the plant's transfer to new regions. Thus the biocoenosis gradually grows. With insects feeding on maize, for example, the majority have been derived from the fauna affecting various wild grasses. According to Neiswander these include at least 154 of the 166 species known to be generally abundant, while only seven can be said definitely to have transferred their feeding to corn from plants other than grasses. Grass-feeders are derived from a wide variety of wild grasses. The attraction of insects to newly introduced plants is of very general occurrence; recently Hefley noted in the case of two introduced plants (tamarisk and sweet-clover) in Oklahoma that some thirteen insects are found to feed on them. This is not surprising in the case of the sweet clover, as it has many close relatives in that region. The tamarisk, however, does not. In Australia where introduced plants and animals of all sorts become established very readily, the development of new associations between plants and insects has been very noticeable.

These enormously complex biocoenoses, each centering about a particular plant, have obviously developed over long periods, and it is possible to envisage several factors which have taken part in determining their final form. Shifts or mutations in the instincts concerned with the selection of specific foods have undoubtedly taken place as we know of the occurrence of these from our own experience, and we have very good evidence of similar occurrences in the geological past. Speciation among plants and the insects feeding upon them has been going on for long periods without any sudden changes in food habits. This is evident from the groups of related insects that feed on related plants. There are numerous clear-cut cases which point to such concurrent evolution. Perhaps the most outstanding example of this is the large family of gall-wasps (Cynipidae) which are almost restricted to various oaks.

These will be dealt with in more detail in the following chapter. Genera of insects whose species are restricted to some particular genus of plants are also understandable on this basis. The migration or introduction of insects into regions where they encounter new plants to their liking may add to their list of food plants. It would appear that such cases cannot be considered as mutations of instinct although an occasional one may be of that nature. It is easily apparent that the migrations of insects and the slower migration of plants whereby they have had opportunities to contact one another must be regarded as a potent aid in building up these plant faunae. An interesting case in this connection showing that the actual contact may lead to strange modifications in food plants has been described by Bondar. A Brazilian longicorn beetle, *Sphallenum setosum*, bores as a larva in the stems of a leguminous vine (*Canavalia*) that frequently twines about the trunks of trees. If the *Canavalia* dies before the year-long development of the beetle is completed the larva continues its feeding in the tree. Both the guava and a species of *Anona* are known to serve thus as an acceptable secondary food plant.

The geographical range of their food plants is, of course, a prime factor in determining the distribution of vegetarian insects. Natural barriers affect both, but not necessarily in the same degree. The seeds of many plants are readily disseminated by wind, water, and other natural agencies. The same is true of certain insects, but by no means does it follow that the insects dependent upon such plants are likewise of the kinds best adapted to the same methods of spread. If they are not, the plants will leave their enemies behind. It happens, however, that many insects migrate readily and have been able to follow their food plants into distant regions. This has happened commonly during geological time and of course also very recently through accidental transfer by human agency. Often when new plants appear in a region they prove attractive to insects already there. A number of instances of this sort are known; some of these relating to Australia have recently been noted by French. All of these factors have thus worked together in integrating great numbers of plant faunae, of which no single one can be considered as an independent system.

In contrasting the polyphagous insects with those of monophagous or restricted oligophagous habits, the question may be asked as to which are the more primitive. Considering the habits of most vegetarian animals, including ourselves, it seems obvious that restricted diets represent specializations, and this is no doubt true, but the occurrence of such restricted habits among primitive insects like grasshoppers shows that all of them are by no means necessarily recent developments. We have, of course, no means of knowing what animals fed on before we ourselves came on the scene, and this leaves a gap of many millions of years about which we may only speculate, albeit with some assurance of reasonable accuracy. All in all, it would seem that restricted diets made their appearance as gradual modifications of those involving more varied food plants.

Although the number of food plants utilized by many polyphagous insects is extensive and may in some instances amount to a hundred or more, there always appears to be one, or at most a few species that are more generally preferred. Thus the caterpillars of the gipsy moth so abundant in New England clearly prefer the foliage of oak trees, although they occur on numerous other deciduous trees and shrubs, especially if oaks are absent or have suffered defoliation. The lack of their preferred food and consequent hunger leads them to feed on less acceptable species, and finally as food grows scarcer they will show less and less choice, sometimes even turning to the needles of white pine which will serve the larger larvae, although the newly hatched brood are unable to utilize pine foliage. This is a general phenomenon among insects and is, of course, something common to most animals of oligophagous habits. We ourselves will recognize such a situation as essentially similar to the eating of raw fish or entrails by the starving survivors of a catastrophe at sea. Most of us have never been really so hungry as that, but it is only a matter of degree.

A monophagous insect will deliberately starve to death in the absence of its proper food plant, and most oligophagous species with a highly restricted dietary will do the same. We cannot appreciate such instinctive intolerance. Probably the nearest human approach is the teetotaler who would die of thirst in the midst of aqueous liquids defiled by the taint of one-half per cent alcohol. This phenomenon, however, does not apply to our species as a whole, but to only a sort of mutation or clone which propagates itself through society and not through its own offspring. The same is true of restricted food habits in phytophagous groups of human beings whose diet excludes any animal beyond the egg stage, except a few noncommittal shellfish, bean-weevil larvae or an occasional green caterpillar in the salad. Assuming that these individuals manage to accumulate a proper assortment of amino acids, including the essential sperm-producing arginine for papa, such strains of mankind may reproduce their kind, although they do not seem to be on the increase. As has already been mentioned, it is known that some purely vegetarian insects are derived from carnivorous ancestors.

Although the choice of food represents the preference of the larval or nymphal stages of the insects in question, it involves also the instincts of the imaginal insects as well. The gravid female almost invariably deposits her eggs on the food plant upon which the young will feed, and her instinct in selecting these plants should coincide with the preference of her prospective offspring (Pl. X). This it does, invariably, and the young on hatching find their first meal ready without the necessity of seeking it out by their own efforts. Thus the instinct to oviposit on a particular plant and to feed on it are perfectly attuned from one generation to the next and the next and the next by countless repetition. There is one striking situation in this sequence. The female returns to the plant on which she herself fed in her youth, although her adult food may have been of an entirely different type as happens with the moth that feeds on nectar after attaining its growth on spinach or something equally unappetizing. She then returns to the scene of her early gluttony to prepare for the offspring that will follow in her footsteps. Returning to the scene of the crime is a phenomenon well known to human minions of the law and may be put to practical use in apprehending the non-professional criminal who has become strongly conditioned by the exceptionally vivid mental picture of his misdoings. In the case of a species that has fed for generations on a particular plant this may be regarded as a sort of ancestral memory, and such an interpretation has been advanced by several entomologists. Among insects with fixed monophagous habits or those restricted to a series of very similar food plants, it is wholly reasonable to postulate the occurrence of some such conditioning. Experimental work has, however, failed to demonstrate this clearly in more than a few instances. There are, moreover, some insects that alternate between two entirely different host plants during succeeding generations. We have already mentioned the occurrence of this phenomenon among certain aphids, and there are cases known of moths that behave in the same way. Thus, in *Tephroclystis virgaureata*, the caterpillars of the spring generation feed on Compositae (*Solidago* and *Senecio*) while those of the summer brood attack Rosaceae (*Crataegus* and *Prunus*). In such cases the process of conditioning and the vagaries of ancestral memory have not been reduced to any simple formula.

It must be admitted that some polyphagous insects are composed of strains or races which show a consistent predilection for certain plants acceptable to their species as a whole. This idea was first advanced in 1864 by the American entomologist Walsh who cited a number of observations which support it. Later several others (Cockerell, Hopkins, Cameron) more or less independently came to the same conclusion, and there is a considerable literature relating to such strains, phytophagic races or physiological species as they have been variously called. There are numerous well-known examples of insects which are indistinguishable, or practically so, that select the food plant on which their parents were reared in preference to those on which other members of their own species commonly feed. When such biological races are freely interfertile, show no tendency to selective mating, and coincide in their seasonal and geographical distribution, we must conclude that each individual retains an hereditary attachment to its particular food plant.

Many of the cases placed in this category do not fulfill all these requirements fully, especially after their behavior is carefully investigated by experimental methods, but some unquestionably show such tendencies. The small anthomyiid fly *Pegomyia hyoscyami* exists as two biological races one of which attacks Chenopodiaceae while the other is restricted to Solanaceae. Here the relation is complicated in that members of the latter race generally prefer *Hyoscyamus*, although in its absence they will affect *Atropa belladonna*. These interesting leaf-mining flies have been fully considered by the British entomologists, Cameron and Thorpe. In America there is a common trypetid fly, *Rhagoletis pomonella* which infests the fruits of cultivated apples in countless numbers. This fly also commonly breeds in blueberries, the blueberry strain of flies being smaller, but not morphologically different. According to Woods (15) neither of the two strains can be transferred from apple to blueberry or *vice versa*. Later studies by Pickett and Neary (40) show, however, that the strain from apples may produce at least some progeny able to develop in blueberry, hawthorn, pear, plum, mountain ash, snowberry and cherry, most of which are not normal hosts. Also, the progeny of a cross between the blueberry and apple strains proved fertile, so some migration between several food plants is possible although clearly not frequent in nature.

Here again it is apparent that the fixity of these strains of *Rhagoletis* is not absolute, but sufficiently definite to serve as a quite efficient means of maintaining their integrity. Such strains where hybridity is restricted can be considered as incipient species since they represent populations kept isolated by preference for different food plants, sometimes associated with a modification of their seasonal life history.[1]

Similar instances have been reported by many observers. Among Orthoptera, in the snowy tree cricket, *Oecanthus niveus*, Fulton has noted two races; they are structurally identical but one occurs on deciduous trees and the other on bushes like raspberry and rose. Here the restriction appears to be ecological rather than purely trophic. However, among longicorn beetles whose larvae bore in the wood of various trees, it appears that certain species with several food plants do not always select the one in which they were reared, although there is a tendency, variable in extent among different species, to do this when given a choice. It was from studies of these beetles that Hopkins ('16) proposed his host-selection principle, essentially similar to the matters just discussed. Extensive experiments by Craighead which followed, showed some indication of hereditary preferences, but the results were variable, and somewhat inconclusive. Similar experiments by Twinn on our common cabbage butterfly, which exhibits a quantitative variation in its choice among various acceptable food-plants, show likewise no definite tendency to select the parental food-plant. However, this is exactly what might be expected if such tendencies leading to segregation, separation, and isolation are actually evolutionary developments representing speciation in an earlier stage than that of species including several partially developed hereditary preferences which finally culminate in biological strains or races. References to many examples which may be fitted into such a sequence will be found in the literature cited at the end of this chapter.

The development of such biological races as outlined above ap-

---

[1] Since this chapter was written, a very clear account of races in the European corn-borer has been published by Arbuthnot (Tech. Bull. U. S. Dept. Agric., No. 869, 20 pp. (1944). By crossing individuals of a single and multiple-generation strain of this species he has established the fact that there are two stocks which show a preference for mating with their own kind. Samples of one field population from Connecticut were found to represent a homozygous multiple-generation strain while one from Ohio proved to be impure. From the latter he was able to isolate a homozygous single-generation strain, but attempts to isolate a homozygous multiple-generation were not successful, as some of the factors determining the single-brooded race are evidently recessive.

pears to be a gradual process as we can readily find examples representing all stages from general uniformity to distinct racial segregation.

It is interesting to note that one other group of invertebrate animals, the plant-infesting nematode worms, seem to have developed races or strains very much like those among the insects. Steiner believes that some of these, *e.g.*, species of *Heterodera* and *Tylenchus*, occur as strains that attack the species of host plant in which their parents lived, although if forced by the threat of starvation they will feed on others. Moreover, he thinks that their preference for the parental food plant builds up during successive generations. In this connection it must be remembered that nematodes are much more difficult to deal with in this respect as they lack the clear-cut morphological characters on which the accurate determination of taxonomic relationships must be based.

It has already been mentioned that there are numerous well-known instances where the food habits of certain groups have suddenly shifted from predatism or parasitism to vegetarianism or *vice versa*. This makes it probable that similar shifts in food habits may have taken place within vegetarian groups. Butterflies of the genus *Papilio* furnish an example of this sort. There are several groups of species among these swallow-tails which show strikingly distinctive food habits of obviously long standing. Thus one widespread series feeds on the poisonous *Aristolochia* and another on the foliage of oranges, lemons, and other species of *Citrus*. As both series are widely distributed, it is evident that their food habits have been retained during the long period required for spread and speciation.

Examples of genera containing several species with distinctive but neither uniform nor correlated food preferences are numerous also. Thus among moths of the genus *Papaipema* there are a number of species, most of them with very constant preferences, but differing widely from one another. Among the North American species only one is a general feeder, while the food plants of the others include some twenty plant families ranging from the primitive ferns to the highly specialized Compositae. Here, just as we find in the evolution of morphological structures, there has been a rapid evolution in one group which contrasts very strongly with the great conservatism characteristic of others. Extensive genera commonly exhibit a wide range of food plants, even though the species within them may have very restricted diets. Among scale insects

(Coccidae), Teodoro states on the basis of wide familiarity, that distinctly polyphagous species tend to occur in each genus, but in association with monophagous or highly restricted oligophagous ones. Isolation due to a change in food habits may readily act in the same way as geographical separation since mating is very frequently accomplished on the food plant, especially in the case of many slow moving beetles and weevils where extensive genera with restricted vegetarian habits are commonly encountered.

From the foregoing discussion, certain salient features stand out very clearly. Insects are able to recognize with unerring accuracy the plants on which they feed. These are sought out by the gravid female who lays her eggs upon them, and the young on hatching find themselves on the proper food plant since their appetites are in agreement with the instinctive selection of the mother. Once such a sequence is established natural selection plays an active part in maintaining it, since any failure to select an acceptable plant spells immediate disaster for the young, who are generally unable to forage for themselves. Any aberration of instinct resulting in oviposition on plants not incompatible with the tastes of the larva might easily become characteristic of a biological race or species, and thus shifts in food habits could result directly from mutations in the egg-laying instincts of the adult female. There is reason to believe that such transformations have occurred, but not frequently. The ability of insects to recognize specific food plants is almost uncanny and caused much wonderment and speculation on the part of those who first noticed it. It was earlier assumed that the insects were possessed of a sort of botanical instinct or "sixth sense" which enabled them to recognize the taxonomic affinities of the plants. It was evident also that their acumen in this respect often transcended the insight of some of the earlier systematic botanists themselves, and, such being the case, it appeared indecent to probe further into the esoteric principles involved.

As adult insects are endowed with an extremely delicate sense of odor perception, it is natural to suppose that olfaction serves as the prime factor in enabling them to differentiate accurately between various and closely related plants. The sense of smell is not so highly developed in the larval forms of insects, since the antennae which serve mainly for this purpose are greatly reduced in size and complexity (Fig. 18). Furthermore, the larvae react mainly to the taste of their food and their gustatory organs are not particularly sensitive. Nevertheless, experiment shows that they are highly discriminative, and that the larvae react also to texture, pilosity, and other characteristics recognizable as stimuli to their tactile organs. These as well as the larval taste organs are located on the mouthparts.

Since the beginning of the present century the recognition of food plants has been very generally attributed to a "chemical sense," and until more recently without much closer inquiry into the details of its operation, just as the early cosmeticians concocted their perfumeries by sniffing at mixtures of floral essences and try-

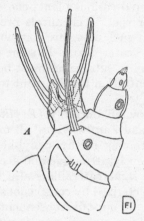 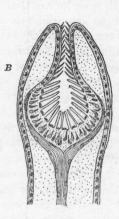

FIG. 18. Sense organs in Lepidoptera. *A*, maxilla of a caterpillar showing the simple nature of the taste organs, present as three sensilla, at the tip of the maxilla and at the side between the four spines. *B*, horizontal section through the labial palpus of an adult butterfly, showing the complex sensory pit lined with a large number of sensilla. *A* redrawn from Dethier, *B*, diagrammatic, from vom Rath.

ing to psychoanalyze their own reactions. Even on such a basis it becomes possible to interpret these food preferences. For example, there are certain butterflies, notably our common cabbage butterfly, *Ascia* (*Pontia*) *rapae* that feed on the leaves of cabbage, mustard, turnips, kohl-rabi, etc., all related members of the plant family Brassicaceae, well known to secrete the highly odorous "mustard oils" that render them palatable to certain "biological races" of the human species. The same oils occur rarely in other families (Capparidaceae, Resedaceae, Geraniaceae) and these butterflies sometimes feed on such plants. A common example is the garden "nasturtium" (*Tropaeolum*) which is a geraniaceous plant, not the real nasturtium of the botanists, although having the same pungent odor and taste. This confusion of different plants by butterflies and likewise by ourselves is clearly due in both instances to the similar essential oils. Further examples of a similar nature are cited in two earlier papers (Brues, '20, '24) should the reader wish to pursue this matter further.

More recently Dethier has considerably extended our knowledge of the potency of odorous chemical substances in determining the selection of food plants by caterpillars. As the result of numerous trials with such larvae that were offered various plants chosen with reference to essential oils and other specific chemicals present in their leaves, it is clearly evident that such odors are commonly the determining factor in influencing the caterpillars. In some cases it even appears that two or more chemical substances must be mixed in the proper proportions and concentration for the plant to be acceptable.

Thus, our North American swallowtail butterfly (*Papilio ajax*) eats some species of Citrus, Ruta, and Zanthoxylum, all members of the plant family Rutaceae, containing methyl-nonyl-ketone in their leaves. This substance is obviously attractive and acceptable to their appetites. However, these same caterpillars also feed on several Umbelliferae, generally characterized by conspicuous flavors (coriander, carrot, caraway, anise). It appears consequently that the larvae are conditioned to general types of chemicals, and it has been shown that they can distinguish at least eight distinct chemicals separately, without confusion. The dietary of this larva may be said therefore to result from an appetite to accept plants containing any of these substances which are also attractive to the butterfly as a substratum for the deposition of her eggs.

Where a great variety of plants are regularly eaten by highly polyphagous caterpillars, Dethier believes that they are accepted because no repellent substances are encountered and that selection is not due to specific attractants to which the insects react. This may be the basis of polyphagy in its most primitive form, but we must admit in the case of general feeders like the caterpillars of the gipsy moth that their marked preference for oak and beech in comparison to many other acceptable plants involves an actual predilection for the foliage of these particular trees.

Thus the botanical instinct of insects may be attributed, at least partially, to a chemical sense in recognizing odors by olfactory organs which in the adult insect are excessively sensitive and capable of acute discrimination. These have made it possible for insects to develop oligophagous and monophagous habits of a very precise type, although they are only accessory to the basic instincts of food selection and cannot serve to explain the persistent attachment to specific substances so commonly encountered.

Although the attractiveness of certain chemicals commonly leads insects to their proper niche in the vast biological environment, it may also lead them hopelessly astray. With chemists busily engaged in extracting the innermost secrets of the most diverse organisms and reproducing innumerable natural substances by synthetic methods, many compounds have been made available for biological experimentation. Entomologists have not been slow to try a great variety of materials as baits or lures for noxious insects. Some of the results are quite unexpected and highly instructive. As already mentioned, the essential oils from food plants exercise the expected attractiveness of the plants themselves. However, certain radically different substances recall the familiar behavior of the domesticated cat when offered a sprig of catnip. Thus the Japanese beetle (*Popillia japonica*), a polyphagous species affecting a wide range of plants, reacts with such irresistible ardor to a mixture of two aromatic vegetable liquids, geraniol and eugenol, that practically all beetles in the vicinity may be caught in a trap thus baited. One such trap actually caught 13,000 beetles in a period of eight hours. These two substances are in no way related chemically and occur in a number of diverse plants; furthermore they are not general attractants since native beetles of allied genera and the Asiatic garden beetle are entirely unresponsive to their presence. In connection with this is the surprising fact, noted by Ballou, that cultivated geraniums (*Pelargonium*), the natural source of geraniol, are highly poisonous to the Japanese beetle and cause a paralysis leading to death in beetles that have fed on the leaves. Still more unexpected is the success of kerosene oil or ammonia as a bait in traps designed to eradicate certain fruit flies whose larvae develop in the pulp of various fruits. It would be a wild stretch of the imagination to interpret this behavior on the basis of natural food, ancestral memory, atavism, or what-not.

Many unusual types of food, vegetable and otherwise, are of course acceptable or regularly utilized by various insects. Some of these are referred to in Chapter II and elsewhere in the present account.

On the basis of the matters just discussed, it is easy to see how mutations have occurred in the selection of food plants, entailing sudden shifts to new diets which are then retained with great tenacity. Those dependent upon the presence of identical substances in several different plants are in a sense predictable, or at least they are not unreasonable in our own eyes; this especially in the days of substitute foods when a translation of *Ersatz* has become a patriotic necessity. Shifts to plants with no apparent similarity in odor or taste are at present beyond the pale of understanding, except that they appear to be permanent changes in instinct, closely resembling mutations in structure. The latter are familiar to experimental biologists, but their source remains inscrutable.

In the nature of things, plants usually show no immediate reactions to the feeding of insects, except to weaken and succumb following destruction of their foliage, stems, roots, or whatever portions may be eaten away or ruined functionally. There are certain difficulties imposed on insect feeding that may come into action promptly, but these are generally the result of normal functions whose primary object bears no relation to insects. Bark beetles boring their way into trees to excavate their brood tunnels may be driven away by the copious flow of sap or resin exuding from strong, healthy trees. Such beetles are forced to confine their depredations to sickly or moribund hosts unless their great abundance may so divide the flood that the beetles are not drowned out. Frequently the regeneration of tissues that have been destroyed takes place, especially in roots that have been severed or in the case of injury to the bark of woody plants. Rarely regenerative changes involving hypertrophy may actually destroy the insect. Thus, many years ago, Hinds noted the destruction of eggs of the Mexican cotton boll-weevil resulting from cell proliferation induced by the egg-puncture of the weevil in the growing boll.

It has been noticed that certain genetic strains of agricultural plants vary in their susceptibility to attack by specific insects, but so many factors are involved that it would be futile to make any generalized statements, especially as we cannot set aside the instinctive behavior of the insects in any case. The statement is often made that certain varieties of their food plants are "more resistant" to the feeding of insects. This rather ambiguous expression refers to the fact that some strains escape with less damage than others, or that specific insects develop or multiply more rapidly on certain strains. Such immunity is a common occurrence with fungous diseases, and has been demonstrated in the case of numerous species of insects. It may depend upon the toughness of leaves, their hairiness or actually the available food materials in the plant, as has been shown by the variable rate of reproduction of aphids on different varieties of beans. In the same way Evans has demonstrated that seasonal changes in the composition of cabbage plants affect not only the rate of growth of the cabbage butterfly but also the final weight of the pupa. Even diurnal variations in the nutrient content of mulberry leaves affect the weight of silkworm cocoons spun by caterpillars fed on leaves picked at different times during the day.

Another type of reaction on the part of plants to the presence of insects results in the development of malformations of very specific and distinctive form and structure. Known as cecidia or galls these growths show so many interesting features that they will be considered separately in the following chapter. They develop only when the insects are actually within the tissues or more rarely when they feed continuously with their mouthparts inserted into the plant.

Many plants contain powerful drugs that are exceedingly poisonous to animal life, and insects form no exception. These are usually organic bases or alkaloids, peculiar to the plants that produce them, although some occur sporadically among unrelated plants and others may be elaborated by several allied species. On account of their toxicity to insects, some are very generally used as insecticides. The tobacco alkaloid, nicotine, is one of these, effective with a wide variety of insects, yet there are numerous others that depend on tobacco as a food plant. Like the practised human smoker, they possess an immunity to its nauseating effects and are able to complete their entire development without resort to materials of more generally accepted nutritional status. Some of these tobacco insects are, moreover, polyphagous species and in such cases immunity is obviously innate and not acquired as in the case of those of us who must continually indulge in Lady Nicotine to appreciate only her virtues. Evidence that immunity to nicotine is not equal among those insects that feed on the foliage of tobacco is furnished by the behavior of the Colorado potato beetle whose larvae feed ordinarily on various species of Solanum. In a field planted with numerous species of Nicotiana, including commercial tobaccos, the beetles will extend their feeding to those of low nicotine content, but injury to plants more strongly fortified with the alkaloid is not noticed. An exactly similar condition prevails in the case of the potato flea-beetle, *Psylliodes*, which sometimes feeds on tobacco. A clear-cut difference between closely related species is seen in moths of the genus *Ephestia*. The Mediterranean flour moth, *Ephestia kuehniella* is a cosmopolitan pest of granaries and pantries where the larvae feed on a great variety of processed cereal products, while the tobacco moth, *Ephestia elutella*, also known as the cacao moth, has become a notorious enemy of stored tobacco.

Another powerful insecticide is rotenone, a poisonous ketone found in combination with certain toxic resins in derris root, derived principally from bean-like plants of the genus *Derris*. Known to aboriginal man as a powerful fish and arrow poison, derris root has become a widely used insecticide, standardized on the basis of its rotenone content, but derris has various insect enemies that feed in the roots, stems, and eat the foliage. These include twelve species of beetles that feed in the dried roots. That these actually digest the contained rotenone is proved by the absence of any insecticidal properties in extracts made from their excreta. About an equal number of lepidopterous caterpillars are known to feed on the leaves of the growing plants. In the American tropics the roots of several other unrelated plants known as cubé contain rotenone also. Another insect poison is pyrethrum, the source of Persian insect powder and the active ingredient of the familiar "Flit." The pyrethrum plant has several insect enemies, one of which is a longicorn beetle, *Phytoecia virgula*. The larvae live in the stems, but occur only in plants growing on poor soil where they do not develop the poisonous principle found in healthy plants. Likewise other drug plants have insects living at their expense. The seeds of *Colchicum* from which the extremely poisonous colchicine is derived, serve as food for the byrrhid beetle, *Simplocaria semistriata*. Even poison ivy is not immune; an aphid, a common leaf-beetle and a leaf-miner feed on the leaves, and a midge forms galls on its roots in addition to a bark-beetle and one or two caterpillars that injure the stems and foliage. Insect enemies of this plant are perhaps not so unexpected as the irritant is an oil which exerts its effects through the skin rather than as an alimentary poison, and the integument of insects is not readily permeable to such substances.

The influence of insect feeding as checks on the growth and reproduction of plants has already been mentioned as a great detriment to floras in their natural condition, and to agriculture, horticulture, and forestry as practiced in the world today. We promised not to deal with these practical matters, as they lie out-

side the scope of the present book, but many of the examples cited
have been drawn from this economically important series for the
reason that their habits are far better known than those of their less
conspicuous relatives.

There is one phase of the economic picture that deserves some
notice, not that it is essentially different, but because it is less
widely publicized. This relates to the utilization of phytophagous
insects in the reduction or control of noxious weeds. In this capacity
we can regard the insects as friendly beings, not as drags on the
merry-go-round of human progress. This cheers up the poor entomologist who must otherwise spend a goodly part of his time
damning the ubiquitous little animals that made his profession
possible. It should also be good for the public morale for us to learn
that something really good is going on right under our noses,
initiated without legal procedure and receiving only an occasional
prod from some governmental agency to speed its progress.

One or two examples will suffice. Undoubtedly, the most important venture has been the control of the prickly-pear cactus in
Australia. The true cacti, or family Cactaceae, are all native to
the New World, but have been imported widely into many parts of
the eastern hemisphere as pot or garden plants, for forage, or for no
particular reason, or even accidentally. They cannot survive in
frigid climates but have become extensively naturalized throughout
tropical and subtropical regions, desert and otherwise. The most
abundant and prolific of these immigrants are species of *Opuntia*
or "prickly-pear." By the early years of the present century several
species had become established over considerable parts of the coastal
regions of Australia, spreading inland into immense areas of wheat,
range, and marginal agricultural land and choking out other plant
life. In 1928 some 75,000 square miles were infested in Queensland
alone. Plowing, burning, spraying or feeding to animals was of no
avail in stemming the tide of invasion and numerous American insects that feed on *Opuntia* were introduced as an aid in destroying
the plants. These included a considerable variety, as these common
cacti support a rich fauna of native American insects of several
orders, scale insects (cochineal), plant bugs, beetles, moths, etc.
Many were of some service, but one of the later importations, a
small moth (*Cactoblastis cactorum*) whose larvae mine in the paddle-
like stems, proved so efficient that it became established by 1931 on
practically every acre of the infested area in Queensland and New
South Wales. Collapse of the primary "pear" followed and although

this does not completely destroy the plants, they are much weakened and succumb to repeated attacks by the insects. At present wide areas have been reclaimed for agricultural production. This moth is practically restricted to cactus and could, like the cochineal scales and others, be introduced without fear of endangering other elements of the Australian flora. Other tropical countries like India and South Africa have successfully applied the same methods. Similar, less completely satisfactory control has been established in the same areas and elsewhere in eradicating other weeds such as gorse, St.-John's-wort, ragwort, and lantana. This method of natural control is, of course, fraught with some danger, as the reactions of insects to unfamiliar plants cannot be predicted with certainty. With weeds that are botanically related to useful plants the danger is of course greater, and extensive experimentation is a necessary prelude to such colonization. Otherwise serious economic losses may occur through the destruction of crop plants or other desirable members of the flora that may prove to be suitable food plants for the imported insects.

The foregoing account includes only a brief outline of the food relations of the phytophagous insects. To serve the purpose of further inquiry a list of references is appended. It is necessarily very incomplete, but will serve to extend the topics that have been presented, as well as to introduce the reader to others upon which we have not touched.

REFERENCES

DISTRIBUTION OF PHYTOPHAGY AMONG INSECTS. LEAVES, ROOTS, STEMS, ETC., AS FOOD

Alexander, C. P.

  1920 The Crane-flies of New York; Part II. Biology and Phylogeny. Mem. Cornell Univ. Agric. Expt. Sta., No. 38, pp. 695–1133, pls. 12–97.

Altson, A. M.

1923   On the Method of Oviposition and the Egg of *Lyctus brunneus*. Journ. Linn. Soc. Zool., vol. 35, pp. 217–227, 2 figs., 1 pl.

Back, E. A.

1940   Bookworms. Ann. Rept. Smithsonian Inst. for 1939, pp. 365–374, 18 pls.

Ball, E. D.

1932   The Food Plants of the Leaf Hoppers. Ann. Entom. Soc. America, vol. 25, pp. 497–501.

Balachowsky, A., and L. Mesnil

1935   Les insectes nuisibles aux plantes cultivées. Minis. Agric. Paris, 1921 pp., 8 pls., 1369 figs. Busson, Paris.

Barber, F. S.

1913   Notes on a Wood-boring Syrphid. Proc. Entom. Soc. Washington, vol. 15, pp. 151–152.

Barbey, A.

1913   Traité d'entomologie forestière. 617 pp., 350 figs., 8 pls., Paris and Nancy.

Barnes, H. F.

1933   Some Biological and Economic Aspects of the Gall Midges. Sci. Progr., vol. 29, pp. 73–86.

1933   A Cambium Miner of Basket Willows. Ann. Appl. Biol., vol. 20, pp. 458–519, 2 pls., 12 figs.

Bedwell, E. C.

1931   Dermestid Beetles Attacking Wood. Entom. Monthly Mag., vol. 67, pp. 93–94.

Bequaert, J.

1922   Ants in their Diverse Relations to the Plant World. (Ants of the American Mus. Congo Exped.). Bull. American Mus. Nat. Hist., vol. 45, pp. 333–530, 4 pls., 24 figs.

1940   Notes on the Distribution of *Pseudomasaris* and on the Foodplants of the Masaridinae and Gayellinae. Bull. Brooklyn Entom. Soc., vol. 35, pp. 37–45.

Blackman, M. W., and H. H. Stage

1924   On the Succession of Insects Living in the Bark and Wood of Dying, Dead and Decaying Hickory. Tech. Publ. New York State Coll. Forestry, pp. 1–269, 14 pls.

Blood, H. L., B. L. Richards, and F. B. Wann

1933   Studies of Psyllid Yellows of Tomato. Phytopathology, vol. 23, p. 930.

Börner, C.

1908   Eine monographische Studie über die Chermiden. Arb. Biol. Anst. f. Land-u. Forstw., Berlin, vol. 6, pp. 81–320.

Boyce, A. M.

1927   A Study of the Biology of the Parsley Stalk Weevil, *Listronotus latiusculus*. Journ. Econ. Entom., vol. 20, pp. 814–821.

Brindley, H. H.

1920   Further Notes on the Food Plants of the Common Earwig (*Forficula auricularia*). Proc. Cambridge Phil. Soc., vol. 30, pp. 50–55.

Brittain, W. H., and D. E. Newton

1933 A Study of the Relative Constancy of Hive Bees and Wild Bees in Pollen Gathering. Canadian Journ. Res., vol. 9, pp. 334–339.

1934 Further Observations on Pollen Constancy in Bees. Canadian Journ. Res., vol. 10, pp. 255–263.

Brues, C. T.

1927 Observations on Wood-boring Insects, their Parasites and Other Associated Insects. Psyche, vol. 34, pp. 73–90.

1930 The Food of Insects Viewed from the Biological and Human Standpoint. Psyche, vol. 37, pp. 1–14.

Buhr, H.

1932– Mecklenburgische Minen. I. Agromyzidenminen. Stettiner Entom. Zeitg.,
33 vol. 93, pp. 57–115. II. Coleopteren- Tenthrediniden- und Dipteren-Minen. *Ibid.*, vol. 94, pp. 47–96.

Butler, E. A.

1918 On the Association between the Hemiptera-Heteroptera and Vegetation. Entom. Monthly Mag., vol. 54, pp. 132–136.

1923 A Biology of the British Hemiptera-Homoptera. 326 pp., illus., London, H. F. and G. Witherby.

Cameron, A. E.

1918 Life History of the Leaf-eating Crane-fly, *Cylindrotoma splendens*. Ann. Entom. Soc. America, vol. 11, pp. 67–89.

1930 Two Species of Anthomyiid Diptera Attacking Bracken. Scottish Naturalist, No. 185, pp. 137–141.

Caudell, A. N.

1904 An Orthopterous Leaf Roller. Proc. Entom. Soc. Washington, vol. 6, pp. 46–49.

Cavara, F.

1922 Danneggiamenti delle Termitidi a pianti diverse. Rend. Accad. Sci. Fis. e. Math. (3a), vol. 27, pp. 190–194.

Chaine, J.

1911– Termites et plantes vivantes. C. R. Soc. Biol., Paris, vol. 71, pp. 678–680
12 (1911); *ibid.*, vol. 72, pp. 113–115 (1912).

Chamberlin, W. J.

1942 The Bark and Timber Beetles of North America. 513 pp., 5 pls., 321 figs. Corvallis, Idaho, Oregon State Coll.

Chambers, V. T.

1877– Notes Upon the American Species of *Lithocolletis*. Psyche, vol. 2, pp. 81–87,
78 137–153.

Christy, M.

1919 Hornets, Wasps and Flies Sucking the Sap of Trees. Essex Naturalist, vol. 19, pp. 10–14.

Chrystal, R. N.

1932 An Oecophorid Moth, *Borkhausenia pseudospretella*, Attacking Book Bindings. Entom. Monthly Mag., vol. 68, pp. 9–10.

Clarke, S. H.

1928 On the Relationship between Vessel Size and *Lyctus* Attack in Timber. Forestry, vol. 2, pp. 47–52, 1 pl.

Cockerell, T. D. A.

1897 The Food Plants of Scale Insects (Coccidae). Proc. U. S. Nat. Mus., vol. 19, pp. 725–785.

Cockle, J. W.

1921 *Vitula serratilineella,* a Honey-feeding Larva. Proc. Entom. Soc. British Columbia, No. 16, pp. 32–33.

Cortantino, G.

1930 Contributo alla conoscenza della mosca delle frutta (*Ceratitis capitata*). Boll. Lab. Zool. Portici, vol. 23, pp. 237–322, 20 figs.

Cotton, R. T.

1920 Four Rhyncophora Attacking Corn in Storage. Journ. Agric. Res., vol. 20, pp. 605–614, 4 pls.

1920 Rice Weevil, (*Calandra*) *Sitophilus oryzae.* Journ. Agric. Res., vol. 20, pp. 409–422, 1 pl.

1937 Insects and Mites Associated with Stored Grain. Misc. Publ. U. S. Dept. Agric., No. 258, 81 pp.

Craighead, F. C.

1923 A Classification and the Biology of North American Cerambycid Larvae. Bull. Canadian Dept. Agric., No. 27, 238 pp., 44 pls., 8 figs.

Crosby, C. R.

1909 On Certain Seed-infesting Chalcis-flies. Bull. Cornell Univ. Agric. Expt. Sta., No. 265, pp. 367–388.

Crumb, S. E.

1932 The More Important Climbing Cutworms. Bull. Brooklyn Entom. Soc., vol. 27, pp. 73–100, 2 pls.

Davenport, D., and V. G. Dethier

1937 Bibliography of the Described Life Histories of the Rhopalocera of America North of Mexico. Entom. Americana, vol. 17, pp. 155–194.

Davidson, J.

1921 Biological Studies of *Aphis rumicis.* Sci. Proc. Roy. Dublin Soc., vol. 16, pp. 304–322, 3 figs.

1923 Biological Studies of *Aphis rumicis.* The Penetration of Plant Tissues and the Source of the Food Supply of Aphids. Ann. Appl. Biol., vol. 10, pp. 35–54, 4 figs., 2 pls.

1924 Factors which Influence the Appearance of the Sexes in Plant Lice. Science, vol. 59, p. 634.

1925 A List of British Aphids. xi + 176 pp., London, Longmans, Green and Co.

De Gryse, J. J.

1916 The Hypermetamorphism of the Lepidopterous Sap-feeders. Proc. Entom. Soc. Washington, vol. 18, pp. 164–168.

Doering, K. C.

1942 Host Plant Records of Cercopidae in North America. Journ. Kansas Entom. Soc., vol. 15, pp. 65–92.

Eckert, J. E.

1942 The Pollen Required by a Colony of Honeybees. Journ. Econ. Entom., vol. 35, pp. 309–311, 1 fig.

Edwards, H.

1889 A Bibliographic Catalogue of the Described Transformations of North American Lepidoptera. Bull. U. S. Nat. Mus., No. 35, 147 pp.

Eliescu, J.

1932 Beiträge zur Kenntnis der Morphologie, Anatomie und Biologie von *Lophyrus pini*. II. Teil. Zeits. angew. Entom., vol. 19, pp. 188–206.

Emery, C.

1912 Alcune esperienze sulle formiche granivore. Rendic. Acad. Sci., Bologna, vol. 16, pp. 107–117.

Escherich, K.

1914 Die Forstinsekten Mitteleuropas. Vol. 1, xii + 433 pp., 1 pl., 248 figs.
1923 Vol. 2, viii + 633 pp., 335 figs.
1931 Vol. 3, 825 pp., 625 figs., Berlin, Paul Parey.

Faure, J. C.

1932 The Phases of Locusts in South Africa. Bull. Entom. Res., vol. 23, pp. 293–424, 25 pls., 1 map.
1933 The Phases of the Rocky Mountain Locust, *Melanoplus mexicanus*. Journ. Econ. Entom., vol. 26, pp. 706–718.

Felt, E. P.

1905– Insects Affecting Park and Woodland Trees. Mem. New York State Mus.,
06 No. 8, 2 vols., 877 pp.

Felt, E. P., and W. H. Rankin

1932 Insects and Diseases of Ornamental Trees and Shrubs. xix + 507 pp., New York, Macmillan.

Fisher, R. C.

1932 Prevention and Control of Damage by Wood-boring Insects. Forestry, London, vol. 6, pp. 67–74.

Forbes, W. T. M.

1923 The Lepidoptera of New York and Neighboring States. Primitive Forms, Microlepidoptera, Pyraloids, Bombyces. Mem. Cornell Univ. Agric. Expt. Sta., No. 68, 729 pp., 439 figs.

Froggatt, W. W.

1923 Forest Insects of Australia. viii + 171 pp., 32 figs., 46 pls., Sydney, Forestry Comm. of New South Wales.

Frost, S. W.

1923 A Study of the Lead-mining Diptera of North America. Mem. Cornell Univ. Agric. Expt. Sta., No. 78, 228 pp., 13 pls., 2 figs.
1924 The Leaf-mining Habit in the Coleoptera. Ann. Entom. Soc. America, vol. 17, pp. 457–467, 1 pl.
1925 The Leaf-mining Habit in the Hymenoptera. Ann. Entom. Soc. America, vol. 18, pp. 399–414, 2 pls.
1925 Convergent Development in Leaf-mining Insects. Entom. News, vol. 36, pp. 299–305.

Funkhouser, W. D.

1917   Biology of the Membracidae of the Cayuga Lake Basin. Mem. Cornell Univ. Agric. Expt. Sta., No. 11, pp. 177–445.

Gahan, A. B.

1922   List of Phytophagous Chalcidoidea. Proc. Entom. Soc. Washington, vol. 24, pp. 33–58.

Gilledge, C. J.

1929   The Insect Pests of Books. An Annotated Bibliography. Library Assoc. Rec. (2), vol. 7, pp. 240–242.

Gilmer, P. M.

1933   The Entrance of Codling Moth Larvae into Fruit, with Special Reference to the Ingestion of Poison. Journ. Kansas Entom. Soc., vol. 6, pp. 19–25.

Gould, G. E., and H. O. Deay

1940   The Biology of Six Species of Cockroaches which Inhabit Buildings. Bull. Indiana Expt. Sta., No. 451, 31 pp., 14 figs.

Graham, S. A.

1929   Principles of Forest Entomology. xiv + 339 pp., 1 pl., 149 figs., New York, McGraw-Hill.

Grandi, G.

1920   Studio morfologico e biologico della *Blastophaga psenes*. Boll. Lab. Zool. Portici, vol. 14, pp. 63–204.

1925   Biologia, morfologia e adattamento negli insetti dei fichi. Atti. Soc. Ital. Sci. Nat., vol. 63, pp. 288–311.

1930   Monografia del gen. *Philotrypesis*. Boll. Lab. Entom. Bologna, vol. 3, pp. 1–181, 76 figs.

Greene, C. T.

1914   The Cambium Miner in River Birch. Journ. Agric. Res., vol. 5, pp. 471–474.

1917   Two New Cambium Miners. *Ibid.*, vol. 10, pp. 313–317.

Grossman, E. F.

1929   Biology of the Mexican Cotton Boll Weevil. III. The Mechanism of Grub Feeding. Florida Entom., vol. 13, pp. 32–33.

Hargreaves, E.

1915   The Life-history and Habits of the Greenhouse White Fly (*Aleyrodes vaporariorum*). Ann. Appl. Biol., vol. 1, pp. 303–334, 56 figs.

Harvey, F. L.

1895   Mexican Jumping Beans. American Naturalist, vol. 29, pp. 767–769.

Hefley, H. M.

1916   The Relations of Some Native Insects to Introduced Food-plants. Journ. Anim. Ecol., vol. 6, pp. 138–144.

Heikertinger, F.

1924   Monographie der paläarktischen Halticinen, Biologischer Teil. Entom. Blätter, vol. 20, pp. 214–224.

1926   Über ein auffälliges Käferfrassbild (*Otiorrhynchus crataegi.*). Coleopterolog. Rundschau, vol. 12, p. 399.

Hering, M.

1920– Minenstudien. Pts. 1–3, Deutsch. Entom. Zeits.; pts. 4, 6, 7, Zeits. f. Mor-
34   phol. Oekol.; pts. 5, 11, Zeits. f. wiss. Insektenbiol.; pt. 9, Zool. Jahrb., Abth. f.
  Anat.; pt. 10, Zeits. f. angew. Entom.; pts. 12, 13, 14, Zeits. f. Pflanzen-
  krankh.

1923 Durch Insektenlarven erzeugte Blattminen. Biol. Tiere Deutschlands, Teil.
  43, 17 pp., 18 figs.

1924– Das histologische Bild der von Insektenlarven erzeugten Blattminen. Mikro-
25   kosmos, vol. 18, pp. 161–166; 167–180, 9 figs.

1926– Die Oekologie der blattminierenden Insektenlarven. 254 pp., 67 figs., 2 pls.,
  Berlin, Bornträger Bros.

Heymons, R., H. von Lengerken, and M. Bayer

1926 Studien über die Lebenserscheinungen der Silphini. I. *Silpha obscura*. Zeits.
  Morph. Oekol. d. Tiere, vol. 6, pp. 287–332.

Howard, N. F.

1926 Feeding of the Mexican Bean Beetle Larva. Ann. Entom. Soc. America,
  vol. 34, pp. 766–769, 1 fig.

Howard, N. F., and L. L. English

1924 Studies of the Mexican Bean Beetle in the Southeast. Bull. U. S. Dept.
  Agric., No. 1243.

Howland, L. J.

1930 The Nutrition of Mosquito Larvae with Special Reference to their Algal Food.
  Bull. Entom. Res., vol. 21, pp. 431–440.

Hutson, J. C.

1933 The Root-eating Ant (*Dorylus orientalis*). Trop. Agriculturist, vol. 80, pp.
  276–280.

Johnson, C. G.

1937 The Biology of *Leptobrysa rhododendri*, the Rhododendron lacebug. II. Feeding
  Habits and the Histology of the Feeding Lesions Produced in Rhododendron
  Leaves. Ann. Appl. Biol., vol. 24, pp. 342–355, 2 pls.

Karny, H. H.

1934 Biologie der Wasserinsekten. 311 pp., Wagner, Vienna.

Kemner, N. A.

1933 Om insekten och insekstador i herbarier. Bot. Notiser, Lund., No. 1–3, pp.
  439–455.

Kirkaldy, G. W.

1909 Catalogue of the Hemiptera, with Biological References, Food Plants, etc.
  xl + 392 pp., Berlin, F. L. Dames.

Kirchner, O. von

1911 Blumen und Insekten, ihre Anpassungen aneinander und ihre gegeneitige
  Abhängigkeit. Leipzig and Berlin, iv + 436 pp., 2 pls., 159 figs.

Kleine, R.

1934 Die Borkenkäfer (Ipidae) und ihre Standpflanzen. Eine vergleichende Studie.
  I. Teil. Zeits. angew. Entom., vol. 21, pp. 123–181.

Knoll, F.

1922– Insekten und Blumen. Experimentelle Arbeiten zur Verteilung unserer
23 Kenntnisse über die Wechselbeziehungen zwischen Pflanzen und Tieren.
Abh. zool.-bot. Ges. Wien, vol. 12, 1921, pp. 1–117; vol. 12, 1922, pp. 123–
377.

Kraus, E. J., and A. D. Hopkins

1911 A Revision of the Powder-post Beetles, Lytidae, of the U. S. and Europe.
Dept. Agric., Bur. Entom., Tech. Ser., No. 20, pt. 3, pp. 111–138.

Krausse, A.

1917 Frassbilder der Larve von *Macrophya albicincta*. Arch. Naturgesch. Jahrg.
82 A., Heft 4, pp. 48–49.

Künckel d'Herculais, J.

1907 Un Diptère vivipare de la famille des Muscides à larves tantôt parasites,
tantôt végétariennes. C. R. Soc. Biol. Paris, vol. 144, pp. 390–393.

Lahille, F.

1920 La Langosta en la República Argentina. Minist. Agric. Lab. Zool., Buenos
Aires, 172 pp., 1 map, 11 pls., 16 figs.

Lengerken, H. von

1929 Der Blattschnittmethode des Ahornblattrollers. Biol. Zentralbl., vol. 49,
pp. 469–490, 9 figs.

Lesne, P.

1911 Les variations du régime alimentaire chez les Coléoptères xylophages de la
famille des Bostrychides. C. R. Acad. Sci. Paris, vol. 152, pp. 625–628.

1924 Les Coléoptères Bostrychides de l'Afrique tropicale. Encycl. Entom., vol. 3,
301 pp., 173 figs., 38 maps.

Lindinger, L.

1912 Die Schildläuse (Coccidae). 388 pp. Ulmer, Stuttgart, Germany.

Linsley, E. G.

1942 Insect Food-caches as Reservoirs and Original Sources of Some Stored Product Pests. Journ. Econ. Entom., vol. 35, pp. 434–439.

Lizer, C.

1920 Zona permanente ó de refugio invernal de la *Schistocerca paranensis* (Langosta voladora). Boll. Minist. Agric. Nac. Buenos Aires, vol. 24, pp. 26–70,
4 maps.

Lovell, J. H.

1913 The Origin of the Oligotropic Habit Among Bees. Entom. News, vol. 24,
pp. 104–112.

1915 The Origin of Anthophily Among the Coleoptera. Psyche, vol. 22, pp. 67–84.

1915 Preliminary List of the Anthophilous Coleoptera of New England. Psyche,
vol. 22, pp. 109–117.

Lüderwaldt, H.

1910 Die Frassspuren von *Cephaloldia dayrollei*. Zeits. wiss. Insektenbiol., vol. 6,
pp. 61–63.

McAtee, W. L.

1908 Notes on an Orthopterous Leaf Roller. Entom. News, vol. 19, pp. 488–491,
1 pl.

MacGillivray, A. D.

1913   The Immature Stages of Tenthredinoidea. Ann. Rept. Entom. Soc. Ontario, vol. 44, pp. 54–75, 1 pl.

Manson, G. F.

1931   Aphid Galls as a Noctuid Feeding Ground. Canadian Entom., vol. 63, pp. 171–172.

Mansour, K., and J. J. Mansour-Bek

1934   On the Digestion of Wood by Insects. Journ. Expt. Biol., vol. 11, pp. 243–256.

Marvin, G. E.

1933   Nectar Secretion of the Tuliptree or Yellow Poplar. Journ. Econ. Entom., vol. 26, pp. 170–176.

Meijere, J. H. C. de

1905–   Im Innern von Farnkräutern parasitierende Insektenformen. Tijdschr. v.
06   Entom., vol. 48, pp. lvi–lviii.
1940   Über die Larven der in Orchideen minierenden Dipteren. *Idem.*, vol. 83, pp. 122–127.

Merrill, G. B., and J. Chaffin

1923   Scale Insects of Florida. Quart. Bull. Florida State Plant Bd., vol. 7, pp. 178–298, 100 figs.

Metcalf, C. L., and W. P. Flint

1939   Destructive and Useful Insects. 981 pp., McGraw-Hill, New York.

Miller, D.

1920   Insects Inhabiting the Gum Fluid of *Phormicum.* New Zealand Journ. Agric., vol. 21, pp. 335–337, 7 figs.

Miller, J. M.

1914   Insect Damage to the Cones and Seeds of Pacific Coast Conifers. Bull. U. S. Dept. Agric., Bur. Entom., No. 95, 7 pp., 2 pls.

Mitterberger, K.

1919   Die Nahrungspflanzen der heimischen Coleophora-Arten. Arch. Natg. Berlin, Abt. A, vol. 83, Heft 6 (1917), pp. 55–75.

Mordvilko, A. K.

1924   Anholocyclic Aphids and the Glacial Epoch. C. R. Acad. Russie, for 1924, pp. 54–56. [In Russian; abstract in Rev. Appl. Entom.]
1929   Food Plant Catalogue of the Aphididae of Russia. Trud. prikl. Entom., vol. 14, pp. 1–100, 20 figs. [In Russian; abstract in Rev. Appl. Entom.]

Morgan, A. C.

1913   New Genera and Species of Thysanoptera, with Notes on Distribution and Food Plants. Proc. U. S. Nat. Mus., vol. 46, pp. 1–55, 79 figs.

Müller, H.

1883   Die Befruchtung der Blumen durch Insekten. 478 pp. Leipzig.

Needham, J. G., S. W. Frost, and B. H. Tothill

1928   Leaf-mining Insects. viii + 351 pp., 91 figs., 3 pls., Baltimore, Williams and Wilkins.

PLATE IX

A. Larva of a wood-boring Buprestid beetle. The head and prothorax are greatly enlarged to accommodate the powerful muscles that actuate the mandibles. Original.

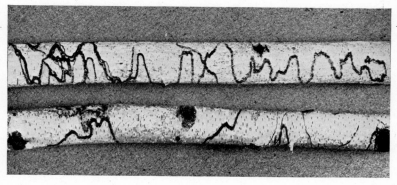

B. Path of feeding burrows of a cambium-mining fly larva, showing on logs of gray birch. Photograph by the author.

PLATE X

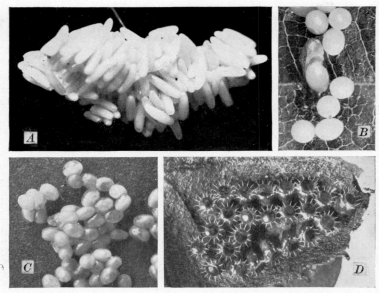

Eggs of various insects. *A*, the housefly (*Musca*); *B*, a leaf beetle (*Monocesta coryli*) on the leaf of the food plant; *C*, the leopard moth (*Zeuzera pyrina*) on bark; *D*, the squash-bug (*Anasa tristis*) on squash leaf. Original photographs.

Needham, J. G., J. R. Traver, and Yiu-Chi Hsu

1935 The Biology of Mayflies. xiv + 759 pp., Comstock Publishing Co., Ithaca, New York.

Needham, P. R.

1938 Trout Streams. x + 233 pp., Comstock Pub. Co., Ithaca, New York.

Neger, F. M.

1910 Neue Beobachtungen an körnersammelnden Ameisen. Biol. Centralbl., vol. 30, pp. 138–150.

Nielsen, J. C.

1906 Zoologische Studien über die Markflecke. Zool. Jahrb. Abth. f. Syst., vol. 23, pp. 725–738.

Nikolski, V. V.

1925 *Locusta migratoria*. Trans. Dept. Appl. Entom., State Inst. Expt. Agron., Leningrad, vol. 12, 330 pp., 37 figs. [In Russian; abstract in Rev. Appl. Entom.]

Norris, M. J.

1936 The Feeding-habits of the Adult Lepidoptera Heteroneura. Trans. Roy. Entom. Soc. London, vol. 85, pp. 61–90.

Oertel, E.

1939 Honey and Pollen Plants of the United States. U. S. Dept. Agric., Cir. 554, pp. 64, 1 fig.

Park, O. W.

1928 Time Factors in Relation to the Acquisition of Food by the Honeybee. Res. Bull. Iowa Agric. Expt. Sta., No. 108, pp. 185–225, 13 figs.

Parker, R. L.

1926 The Collection and Utilization of Pollen by the Honeybee. Mem. Cornell Expt. Sta., No. 98, 55 pp.

Parkin, E. A.

1936 A Study of the Food Relations of the *Lyctus* Powder Post Beetles. Ann. Appl. Biol., vol. 23, pp. 369–400, 1 pl., 4 figs.

Patch, E. M.

1938 Food-plant Catalogue of the Aphids of the World. Bull. Maine Agric. Expt. Sta., No. 393, 431 pp.

Phillips, W. I., and W. T. Emery

1919 Revision of the Chalcid-flies of the Genus *Harmolita*. Proc. U. S. Nat. Mus., vol. 55, pp. 433–471, 10 pls.

Pierce, W. D.

1907 On the Biologies of the Rhyncophora of North America. Ann. Rept. Nebraska Bd. Agric., for 1906–07, pp. 247–319, 8 pls.

1911 Some Factors Influencing the Development of the Boll-weevil. Proc. Entom. Soc. Washington, vol. 13, pp. 111–114.

Poos, F. W., and N. H. Wheeler

1934 On the Hereditary Ability of Certain Insects to Transmit Diseases and to Cause Disease-like Injuries to Plants. Journ. Econ. Entom., vol. 27, pp. 58–69, 7 figs.

Pussard, R.

1933 Contribution à l'étude de la nutrition des Psyllides. Bull. Soc. Entom. France, vol. 37, pp. 292–297.

Reclaire, A., and D. MacGillavry

1932 Naamlijst der in Nederlanden het omliggend Gebied waargenomen Wantsen (Hemiptera-Heteroptera) met Aateekeningen omtrent de voedsel-of verblijfplant en de Levenswijze. Tijdschr. Entom., vol. 85, pp. 59–258.

Reh, L., and others

1925 Sorauer: Handbuch der Pflanzenkrankheiten. Vol. IV. Tierisch Schädlingen an Nutzpflanzen. Pt. 1, xvi + 483 pp., 218 figs., Berlin, Paul Parey.

Riley, C. V.

1892 The Yucca Moth and Yucca Pollination. 3d Ann. Rept. Missouri Botan. Gard., pp. 99–158, 19 pls.

Rogers, J. S.

1926 Notes on the Feeding Habits of Adult Crane Flies. Florida Entom., vol. 10, pp. 5–8.

Rohwer, S. A.

1912 Chalcidids Injurious to Forest Tree Seeds. Bull. U. S. Dept. Agric., Bur. Entom., Tech. Ser., No. 20, pp. 157–163.

Russo, G.

1926 Contributo alla conoscenza delgi Solytidi. Studio morfobiologico del *Chaetoptelius vestitus* e dei suoi simbionti. Boll. Lab. Zool., Portici, vol. 19, pp. 103–260, 48 figs., 1 pl.

Savin, W. M.

1925 A Remarkable Partnership between the Spanish Bayonet and the Yucca Moth. Nat. Hist., New York, vol. 25, pp. 276–281, 6 figs.

Scheidter, F.

1923 Zur Lebensweise unserer Holzwespen. Zeits. Schädlingsbekämpfung, vol. 1, pp. 89–98, 1 pl., 6 figs.

Schilder, F. A.

1928 Die Nahrung der Coccinelliden und ihre Beziehung zur Verwandschaft der Arten. Arbeit. Biol. Reichanst., vol. 16, pp. 213–282.

Schütze, K. T.

1931 Die Biologie der Kleinschmetterlinge unter besonderer Berücksichtigung ihrer Nährpflanzen und Erscheinungszeiten. 235 pp., Frankfurt a. M. Inst. Entom. Verein.

Scudder, S. H.

1889 Classified List of Food Plants of American Butterflies. Psyche, vol. 5, pp. 274–278.

Séguy, E.

1934 Contribution à l'étude des mouches phytophages de l'Europe Occidentale. II. Encycl. Entom., B. II. Dipt. 7, pp. 167–264.

Senior-White, R.

1928 Algae and the Food of Anopheline Larvae. Indian Journ. Med. Res., vol. 15, pp. 969–988.

Shipley, A. E.

1925 Enemies of Books. Tropical Agric., vol. 2, pp. 223–224; 244–246.

Smith, F. F.

1933 The Nature of the Sheath Material in the Feeding Punctures Produced by the Potato Leafhopper and the Alfalfa Hopper. Journ. Agric. Res., vol. 47, pp. 475–485.

Smith, F. F., and F. W. Poos

1931 The Feeding Habits of Some Leaf Hoppers of the Genus *Empoasca*. Journ. Agric. Res., vol. 43, pp. 267–285.

Smith, K. M.

1920 Investigations of the Nature of the Damage to Plant Tissue Resulting from the Feeding of Capsid Bugs. Ann. Appl. Biol., vol. 7, pp. 40–55, 5 figs., 1 pl.

1926 A Comparative Study of the Feeding Methods of Certain Hemiptera and of the Resulting Effects upon the Plant Tissue, with Special Reference to the Potato Plant. Ann. Appl. Biol., vol. 13, pp. 109–139.

Sorhagen, L.

1922 Beiträge zur Biologie europäischer Nepticula-Arten. Arch. Naturgesch., vol. 88 A, No. 3, 60 pp., 4 pls., 7 figs.

Standfuss, M.

1896 Handbuch der paläarktischen Gross-Schmetterlinge. Jena, xii + 392 pp., 8 pls.

Swezey, O. H.

1915 A Leaf-mining Cranefly in Hawaii. Proc. Hawaiian Entom. Soc., vol. 3, pp. 87–89.

1922 Insects Attacking Ferns in the Hawaiian Islands. Proc. Hawaiian Entom. Soc., vol. 5, pp. 57–65.

Tate, H. D.

1937 Method of Penetration, Formation of Stylet Sheaths and Source of Food Supply of Aphids. Iowa State College Journ. Sci., vol. 11, pp. 185–206, 8 pls.

Taylor, R. L.

1928 The Destructive Mexican Book Beetle Comes to Boston. Psyche, vol. 35, pp. 44–50.

Teodoro, G.

1926 Considerazioni sulle cocciniglie parassite e loro piante nutrici. Riv. Biologia, Milan, vol. 8, pp. 629–637.

Theobald, F. V.

1926– The Plant Lice or Aphididae of Great Britain. 3 vols., London.
29

Townsend, C. H. T.

1913 On the History of Cottons and Cotton Weevils. Science, vol. 37, pp. 638–639.

Trägårdh, I.

1930   Studies on the Galleries of Bark-beetles. Bull. Entom. Res., vol. 21, pp. 469–480, 7 figs.

Uvarov, B. P.

1921   A Revision of the Genus *Locusta*, with a New Theory as to the Periodicity and Migrations of Locusts. Bull. Entom. Res., London, vol. 12, pp. 135–163.

1923   Quelques problèmes de la biologie des sauterelles. Ann. Epiphyties, vol. 9, pp. 84–108.

Voukassovitch, P.

1926   Sur deux Diptères parasites; *Siphonella ruficornis* et *Discochaeta cognata*. Bull. Soc. Hist. Nat., Toulouse, vol. 54, pp. 219–222.

Wadley, F. M.

1923   Factors Affecting the Proportion of Alate and Apterous Forms of Aphids. Ann. Entom. Soc. America, vol. 16, pp. 279–303.

Walkden, H. H., and H. R. Bryson

1938   Notes on an Interesting Food Habit of False Wireworm Adults. Journ. Kansas Entom. Soc., vol. 11, pp. 10–13.

Watt, M. N.

1920–   Leaf-mining Insects of New Zealand. Parts I to IV. Trans. New Zealand
24   Inst., vol. 52, pp. 439–466; vol. 53, pp. 197–219; vol. 54, pp. 465–489; vol. 55, pp. 327–341; 684–687.

Weiss, H. B., and R. H. Carruthers

1936   The More Important Insect Enemies of Books. Bull. New York Pub. Lib., vol. 40, pp. 739–752, 827–841, 985–995, 1049–1063.

Wheeler, W. M.

1906   An Ethological Study of Certain Maladjustments in the Relations of Ants to Plants. Bull. American Mus. Nat. Hist., vol. 22, pp. 403–418, 6 pls., 1 fig.

1910   Ants: Their Structure, Development and Behavior. xxv + 663 pp., 286 figs., Columbia Univ. Press, New York.

Wilson, G. F.

1926   Insect Visitors to Sap-exudations of Trees. Trans. Entom. Soc. London, vol. 74, pp. 243–253, 3 pls., 1 fig.

Wilson, H. F., and R. A. Vickery

1920   A Species List of the Aphididae of the World and their Recorded Food-plants. Trans. Wisconsin Acad. Sci. Arts and Letters, vol. 19, pp. 22–355.

Wilson, S. E.

1932   "Powder-post" Beetles. Nature, vol. 130, pp. 22–23.

Zukowsky, B.

1915   Aphoristische Skizze über die bisher bekannt gewordenen Futterpflanzen der paläarktischen Aegeriidae. Internat. Entom. Zeitschr., vol. 9, pp. 77–79.

PLANT FAUNAE. EFFECTS OF INSECT FEEDING ON PLANTS

Balachowsky, A.

1932 La théorie des "places vides" et le peuplement des végétaux par les coccidae. Soc. Entom. France, Livre du Centenaire, pp. 517–522.

Banks, N.

1912 At the Ceanothus in Virginia. Entom. News, vol. 23, pp. 102–110.

Barrett, R. E.

1932 An Annotated List of the Insects and Arachnids Affecting Various Species of Walnuts or Members of the Genus *Juglans*. Univ. California Pubs. Entom., vol. 5, pp. 275–309.

Bodenheimer, F. S.

1925 Über die Ausnutzung des durch Pflanzenneueinführungen entstandenen freien Nahrungraums durch einheimische Insekten. Biol. Zentralbl., vol. 44, pp. 671–675.

Brown, A. W. A.

1941 Foliage Insects of Spruce in Canada. Tech. Bull. Dept. Agric., Canada, vol. 31, 29 pp., 1 pl.

Candura, G. S.

1932 Studi e ricerci sugli insetti viventi nelle Paste Alimentari. Boll. Soc. Nat. Napoli, vol. 44, pp. 159–203.

Claassen, P. W.

1921 Typha Insects: Their Ecological Relationships. Mem. Cornell Agric. Expt. Sta., No. 47, pp. 457–531, 11 pls.

Clausen, C. P.

1933 The Citrus Insects of Tropical Asia. Circ. U. S. Dept. Agric., No. 266, 35 pp.
1927 The Citrus Insects of Japan. Tech. Bull. U. S. Dept. Agric., No. 15, 15 pp., 5 figs.

Clément, A. L.

1916 Les insectes du saule. La Vie Agric. et Rur, Paris, vol. 6, pp. 99–103, 10 figs.

Cockerell, T. D. A.

1882 Notes on Plant Faunae. Insect Life, vol. 5, pp. 117–121.
1914 Bees Visiting *Helianthus*. Canadian Entom., vol. 46, pp. 409–415.
1916 Sunflower Insects in California and South Africa. Canadian Entom., vol. 48, pp. 76–79.
1917 Sunflower Insects in Virginia and Connecticut. Canadian Entom., vol. 49, p. 212.

Copeland, E. B.

1931 The Coconut. 233 pp., 22 pls., 2 figs., London, Macmillan.

Coulon, L.

1914– Les Insectes du chêne. Bull. Soc. d'Etude Sci. Nat., Elbeuf, vol. 32, pp. 159–
17 179; vol. 32–34, pp. 61–80; vol. 35, pp. 53–80.
1929– Les insectes des arbes résineux. Bull. Soc. Sci. Nat., Elbeuf, vol. 48, pp. 87–
32 112 (1929); vol. 49, pp. 87–94 (1930); vol. 50, pp. 77–78 (1932).

Dickerson, E. L., and H. B. Weiss

1920 Insects of the Evening Primroses in New Jersey. Journ. New York Entom. Soc., vol. 28, pp. 32–74, 3 pls.

Dingler, M.

1934 Die Tierwelt des Spargelfeldes. Zeits. angew. Entom., vol. 21, pp. 291–328, 6 figs.

Durrant, J. H.

1921 Insects Associated with Grain. Rept. Grain Pests Comm., Roy. Soc. London, No. 9, pp. 33–52.

Forbes, S. A.

1909 The General Entomological Ecology of the Indian Corn Plant. American Naturalist, vol. 43, pp. 286–301.

French, C.

1933 New Records of Plants Attacked by Insects. Victorian Natural., vol. 49, p. 264, p. 296; vol. 50, pp. 91, 119, 147, 190.

Goidanich, A.

1928 Contributi alla conoscenza dell' entomofauna della canapa. I. Prospetto generale. Boll. Lab. Entom. R. Ist. Sup. Agrar. Bologna, vol. 1, pp. 37–64.

Gondé, H.

1930 Les parasites du blé. Bull. Assoc. Nat. Vallée du Loing, vol. 6, pp. 59–87.

Graenicher, S.

1914 Wisconsin Bees of the Genus *Perdita*. Geographical Distribution and Relations to Flowers. Canadian Entom., vol. 46, pp. 51–57.

Harris, J. A.

1905 The Influence of Apidae upon the Geographical Distribution of Certain Floral Types. Canadian Entom., vol. 37, pp. 353–357, 373–380, 393–398.

Hoffmann, C. H.

1940 Additions to Annotated List of Insects Reared from Elm Bark and Wood. Bull. Brooklyn Entom. Soc., vol. 35, pp. 54–63.

1942 Annotated List of Elm Insects in the U. S. Misc. Publ. U. S. Dept. Agric., No. 466, 20 pp.

Holloway, T. E.

1930 Sugar Cane Insects of North America and the West Indies (excluding Cuba). A Bibliography and List of Known Parasites, Insect Predators and Diseases. Proc. Third Congr. Intern. Soc. Sug. Cane Technol. Soerabaia, 1929, pp. 184–216.

Hori, M., and K. Oshima

1924 The Insect Fauna of the Genus *Populus* in Japan and its Relation to Horticulture and Agriculture. Hokkaida Nokaiho, vol. 24, pp. 1–34.

Hunter, W. D., F. C. Pratt, and J. D. Mitchell

1912 The Principal Cactus Insects of the United States. Bull. U. S. Dept. Agric., Bur. Entom., No. 113, 71 pp., 7 pls., 8 figs.

Meikle, A. A.

1937   The Insects Associated with Bracken. Agric. Prog. (Agric. Ed. Assoc., Great
Britain), vol. 14, pp. 58–60.

Miller, N. C. E.

1941   Insects Associated with Cocoa (*Theobroma cacao*) in Malaya. Bull. Entom.
Res., vol. 32, pp. 1–15.

Miyoshi, K.

1926   List of Injurious Insects of Citrus from Japan. Insect World, vol. 30, pp.
303–308, 338–343. [In Japanese.]

Neiswander, C. R.

1931   The Sources of American Corn Insects. Bull. Ohio Agric. Expt. Sta., No. 473,
98 pp., 3 figs.

Okajima, G.

1929   On the Distribution in Japan of Insects Injurious to the Rice Plant. Proc.
Third Pan-Pacific Sci. Congr., Tokyo, 1926, pp. 2050–2067.

Pechuman, L. L.

1937   Annotated List of Insects Found in the Bark and Wood of *Ulmus americana*
in New York State. Bull. Brooklyn Entom. Soc., vol. 32, pp. 8–21.

Picard, F.

1919   La fauna entomologique du figuier. Ann. Service Epiphyties, vol. 6, pp. 35–
174.
1924   Les origines de la fauna de la vigne. Feuille Natural., ann. 45, pp. 25–27.

Quayle, H. J.

1938   Insects of Citrus and other Subtropical Fruits. ix + 583 pp. Comstock Pub.
Co., Ithaca, New York.

Ramachandra Rao, Y.

1920   Lantana Insects in India. Mem. Dept. Agric. India, Pusa, Entom. Ser., vol.
5, pp. 239–314, 14 pls., 3 figs.

Ramsey, Helen

1941   Fauna of Pine Bark. Journ. Elisha Mitchell Sci. Soc., vol. 57, pp. 91–97.

Rosewall, A. U.

1922   Insects of the Yellow Thistle. Entom. News, vol. 33, pp. 176–180.

Schwitzgebel, R. B., and D. A. Wilbur

1943   Diptera Associated with Ironweed, *Veronica interior* in Kansas. Journ.
Kansas Entom. Soc., vol. 16, pp. 4–13.

Sonan, J.

1924   Insect Pests of Tea in Formosa. Rept. Dept. Agric. Res. Inst., Taihoku,
No. 12, pp. 1–132, 3 pls.

Stellwaag, F.

1924   *Tinea cloacella* und *Tinea granella*. Zeits. angew. Entom., vol. 10, pp. 181–
188.
1924   Die Tierwelt tiefer Weinkeller. Wein u. Rebe, pp. 277–297. Abstr. in Rev.
Appl. Entom., vol. 12, p. 292.

Stephan, J.

  1923 Moos- und Flechtfresser unter den Raupen. Entom. Jahrb., vol. 32, pp. 95–100.

Suzuki, B.

  1930 Illustrations of Mulberry Insects. 122 pp., 20 pls., Tokyo, Maruyama-Sha.

Swezey, O. H.

  1925 The Insect Fauna of Trees and Plants as an Index of their Endemicity and Relative Antiquity in the Hawaiian Islands. Proc. Hawaiian Entom. Soc., vol. 6, pp. 195–209.

Tryon, H.

  1911 The Insect Enemies of the Prickly Pear. Queensland Agric. Journ., vol. 27, pp. 80–83.

Van Emden, F.

  1929 Über die Rolle der Feuchtigkeit im Leben der Speicherschädlinge. Anz. f. Schädlingsk., vol. 5, pp. 58–60.

Van Hall, C. J. J.

  1932 Cacao. 2nd. Edit., 514 pp., London, Macmillan.

Vayssière, P.

  1930 Les insectes nuisible au Cotonnier dans les colonies françaises. Faune Colon. françaises, vol. 4, pp. 193–439, 18 pls., 57 figs.

Weiss, H. B., and E. L. Dickerson

  1922 Notes on Milkweed Insects in New Jersey. Journ. New York Entom. Soc., vol. 24, pp. 123–145 (1921).

Wellhouse, W. H.

  1922 The Insect Fauna of the Genus *Crataegus*. Mem. Cornell Agric. Expt. Sta., No. 56, pp. 1041–1136, 23 figs., 3 pls.

Williams, F. X.

  1931 Handbook of Insects and other Invertebrates of Hawaiian Sugar Cane Fields. Hawaiian Sugar Planters' Expt. Sta., 400 pp., 41 pls., 190 figs.

Yokoyama, K.

  1925 List of Insects Injurious to Mulberry and *Bombyx mori* in Japan. Insect World, vol. 29, pp. 187–195; 227–233; 265–270; 296–301. [In Japanese.]

Zacher, F.

  1933 Die Biocoenose der Getreidespeicher und Mühlen. Trav. Congr. Intern. Entom., 1932, vol. 5, pp. 699–703.

CONTROL OF NOXIOUS WEEDS BY INSECTS

Alexander, W. B.

  1919 The Prickly Pear in Australia. Institute Science and Industry, Melbourne, Bull. No. 12, 48 pp., 16 figs., 1 map.

Beeson, C. F. C.

  1934 Prickly Pear and Cochineal Insects. Indian Forestry, vol. 60, pp. 203–205.

Bodenheimer, F. S.

1932 Über die Ausrottung von *Opuntia* spp. durch *Dactylopius* spp. auf Grund einiger Beobachtungen auf Ceylon. Zentralbl. f. Bakter. (2) Vol. 86, pp. 155–160.

Cameron, E.

1935 A Study of the Natural Control of Ragwort (*Senecio jacobaea*). Journ. Ecology, vol. 23, pp. 265–322.

Chater, E. H.

1931 A Contribution to the Study of the Natural Control of Gorse. Bull. Entom. Res., vol. 22, pp. 225–235.

Cockerell, T. D. A.

1929 Biological Control of the Prickly Pear. Science, vol. 69, pp. 328–329.

Currie, G. A., and R. V. Fyfe

1938 The Fate of Certain European Insects Introduced into Australia for the Control of Weeds. Journ. Counc. Sci. Industr. Res., vol. 11, pp. 289–301.

Currie, G. A., and S. Garthside

1932 The Possibility of the Entomological Control of St. John's Wort in Australia. Progress Report, Counc. Sci. and Indus. Res., Australia, No. 29, pp. 1–28.

Dodd, A. P.

1926 The Campaign against Prickly-Pear in Australia. Nature, vol. 117, pp. 625–626.

1929 The Progress of Biological Control of Prickly-Pear in Australia. 44 pp., 12 pls., Commonwealth Prickly-Pear Board, Brisbane.

1940 The Biological Campaign against Prickly-pear. 177 pp., Commonwealth Prickly-pear Board, Brisbane.

Fyfe, R. V.

1937 The Lantana Bug (*Teleonemia lantanae*). Journ. Counc. Sci. Industr. Res., Australia, vol. 10, pp. 181–186.

Hamlin, J. C.

1924 Biological Control of Prickly-pear in Australia: Contributing Efforts in North America. Journ. Econ. Entom., vol. 17, pp. 447–460, 2 pls.

1926 Biological Notes on Important *Opuntia* Insects of the United States. Pan Pacific Entom., vol. 2, pp. 97–105.

Hutson, J. C.

1926 Prickly-pear and Cochineal Insects. Tropical Agriculturist, vol. 47, pp. 290–292.

Imms, A. D.

1929 Remarks on the Problem of Biological Control of Noxious Weeds. Trans. Fourth Intern. Congr. Entom., 1928, pp. 10–17.

1941 The Prickly-pear Problem in Australia. Nature, vol. 148, pp. 303–305.

Jacques, C.

1933 Le *Cactoblastis cactorum*. Rev. Agric. Nouvelle-Calédonie, pp. 1085–1094.

Jepson, F. P.

1930 Present Position in Regard to Control of Prickly-pear in Ceylon by the Introduced Cochineal Insect, *Dactylopius tomentosus*. Trop. Agric., Ceylon, vol. 75, pp. 63–72.

Kelley, S. G.

1931 The Control of Noogoara and Bathurst Burr by Insects. Journ. Counc. Sci. Ind., vol. 4, pp. 161–172.

Kuhnikannan, K.

1930 Control of Cactus in Mysore by Means of Insects. Journ. Mysore Agric. Expt. Un., vol. 11, pp. 95–98.

Marcovitch, S.

1917 Insects Attacking Weeds in Minnesota. 16th Rept. Minnesota State Entom. for 1915–16, pp. 135–152, 4 pls.

Miller, D.

1929 Control of Ragwort through Insects. New Zealand Journ. Agric., vol. 39, pp. 9–17.

Miller, D., A. F. Clark, and L. J. Dumbleton

1936 Biological Control of Noxious Insects and Weeds in New Zealand. New Zealand Journ. Sci. and Technol., vol. 18, pp. 579–593.

Perkins, R. C. L., and O. H. Swezey

1924 The Introduction into Hawaii of Insects that Attack Lantana. Expt. Sta. Hawaiian Sugar Planters' Assoc., Entom. Serv., Bull. 16, 83 pp.

Pettey, F. W.

1934 *Cactoblastis cactorum:* Government's Policy with Regard to Distribution of the Insect. Farming in S. Africa, vol. 9, (97), pp. 138–139, 150.

Ramakrishna Ayyar, T. V.

1931 Control of the Prickly-pear by the Cochineal Insect. Nature, vol. 128, p. 837.

1931 The Coccidae of the Prickly-pear in South India and their Economic Importance. Agric. Live Stock, India, vol. 1, pp. 229–237, 3 pls.

Simmonds, H. W.

1929 Lantana Bug, *Teleonemia lantanae.* Agric., Journ. Dept. Agric. Fiji, vol. 1, pp. 16–21.

1932 Weeds in Relation to Agriculture. *Ibid.*, vol. 5, pp. 58–62.

1933 The Biological Control of the Weed, *Clidema hirta* in Fiji. Bull. Entom. Res., vol. 24, pp. 345–348.

Sweetman, H. L.

1935 Successful Examples of Biological Control of Pest Insects and Plants. *Ibid.*, vol. 26, pp. 373–377.

1936 The Biological Control of Insects. Chap. 14, pp. 359–383. Comstock Pub. Co., Ithaca, New York.

Tillyard, R. J.

1927 Biological Control of St. John's Wort. New Zealand Journ. Agric., vol. 35, pp. 42–45.

1927 Summary of the Present Position as Regards Biological Control of Noxious Weeds. *Ibid.*, vol. 24, pp. 85–90.

1929 The Biological Control of Noxious Weeds. Trans. Fourth Intern. Congr. Entom., 1928, vol. 2, pp. 4–9, Ithaca, New York.

1930 The Biological Control of Noxious Weeds. Pap. Proc. Roy. Soc. Tasmania, 1929, pp. 51–86, 8 pls., 1 fig.

Tooke, F. G. C.

1930   Insects in Relation to Prickly Pear Control.   South African Journ. Nat. Hist., vol. 6, pp. 386–393.

Van der Goot, P.

1940   Biological Control of Prickly-pear in the Palu Valley (N. Celebes).   Landbouw (Buitenzorg), vol. 16, pp. 413–429.

Warren, E.

1914   The Prickly Pear Pest.   Agric. Journ. Union of South Africa, vol. 7, pp. 387–391, 2 figs.

SPECIALIZATION IN FOOD-HABITS.   POLYPHAGY, OLIGOPHAGY AND MONOPHAGY.   SELECTION OF FOOD PLANTS.   INSTINCTIVE BEHAVIOR. RACIAL SEGREGATION

Abbott, C. E.

1936   The Physiology of Insect Senses.   Entom. Americana, vol. 16, pp. 225–280.

Allan, P. B. M.

1943   Substitute Food Plants.   Entom. Rec. Journ. Variation, vol. 55, pp. 1–3.

Ball, E. D.

1934   Food Plants of Some Arizona Grasshoppers.   Journ. Econ. Entom., vol. 29, pp. 679–684.

Bedford, H. W.

1931   The Weed "Hambuk" (*Abutilon* spp.) and the Part it Plays in the Conservation of Parasites of the Various Species of Bollworm which Attack Cotton in the Sudan.   Bull. Wellcome Trop. Res. Lab., Entom. Sec., No. 34, pp. 39–43.

Bondar, G.

1928   Uma broca polyphaga, *Sphallenum setosum*.   Chacarus e Quintaes, vol. 38, pp. 33–34, 2 figs.

Bird, H.

1934   Decline of the Noctuid Genus *Papaipema*.   Ann. Entom. Soc. America, vol. 27, pp. 551–556.

Blaschke, P.

1914   Die Raupen Europas mit ihren Futterpflanzen.   Annaberg, xxix + 349 pp., 34 pls.

Börner, C.

1932   Apfel- und Dorn-Blutlaus.   Anz. f. Schädlingsk, vol. 8, pp. 52–54.

Bos, J. R.

1887   Futteränderung bei Insekten.   Biol. Centralbl., vol. 7, pp. 322–331.

Bridwell, J. C.

1929   Description of a Bruchid Immigrant into Hawaii Breeding in the Seeds of Convolvulaceae.   Proc. Entom. Soc. Washington, vol. 31, pp. 112–114.

1931   Bruchidae Infesting Seeds of Compositae with Descriptions of New Genera and Species.   *Ibid.*, vol. 33, pp. 37–42.

Browne, A. C.

1932 A New Host for Elm Leaf Beetle. Monthly Bull. Dept. Agric., California, vol. 21, p. 347.

Brues, C. T.

1920 The Selection of Food-plants by Insects, with Special Reference to Lepidopterous Larvae. American Naturalist, vol. 54, pp. 313–322.

1924 The Specificity of Food-plants in the Evolution of Phytophagous Insects. *Ibid.*, vol. 58, pp. 127–144.

1936 Aberrant Feeding Behavior Among Insects and its Bearing on the Development of Specialized Food Habits. Quart. Rev. Biol., vol. 11, pp. 305–319.

1940 Food Preferences of the Colorado Potato Beetle (*Leptinotarsa decemlineata*). Psyche, vol. 47, pp. 38–43.

Brunson, A. M., and R. H. Painter

1938 Differential Feeding of Grasshoppers on Corn and Sorghums. Journ. American Soc. Agron., vol. 30, pp. 334–346.

Burgess, A. F., and D. M. Rogers

1913 Results of Experiments in Controlling the Gipsy Moth by Removing its Favorite Food Plants. Journ. Econ. Entom., vol. 6, pp. 75–79.

Cameron, A. E.

1914 A Contribution to a Knowledge of the Belladonna Leaf-miner, *Pegomyia hyoscyami*, its Life-History and Biology. Ann. Appl. Biol., vol. 1, pp. 43–76, 2 pls., 4 figs.

1916 Some Experiments on the Breeding of the Mangold Fly (*Pegomyia hyoscyami*) and the Dock Fly (*P. bicolor*). Bull. Entom. Res., vol. 7, pp. 87–92, 2 figs.

Campbell, F. L.

1927 Notes on Silkworm Nutrition. Journ. Econ. Entom., vol. 20, pp. 88–90.

Cartwright, W. B.

1922 Host Plant Selection by Hessian Fly (*Phytophaga destructor*). *Ibid.*, vol. 15, pp. 360–363, 2 pls., 5 tables.

Coad, B. R.

1914 Feeding Habits of the Boll Weevil on Plants other than Cotton. Journ. Agric. Res., vol. 2, pp. 235–245.

Cockerell, T. D. A.

1897 Physiological Species. Entom. News, vol. 8, pp. 234–236.

Comstock, J. H.

1882 The Apple Maggot. Rept. U. S. Entom. Comm. 1881–82, pp. 195–198, 1 pl.

Coolidge, K. R.

1910 A California Orange Dog. Pomona Journ. Entom., vol. 2, pp. 333–334.

Craighead, F. C.

1921 Hopkins' Host-selection Principle as Related to Certain Cerambycid Beetles. Journ. Agric. Res., vol. 22, pp. 189–220.

1923 The Host Selection Principle as Advanced by Walsh. Canadian Entom., vol. 55, pp. 76–79.

Criddle, N.

1933 Notes on the Habits of Injurious Grasshoppers in Manitoba. *Ibid.*, vol. 65, pp. 97–102.

Curran, C. H.

1924 *Rhagoletis pomonella* and Two Allied Species. Ann. Rept. Entom. Soc. Ontario, No. 54, pp. 56–57.

Currie, G. A.

1932 Oviposition Stimuli of the Burr-seed Fly, *Euaresta aequalis*. Bull. Entom. Res., vol. 23, pp. 191–193.

Davidson, J.

1927 The Biological and Ecological Aspect of Migration in Aphids. Sci. Prog., vol. 21, pp. 641–658, 7 figs.; vol. 22, pp. 57–69.

Davidson, J., and R. A. Fisher

1922 Biological Studies on *Aphis rumicis*. Reproduction on Varieties of *Vicia faba*. Ann. Appl. Biol., vol. 9, pp. 135–145, 1 fig.

Davidson, N. M.

1918 Alternation of Hosts in Economic Aphids. Journ. Econ. Entom., vol. 11, pp. 289–293.

DeLong, D. M.

1926 Food Plant and Habitat Notes on some North American Species of *Phlepsius*. Ohio Journ. Sci., vol. 26, pp. 69–72.

Dethier, V. G.

1937 Gustation and Olfaction in Lepidopterous Larvae. Biol. Bull., vol. 72, pp. 7–23, 4 figs.

1939 Taste Thresholds in Lepidopterous Larvae. *Ibid.*, vol. 76, pp. 325–329.

1941 The Function of the Antennal Receptors in Lepidopterous Larvae. *Ibid.*, vol. 80, pp. 403–414.

1941 Chemical Factors Influencing the Choice of Food Plants by *Papilio* Larvae. American Naturalist, vol. 75, pp. 61–73.

1943 Testing Attractants and Repellents. Pub. American Assoc. Adv. Sci., No. 20, pp. 167–172.

Dickson, R. C.

1940 Inheritance of Resistance to Hydrocyanic Acid Fumigation. Hilgardia, vol. 13, pp. 515–521.

Dobzhansky, T.

1937 Genetics and the Origin of Species. xvi + 364 pp., New York, Columbia Univ. Press.

Ebeling, W.

1938 Host-determined Morphological Variations in *Lecanium corni*. Hilgardia, vol. 11, pp. 611–631.

Ehinger, K.

1916 Ein neuer abgeleiteter Deilephila-Hybride. Intern. Entom. Zeit., Guben, Jahrg. 10, pp. 91–92.

Eliescu, G.

1932 Beiträge zur Kenntnis der Morphologie, Anatomie und Biologie von *Lophyrus pini*. I Teil. Zeits. angew. Entom., vol. 19, pp. 22–67.

Evans, A. C.

1938 Physiological Relations between Insects and their Host Plants. Ann. Appl. Biol., vol. 25, pp. 558–572.

Field, W. L. W.

1910 The Offspring of a Captured Female of *Basilarchia proserpina*. Psyche, vol. 17, pp. 87–89.

Fölsch, W.

1926 Über die Mehlmotte, *Ephestia Kühniella*. Anz. Schädlingsk., vol. 2, pp. 98–99.

Forbes, S. A.

1907 On the Life-history, Habits and Economic Relations of the White Grubs and Maybeetles. Bull. Illinois Agric. Expt. Sta., No. 116, pp. 447–480.

Foster, S. W.

1910 On the Nut-feeding Habits of the Codling Moth. Bull. U. S. Dept. Agric., Bur. Entom., No. 80, pt. 5, pp. 67–70, 2 pls.

Franssen, C.

1931 Nogmaals de "zwarte Bladluizen". Entom. Ber., vol. 8, pp. 306–309.

French, C.

1913 A New Insect Pest of Roses; the Vine Curculio. Journ. Agric., Victoria, pp. 240–241.

Frost, S. W.

1933 Baits for the Oriental Fruit Moth. Bull. Pennsylvania Expt. Sta., No. 301, 35 pp., 11 figs.

Fuller, C.

1924 White Ant Experiments: Tests of the Resistance of Timbers. Entom. Mem., Union South Africa, No. 2, pp. 81–104, 4 figs.

Fulton, B. B.

1915 The Tree Crickets of New York: Life History and Bionomics. Tech. Bull. New York Agric. Expt. Sta., No. 42, 47 pp.

1925 Physiological Variation in the Snowy Tree-cricket, *Oecanthus niveus*. Ann. Entom. Soc. America, vol. 18, pp. 363–383, 6 figs.

1933 Inheritance of Song in Hybrids of the Subspecies of *Nemobius fasciatus*. Ann. Entom. Soc. America, vol. 26, pp. 368–376.

Funkhouser, W. D.

1917 Biology of the Membracidae of the Cayuga Lake Basin. Mem. Cornell Univ. Expt. Sta., No. 11, pp. 177–445, 43 figs.

Gaines, R. C.

1933 Progress Report on the Development of the Boll Weevil on Plants other than Cotton. Journ. Econ. Entom., vol. 26, pp. 940–943.

Galvagni, E.

1920 Über die Generationsverschiedenheit bei der Geometride *Mesotype virgata*. Verh. zool.-bot. Ges., Wien, vol. 70, pp. 82–83.

Goeschen, F.

1913 *Salix babylonica* als Futter für Hybriden der Schwärmengattung *Celerio*. Zeits. wiss. Insektenbiol., vol. 9, pp. 72–73.

Goff, C. C., and A. N. Tissot

1932 The Melon Aphid, *Aphis gossypii*. Bull. Florida Agric. Expt. Sta., No. 252, 23 pp., 14 figs.

Gonzales, S. S.

1939 A New Phytophagous Lady Beetle in the Philippines (*Plagiodera*). Philippine Journ. Agric., vol. 10, pp. 415–417.

Graham, S. A.

1929 Host Selection by the Spruce Budworm. Pap. Michigan Acad. Sci., vol. 9, pp. 517–523.

Hamlin, J. C.

1932 An Inquiry into the Stability and Restriction of Feeding Habits of Certain Cactus Insects. Ann. Entom. Soc. America, vol. 25, pp. 89–120, 3 figs.

Hamm, A. H.

1923 A Substitute Food for *Melitaea aurinia*. Entom. Monthly Mag. (3), vol. 9, p. 183.

Harrison, J. W. H.

1926– On the Inheritance of Food Habits in the Hybrids between the Geometrid
27 Moths (*Poecilopsis pomonaria* and *P. isabellae*). Proc. Univ. Durham Phil. Soc., vol. 7, pp. 194–201.

1927 Experiments on the Egg-laying Instinct of the Sawfly, *Pontania salicis* and their Bearing on the Inheritance of Acquired Characters; with some Remarks on a New Principle of Evolution. Proc. Roy. Soc. London, vol. 101 (B), pp. 115–126.

Harrison, J. W. H., and F. C. Garrett

1926 The British Races of *Aricia medon*, with Special Reference to the Areas in which they Overlap. Trans. Nat. Hist. Soc., Newcastle, vol. 6, pp. 89–106.

Harvey, F. L.

1890 The Apple Maggot. Rept. Maine Agric. Expt. Sta., 1889, pp. 190–241, 4 pls.

Haseman, L., and R. L. Meffert

1933 Are We Developing Strains of Codling Moths Resistant to Arsenic? Bull. Missouri Expt. Sta., Res., No. 202, 11 pls.

Hegner, R. W.

1910 The Food of *Calligrapha bigsbyana*, a Chrysomelid Beetle. Psyche, vol. 17, p. 160.

Hering, M.

1926 Die Oligophagie der blattminierenden Insekten in ihrer Bedeutung für die Klärung phyto-phyletischer Probleme. Verhandl. IIIte Intern. Entom. Kongr., Zurich, pp. 216–230.

Hodgson, B. E.

1928 The Host Plants of the European Corn Borer in New England. Tech. Bull. U. S. Dept. Agric., No. 77, 63 pp.

Hodson, W. E.

1926 Observations on the Biology of *Tylenchus dipsaci*, and the Occurrence of Biologic Strains of the Nematode. Ann. Appl. Biol., vol. 13, pp. 219–228.

Hopkins, A. D.

1910 Contributions toward a Monograph of the Bark-weevils of the Genus *Pissodes*. Bull. U. S. Dept. Agric., Bur. Entom., Tech. Ser., No. 20, pp. 1–68, 22 pls., 9 figs.

1916 Economic Investigations of the Scolytid Bark and Timber Beetles of North America. U. S. Dept. Agric. Program of Work, 1917, p. 353.

Hoskins, W. M., and R. Craig

1934 The Olfactory Responses of Flies in a New Type of Insect Olfactometer. Journ. Econ. Entom., vol. 27, pp. 1029–1036.

Hottes, F. C.

1928 Borderline Aphid Studies. Proc. Biol. Soc. Washington, vol. 41, pp. 133–138.

Hough, W. S.

1929 Studies of the Relative Resistance to Arsenical Poisoning of Different Strains of Codling-moth Larvae. Journ. Agric. Res., vol. 38, pp. 245–256.

1943 Development and Characteristics of Vigorous or Resistant Strains of Codling Moth. Tech. Bull. Virginia Agric. Expt. Sta., No. 91, 32 pp.

Hubbell, T. H., and F. W. Walker

1928 A New Shrub-inhabiting Species of *Schistocerca* from Central Florida. Occas. Pap. Mus. Zool., Univ. Michigan, No. 197, 10 pp., 1 pl.

Huber, L. L.

1927 A Taxonomic Ecological Review of the North American Chalcid Flies of the Genus *Callimome*. Proc. U. S. Nat. Mus., vol. 70, Art. 14, 114 pp., 4 pls.

Hurd, W. E.

1920 Influence of the Wind on the Movements of Insects. Monthly Weather Rev., Washington, D. C., vol. 48, pp. 94–98.

Illingworth, J. F.

1912 A Study of the Biology of the Apple Maggot (*Rhagoletis pomonella*), together with an Investigation of Methods of Control. Bull. Cornell Agric. Expt. Sta., No. 324, pp. 129–187, 44 figs.

Imms, A. D.

1937 Host Selection and Biological Races. Recent Advances in Entomology, pp. 286–299, Blakiston, Philadelphia.

Imms, A. D., and M. A. Husain

1920 Field Experiments on the Chemotropic Responses of Insects. Ann. Appl. Biol., vol. 6, pp. 269–292.

Isely, F. B.

1938 The Relations of Texas Acrididae to Plants and Soils. Ecological Monog., vol. 8, pp. 551–604.

Jancke, O., and L. Lange

1930 Über den Befall von herbarisierten Pflanzen durch den Brotkäfer (*Sitodrepa panicea*). Zeits. angew. Entom., vol. 17, pp. 386–403, 5 figs.

Joannis, J. de

1913 Remarque sur un cas collectif de mimétisme chez les lépidoptères. Bull. Soc. Entom., France, pp. 135–139.

Kleine, R.

1915 *Chrysomela fastuosa* und ihre Nahrungspflanzen. Entom. Blätter, 1914, p. 110; 1915, p. 56.
1916 Geschmacksverirrung. Intern. Entom. Zeitschr., Guben, Jahrg. 10, p. 105.

Klos, R.

1901 Zur Lebensgeschichte von *Tephroclystis virgaureata*. Verh. K.K. zool.-bot. Ges. Wien, vol. 51, p. 785.

Lal, K. B.

1933 Biological Races in *Psylla mali*. Nature, vol. 132, p. 934.
1934 *Psylla pergrina*, the Hawthorn Race of the Apple Sucker, *P. mali*. Ann. Appl. Biol., vol. 21, pp. 641–648.

Lancefield, D. E.

1929 A Genetic Study of Crosses of Two Races or Physiological Species of *Drosophila obscura*. Zeits. ind. Abstamm. u. Vererbungrl., vol. 52, pp. 287–317.

Larson, A. O.

1927 The Host Selection Principle as Applied to *Bruchus quadrimaculatus*. Ann. Entom. Soc. America, vol. 20, pp. 37–79, 1 pl.

Lathrop, F. H., and C. B. Nickels

1931 The Blueberry Maggot from an Ecological Viewpoint. *Ibid.*, vol. 24, pp. 260–281, 3 diags.

Lindgren, D. L.

1938 The Stupefaction of Red Scale, *Aonidiella aurantii*, by Hydrocyanic Acid. Hilgardia, vol. 11, pp. 11–25.

MacGillivray, A. D.

1921 The Coccidae. 502 pp., Urbana, Illinois, The Scarab Co.

McIndoo, N. E.

1914 The Olfactory Sense of Insects. Smithsonian Misc. Coll., vol. 63, No. 9, 63 pp.
1919 The Olfactory Sense of Lepidopterous Larvae. Ann. Entom. Soc. America, vol. 12, pp. 65–84.
1926 An Insect Olfactometer. Journ. Econ. Entom., vol. 19, pp. 545–571, 8 figs.
1926 Senses of the Cotton Boll Weevil. An Attempt to Explain How Plants Attract Insects by Smell. Journ. Agric. Res., vol. 33, pp. 1095–1141.
1928 Responses of Insects to Smell and Taste and their Value in Control. Journ. Econ. Entom., vol. 21, pp. 903–913.

Mail, G.

1931 Food Preferences of Grasshoppers. Journ. Econ. Entom., vol. 24, pp. 767–768.

Marchal, P.

1908 Le Lécanium du Robinia. C. R. Soc. Biol., Paris, vol. 65, pp. 2–5.

1925 La question des races du Phylloxera de la vigne. Ann. Épiphyties, vol. 9, pp. 411–418.

1933 Les aphides de l'orme et leurs migrations. *Ibid.*, vol. 19, pp. 207–329, 3 pls., 52 figs.

Marshall, J.

1935 The Location of Olfactory Receptors in Insects. Trans. Roy. Entom. Soc. London, vol. 83, pp. 49–72.

Mayer, A. G., and C. G. Soule

1906 Some Reactions of Caterpillars and Moths. Journ. Exper. Zool., vol. 3, pp. 415–433.

Meissner, O.

1926 Monophagie und Polyphagie. Intern. Entom. Zeits., Jahrg. 20, pp. 130–132.

Melander, A. L.

1914 Can Insects Become Resistant to Sprays? Journ. Econ. Entom., vol. 7, pp. 162–172.

1915 Varying Susceptibility of the San José Scale to Sprays. *Ibid.*, vol. 8, pp. 475–480.

Meyrick, R.

1927 The Hereditary Choice of Food Plants in the Lepidoptera and its Evolutionary Significance. Nature, vol. 119, p. 388.

Minnich, D. E.

1929 The Chemical Senses of Insects. Quart. Rev. Biol., vol. 4, pp. 100–112.

Mordvilko, A. K.

1933 The Development of Species Among Aphids. Rev. Entom. U. R. S. S., vol. 25, pp. 7–39. [In Russian].

1924 Cases of Heteroecy in the Plant Lice Resulting from Primary Polyphagy. C. R. Acad. Russie, pp. 160–162.

1934 On the Evolution of Aphids. Arch. Naturg., vol. 3, pp. 1–60, 38 figs.

Morgan, A. C., and S. E. Crumb

1928 Notes on the Chemotropic Responses of Certain Insects. Journ. Econ. Entom., vol. 21, pp. 913–920.

Morse, A. P.

1920 Manual of the Orthoptera of New England, including Locusts, Grasshoppers, Crickets and their Allies. Proc. Boston Soc. Nat. Hist., vol. 35, pp. 197–556, 99 figs., 20 pls.

Mosher, F. H.

1915 Food Plants of the Gipsy Moth in America. Bull. U. S. Dept. Agric., Bur. Entom., No. 250, 39 pp.

Mosher, F. H., and R. T. Webber

1914 The Relation of Variation in Number of Larval Stages to Sex Development in the Gipsy Moth. Journ. Econ. Entom., vol. 7, pp. 368–373.

Mote, D. C.

1926 Codling Moth Attacks Cherries. *Ibid.*, vol. 19, pp. 777–778.

O'Kane, W. C.

1914 The Apple Maggot. Bull. New Hampshire Expt. Sta., No. 171, 120 pp., 8 pls., 2 figs.

Packard, C. M., and B. G. Thompson

1921 The Range Crane-flies in California. Circ. U. S. Dept. Agric., No. 172, 8 pp., 5 figs.

Painter, R. H.

1930 The Biological Strains of Hessian Fly. Journ. Econ. Entom., vol. 23, pp. 322–326.

1938 Differential Feeding of Grasshoppers on Corn and Sorghums. Journ. American Soc. Agron., vol. 30, pp. 334–346.

Parker, J. R., R. L. Shotwell, and F. A. Morton

1934 The Use of Oil in Grasshopper Baits. Journ. Econ. Entom., vol. 27, pp. 89–96.

Patch, E. M.

1912 Elm-leaf Curl and Woolly Apple Aphids. Bull. Maine Agric. Expt. Sta., No. 203, pp. 235–258, 3 pls., 8 figs.

1921 Rose Bushes in Relation to Potato Culture. *Ibid.*, No. 303, pp. 321–344, 1 fig.

1925 The Primary Foodplant of the Melon Aphid. Science, vol. 62, p. 510.

Patch, E. M., and W. C. Woods

1922 The Blueberry Maggot. Bull. Maine Agric. Expt. Sta., No. 308, pp. 77–92, 2 figs.

Persing, C. O., et al.

1942 Resistance of Citrus Thrips to Tartar Emetic Treatment. California Citrograwer, vol. 28, pp. 24–25.

Perron, H. L.

1931 Theory in Explanation of the Selection of Certain Trees by the Western Pine Beetle. Journ. Forestry, vol. 29, pp. 696–699.

Pettey, F. W.

1925 Codling Moth in Apricots. Journ. Agric. Union South Africa, vol. 11, pp. 56–65; 137–152.

Pickett, A. D.

1937 Studies on the Genus *Rhagoletis*. Canadian Journ. Res., vol. 15, pp. 53–75.

Pickett, A. D., and M. E. Neary

1940 Further Studies on *Rhagoletis pomenella*. Sci. Agric., vol. 20, pp. 551–556.

Pictet, A.

1902　L'influence des changements de nourriture des chenilles sur le développement de leurs papillons. Arch. Sci. Phys. Nat., vol. 14, pp. 537–540.

1903　Variations des papillons provenant des changements d'alimentation. *Ibid.*, vol. 16, pp. 585–588.

1905　Influence de l'alimentation et de l'humidité sur la variation des papillons. Mem. Soc. Phys. Genève, vol. 35, pp. 45–127, 5 pls.

1926　Localisation dans une région du parc national Suisse, d'une race constante de papillons exclusivement composée d'hybrides. Rev. Suisse. Zool., vol. 33, pp. 399–406.

Porter, B. A.

1928　The Apple Maggot. Bull. U. S. Dept. Agric., Tech. Ser., No. 66, 48 pp.

Quayle, H. J.

1926　The Codling Moth in Walnuts. Bull. California Agric. Expt. Sta., No. 402, 33 pp., 11 figs.

1938　The Development of Resistance to Hydrocyanic Acid in Certain Scale Insects. Hilgardia, vol. 11, pp. 183–210.

Richmond, E. A.

1927　Olfactory Response of the Japanese Beetle. Proc. Entom. Soc. Washington, vol. 29, pp. 36–44.

Roubaud, E.

1929　Biological Researches on *Pyrausta nubilialis*. Intern. Corn Borer Invest., Sci. Rept., vol. 2, pp. 1–21.

Rubtzov, I. A.

1932　The Food-plants of Siberian Acaididae. Bull. Plant Prot. (I. Entom.), No. 3, pp. 13–31. [In Russian with English Summary.]

1932　The Habitats and Conditions of Grasshopper Outbreaks in East Siberia. *Ibid.*, No. 3, pp. 33–130.　[In Russian with English Summary.]

Sacchi, R.

1919　Relation between the Hour at which the Mulberry Leaves are Picked and the Silk-yield of the Silkworm. Informazione Seriche, vol. 6, pp. 61–63. (Abstract in Chem. Abstr.)

Savin, M. B.

1927　Food Preferences of the Black Cricket (*Gryllus assimilis*) with Special Reference to the Damage Done to Fabrics. Entom. News, vol. 38, pp. 4–10; 33–39.

Schrader, F.

1926　Notes on the English and American Races of the Greenhouse Whitefly (*Trialeurodes vaporariorum*). Ann. Appl. Biol., vol. 13, pp. 189–196.

Schwarz, E.

1923　The Reason Why *Catocala* Eggs are Occasionally Deposited on Plants Upon which the Larva cannot Survive. Entom. News, vol. 34, pp. 272–273.

Sladden, D. E.

1934–　Transference of Induced Foot-habit from Parent to Offspring. Proc. Roy. 35　Soc. (B), vol. 114, pp. 441–449; vol. 119, pp. 31–46.

Smith, G. D.

1921 Studies on the Biology of the Mexican Cotton Boll Weevil. Bull. U. S. Dept. Agric., No. 926, 44 pp.

Smith, H. S.

1941 Racial Segregation in Insect Populations and its Significance. Journ. Econ. Entom., vol. 34, pp. 1–13.

Steiner, G.

1925 The Problem of Host Selection and Host Specialization of Certain Plant-infecting Nemas and its Application in the Study of Nemic Pests. Phytopathology, vol. 15, pp. 499–534, 8 figs.

Swezey, O. H.

1941 Food-plant Relations of Scolytidae and Platypodidae in the Hawaiian Islands. Proc. Hawaiian Entom. Soc., vol. 11, pp. 117–126.

Teodoro, G.

1926 Considerazioni sulle Cocciniglie parassite e loro piante nutrici. Rev. Biol., vol. 8, pp. 629–637.

Thompson, W. R., and H. L. Parker

1928 Host Selection in *Pyrausta nubilalis*. Bull. Entom. Res., vol. 18, pp. 359–364.

Thorpe, W. H.

1929 Biological Races in *Hyponomeuta padella*. Journ. Linn. Soc. Zool., vol. 36, pp. 621–634.

1930 Biological Races in Insects and Allied Groups. Biol. Rev., vol. 5, pp. 177–212.

1931 Further Observations on Biological Races in *Hyponomeuta padella*. Journ. Linn. Soc. Zool., vol. 37, pp. 489–492.

1938 Further Experiments on Olfactory Conditioning in a Parasitic Insect, *Nemeritis*. Proc. Roy. Soc. London, (B), vol. 126, pp. 370–397.

Thorpe, W. H., and H. B. Caudle

1938 A Study of the Olfactory Responses of Insect Parasites to the Food Plant of their Host. Parasitology, vol. 30, pp. 523–528.

Thorpe, W. H., and F. G. W. Jones

1937 Olfactory Conditioning in a Parasitic Insect and its Relation to the Problem of Host Selection. Proc. Roy. Soc. London, (B), vol. 124, pp. 56–81.

Trägårdh, I.

1913 On the Chemotropism of Insects and its Significance. Bull. Entom. Res., vol. 4, pp. 113–117.

Trouvelot, B.

1931 Recherches expérimentales sur les déplacements à la marche et au vol de doryphores adultes. Rev. Path. vég. Entom. Agric., vol. 18, pp. 6–8.

Trouvelot, B., Lacotte, Dussy, and Thénard

1933 Observations sur les affinités trophiques existant entre les larves de *Leptinotarsa decemlineata* et les plantes de la famille des Solanées. C. R. Acad. Sci. Paris, vol. 197, pp. 273–275.

1933 Les qualités élémentaires des plantes nourricières du *Leptinotarsa decemlineata* et leur influence sur le comportement de l'insecte. C. R. Acad. Sci. Paris, vol. 197, pp. 355–356.

Trouvelot, B., and J. Thénard

1931   Remarques sur les éléments des végétaux contribuant à limiter ou à empêcher la pullulation du *Leptinotarsa decemlineata* sur de nombreuses espèces ou races végétales.   Rev. Path. Vég. Entom. Agric., vol. 18, pp. 277–285,   2 pls.

Twinn, C. R.

1925   Observations on the Host-selection Habits of *Pieris rapae*.   55th Ann. Rept. Entom. Soc. Ontario 1924, pp. 75–80.

Uichanco, L. B.

1928   Insects in Relation to the Introduced Cultivated Element of the Philippine Flora.   Proc. 3rd. Pan-Pacific Sci. Congr., Tokyo, 1926, vol. 2, pp. 2069–2076.

Vassiliev, E. M.

1913   Plants Serving as Food for Some Herbivorous Insects and the Causes for their Selection.   Stud. Expt. Entom. Sta., All-Russian Sugar Refiners, 1912, pp. 63–66.   [In Russian.   Abstract in Rev. Appl. Entom., vol. 2A, p. 63.]

Vayssière, P.

1921   Dégâts causés par le grillon domestique.   Bull. Soc. Entom. France, p. 248.

Vecchi, A.

1926   Ulteriore esperienze sull'alimentazione del baco da seta con *Maclura aurantiaca*.   Boll. Soc. Entom. Ital., vol. 58, pp. 122–130.

Verschaffelt, E.

1910   The Cause Determining the Selection of Food in some Herbivorous Insects.   Proc. Sec. Sci. K. Akad. Wetensch., Amsterdam, vol. 13, pp. 536–542.

Vickery, R. A.

1910   Contributions to a Knowledge of the Corn Root-aphis.   U. S. Dept. Agric., Bur. Entom., Bull. 85, pt. 6, pp. 97–118.

Wachs, H.

1917   Ein Beitrag zum Problem der Seidenraupenzucht mit Schwarzelfütterung.   Nat. Wochenschr., vol. 32, pp. 729–732.

Walsh, B. D.

1864   On Phytophagic Varieties and Phytophagous Species.   Proc. Entom. Soc. Philadelphia, vol. 3, pp. 403–430.

1865   On the Phytophagic Varieties of Phytophagous Species, with Remarks on the Unity of Coloration in Insect   Proc. Entom. Soc. Philadelphia, vol. 5, pp. 194–216.

Watson, J. R.

1941   Migration and Food Preferences of the Lubberly Locust.   Florida Entom., vol. 24, pp. 40–42.

Webster, F. M.

1887–   The Relation of Ants to the Corn-aphis.   Rept. Comm. Agric., Washington, 88   pp. 148–149.

1907   The Corn-leaf Aphis and Corn-root Aphis.   Circ. Bur. Entom., U. S. Dept. Agric., No. 86, 13 pp.

Wieting, J. O. G., and W. M. Hoskins

1939 The Olfactory Responses of Flies in a New Type of Insect Olfactometer. Journ. Econ. Entom., vol. 32, pp. 24–29.

Williams, F. X.

1913 Notes on Three Sesiidae Affecting the Missouri Gourd (*Curcurbita foetidissima*) in Kansas. Kansas Univ. Sci. Bull., vol. 8, pp. 217–220, 2 pls.

Wirth, W.

1928 Untersuchungen über Reichschwellenwerte von Geruchsstoffen bei Insekten. Biol. Zentralbl., vol. 48, pp. 567–576.

Woglum, R. S.

1925 Observations on Insects Developing Immunity to Insecticides. Journ. Econ. Entom., vol. 18, pp. 593–597.

Wolcott, G. N.

1940 List of Woods Arranged According to their Resistance to the Attack of the Dry-wood Termite, *Cryptotermes brevis*. Caribbean Forester, vol. 1, No. 4, pp. 1–10, 2 figs.

Woods, W. C.

1915 Blueberry Insects in Maine. Bull. Maine Agric. Expt. Sta., No. 224, pp. 249–288, 4 pls., 3 figs.

Zacher, F.

1928 Nahrungswahl und Fortpflanzungsbiologie der Samenkäfer. Anz. f. Schädlingsk., vol. 4, p. 148.

1929 Nahrungswahl und Biologie der Samenkäfer. Verh. deuts. Gesellsch. angew. Entom., 7ten Versamml., 1928, pp. 55–62.

REACTIONS OF PLANTS TO INSECT FEEDING, POISONOUS PLANTS AS FOOD

Baker, A. C.

1922 Feeding Punctures of Insects. Journ. Econ. Entom., vol. 15, p. 312.

Ballou, C. H.

1929 Effects of Geranium on the Japanese Beetle. Journ. Econ. Entom., vol. 22, pp. 289–293, 4 pls.

Barnes, H. F.

1931 Further Results of an Investigation into the Resistance of Basket Willows to Button Gall Formation. Ann. Appl. Biol., vol. 18, pp. 75–82, 2 pls.

Bigger, J. H., J. R. Holbert, et al.

1938 Resistance of Certain Corn Hybrids to Attack of Southern Corn Rootworm. Journ. Econ. Entom., vol. 31, pp. 102–107.

Blunck, H.

1931 *Psylliodes affinis* an Tabak. Anz. f. Schädlingsk., vol. 7, pp. 133–136, 1 fig.

Börner, C.

1915 Über blutlösende Safte im Blattlauskörper und ihr Verhalten gegenüber Pflanzensaften. Mitt. Biol. Reichanst. Land- u. Forst-wirtschaft, Berlin, No. 16, pp. 43–49.

Bovingdon, H. H. S.

1931   Pests of Cured Tobacco. The Tobacco Beetle, *Lasioderma serricorne* and the Cacao Moth, *Ephestia elutella*. Tobacco, August 1st, 1931, 4 pp., 12 figs.

1933   Report on the Infection of Cured Tobacco in London by the Cacao Moth *Ephestia elutella*. Pub. Empire Marketing Board, No. 67, 88 pp.

Boyce, A. M.

1933   Influence of Host Resistance and Temperature during Dormancy upon the Seasonal History of the Walnut Husk Fly, *Rhagoletis completa*. Journ. Econ. Entom., vol. 26, pp. 813–819.

Brown, K. B.

1916   The Specific Effects of Certain Leaf-feeding Coccidae and Aphididae upon the Pines. Ann. Entom. Soc. America, vol. 9, pp. 414–422, 2 pls.

Büsgen, M.

1891   Der Honigtau. Biolog. Centralbl., vol. 11, pp. 193–200.

Cameron, A. E.

1914   A Contribution to a Knowledge of the Belladonna Leaf-miner, *Pegomyia hyoscyami*, its Life-history and Biology. Ann. Appl. Biol., vol. 1, pp. 43–76, 2 pls., 4 figs.

Campos, R. F.

1933   El gigantesco tabano, *Pantophthalmus bellardii*. Su larva affecta al arbol de cacao. Rev. Col., Rocafuerte, vol. 14, pp. 17–19.

Collins, G. N., and J. H. Kempton

1917   Breeding Sweet Corn Resistant to the Corn Earworm. Journ. Agric. Res., vol. 11, pp. 549–572.

Coquillett, D. W.

1895   A Cecidomyiid that Lives on Poison Oak. Insect Life, vol. 7, p. 348.

Dahms, R. G., and F. A. Fenton

1939   Plant Breeding and Selecting for Insect Resistance. Journ. Econ. Entom., vol. 32, pp. 131–134.

Davidson, J., and H. Henson

1929   The Internal Condition of the Host Plant in Relation to Insect Attack, with Special Reference to the Influence of Pyridine. Ann. Appl. Biol., vol. 16, pp. 458–471.

Fiebrig, K.

1906   Eine morphologisch und biologisch interessante Dipterenlarve aus Paraguay. Zeits. Wiss. Insektenbiol., vol. 2, pp. 316–323, 14 figs.

1908   Eine Schaum bildende Käferlarve, *Pachyschelus* sp. Zeits. wiss. Insektenbiol., vol. 4, pp. 333–339; 353–363, 13 figs.

Gater, B. A. R.

1925   Investigations on "Tuba". Malayan Agric. Journ., vol. 8, pp. 312–329.

Haseman, L.

1916   An Investigation of the Supposed Immunity of Some Varieties of Wheat to the Attack of the Hessian Fly. Journ. Econ. Entom., vol. 9, pp. 291–294.

Hinds, W. E.

1906   Proliferation as a Factor in the Natural Control of the Mexican Boll Weevil. Bull. Bur. Entom., U. S. Dept. Agric., No. 59, 45 pp.

Holbert, J. R., W. P. Flint, and J. H. Bigger

1934   Chinch Bug Resistance in Corn: an Inherited Character. Journ. Econ. Entom., vol. 27, pp. 121–124.

Hollick, A.

1921   Loco Weeds. Journ. American Mus. Nat. Hist., vol. 21, pp. 85–91, 7 figs.

Horsfall, J. L.

1923   The Effects of Feeding Punctures of Aphids on Certain Plant Tissues. Bull. Pennsylvania State Coll., No. 182, 22 pp., 7 pls.

Isley, D.

1934   Relationship Between Early Varieties of Cotton and Boll Weevil Injury. Journ. Econ. Entom., vol. 27, pp. 762–766.

Jordan, K. H. C.

1930   Zerfressene Zigarren. Mitt. Gesellsch. Vorratsschutz, vol. 6, p. 19.

Juckenack, A., and C. Griebel

1910   Über den Einfluss strychninhaltiger Nahrung auf Insekten. Zeitschr. Untersuch. Nahrungs- u. Genussmittel, vol. 19, pp. 571–573.

Kalshoven, L. G. E.

1924   Aanteeleningen over eukele Kina-insekten. Meded. Inst. Plantenziekten, Buitenzorg, No. 65, 27 pp., 4 pls.

Kazanetshi, N.

1927   Les ennemies du chrysanthème (pyrèthre) de Dalmatie en rapport avec la dégénération de la plante. C. R. Acad. Agric. France, vol. 13, pp. 1080–1086.

Lagerheim, G.

1900   Zur Frage der Schutzmittel der Pflanzen gegen Raupenfrass. Entom. Tidskr., vol. 21, pp. 209–232.

Lees, A. H.

1926   Insect Attack and the Internal Condition of the Plant. Ann. Appl. Biol., vol. 13, pp. 506–515.

Ludwig, F.

1907   Weiteres zur Biologie von *Helleborus foetidus*. Zeits. wiss. Insektenbiol., vol. 3, pp. 45–50.

McIndoo, N. E., and A. F. Sievers

1924   Plants Tested for or Reported to Possess Insecticidal Properties. Bull. U. S. Dept. Agric., No. 1201, 62 pp.

MacLeod, G. F.

1933   Some Examples of Varietal Resistance of Plants to Insect Attacks. Journ. Econ. Entom., vol. 26, pp. 62–67.

Malloch, W. A.

1923 The Problem of Breeding Nematode-resistant Plants. Phytopathology, vol. 13, pp. 436–450.

Marie, P.

1927 Procédés mécaniques d'auto–défense corticale chez certain conifers contre les attaques des Scolytides. Rev. Path. Vég. Entom. Agric., vol. 13, pp. 167–171.

Miller, N. C. E.

1930 A Major Pest of Derris; *Neolepta biplagiata.* Malayan Agric. Journ., vol. 18, pp. 541–544.

1934 Coleopterous Pests of Stored Derris in Malaya. Sci. Ser. No. 14, Straits Settlements and Fed. Malay States, Dept. Agric., 34 pp.

Monzen, K.

1926 The Woolly Apple Aphis (*Eriosoma lanigerum*) in Japan, with Special Reference to its Life-History and the Susceptibility of the Host-plant. Verhandl. IIIte Intern. Entom. Kongr., Zürich, pp. 249–275.

Morgan, A. C., and S. C. Lyon

1928 Notes on Amyl Salicylate as an Attractant to the Tobacco Hornworm Moth. Journ. Econ. Entom., vol. 21, pp. 189–191.

Mossop, M. C.

1932 Pests of Stored Tobacco in Southern Rhodesia. Rhodesia Agric. Journ., vol. 24, pp. 245–265.

Müller, A.

1926 Die innere Therapie der Pflanzen. Zeits. angew. Entom., vol. 12, supplement, 206 pp., 29 figs.

Painter, R. H.

1936 The Food of Insects and its Relation to Resistance of Plants to Insect Attack. American Naturalist, vol. 70, pp. 547–566.

Parker, J. H., and R. H. Painter

1932 Insect Resistance in Crop Plants. Proc. Intern. Congr. Genetics, vol. 6, pp. 150–152.

Parkin, E. A.

1934 Observations on the Biology of the *Lyctus* Powder Post Beetles, with Special Reference to Oviposition and the Egg. Ann. Appl. Biol., vol. 21, pp. 495–518, 9 figs.

1936 A Study of the Food Relations of the *Lyctus* Powder Post Beetles. Ann. Appl. Biol., vol. 23, pp. 369–400, 1 pl., 4 figs.

Peterson, C. E., and E. S. Haber

1942 The Relation of Leaf Structure to Thrips Resistance in Onions. Proc. American Hort. Soc., vol. 40, pp. 421–422.

Petri, L.

1910 Osservazioni sopra il rapporto fra la composizione chimica della radici della vite e il grado de resistenza alla fillossera. Rendic. R. Acad. Lincei, vol. 19, pp. 27–34.

Phillips, J. S.

1933 The Relation of Ants with Insects Attacking Certain Poisonous Plants in Hawaii. Trans. Entom. Soc. London, vol. 82, pp. 1–3.

Powell, T. E., Jr.

1931 An Ecological Study of the Tobacco Beetle, *Lasioderma serricorne*, with Special Reference to its Life History and Control. Ecological Monographs, vol. 1, pp. 333–393, 20 figs.

Reed, W. D., E. Livingstone, and A. W. Morrill

1933 A Pest of Cured Tobacco, *Ephestia elutella*. Circ. U. S. Dept. Agric., No. 269, 16 pp., 7 figs.

Roach, W. A.

1936 Studies on Possible Causes of Immunity. Ann. Appl. Biol., vol. 24, pp. 206–210.

Roark, R. C.

1919 Plants Used as Insecticides. American Journ. Pharmacy, vol. 91, pp. 25–37; 91–107.

1939 Insect Pests of Derris. Journ. Econ. Entom., vol. 32, pp. 305–309.

Sakai, K.

1933 A Bostrychid Attacking Roots of Derris. Kontyu, vol. 7, pp. 272–273, Tokyo.

Schroeder, J.

1903 Über experimentell erzielte Instinktvariationen. Verh. Deutsch. Zool. Ges., pp. 158–166.

Seamans, H. L., and E. McMillan

1935 Effects of Food Plants on the Development of the Pale Western Cutworm. Journ. Econ. Entom., vol. 28, pp. 421–425.

Searls, E. M.

1932 A Preliminary Report on the Resistance of Certain Legumes to Certain Homopterous Insects. Journ. Econ. Entom., vol. 25, pp. 46–49.

Staniland, L. N.

1925 The Immunity of Apple Sticks from Attacks of Woolly Aphis (*Eriosoma lanigerum*). The Causes of the Relative Resistance of the Stock. Bull. Entom. Res., vol. 15, pp. 157–170.

Tissot, A. N.

1928 A New Aphid from Poison Ivy, *Carolinaia rhois*. Florida Entom., vol. 12, pp. 1–2, 1 fig.

Vansell, G. H., and W. G. Watkins

1933 A Plant Poisonous to Adult Bees. Journ. Econ. Entom., vol. 26, pp. 168–170.

Wilson, G. F.

1931 Insects Associated with the Seeds of Garden Plants. Journ. Roy. Hort. Soc., vol. 56, pp. 31–47, 6 pls.

Wolcott, G. N.

1924 The Comparative Resistance of Woods to the Attack of the Termite, *Cryptotermes brevis*. Bull. Porto Rico Insular Expt. Sta., No. 33, 15 pp.

Worrall, L.

1925 A Jassid-Resistant Cotton. Journ. Dept. Agric. Pretoria, vol. 10, pp. 487–491, 3 figs.

Zacher, F.

1921 Tierische Schädlinge an Heil- und Giftpflanzen und ihre Bedeutung für den Arzneipflanzenbau. Ber. Deutsch. Pharmazet. Ges., Berlin, vol. 31, pp. 55–65, 6 figs.

CHAPTER IV

GALL INSECTS

Tam ferax itaque muscarum familia è Gallis erumpentium, sic dictante Naturâ, terebrae, seu duplicis limae usu vulnerat, & perfodet molles plantarum partes; itâ ut ex diversa ipsarum natura, & compage variae pariter emergant morbosae excrescentiae, & tumores. Ex infuso namque liquore, à terebrae extremo effluente, qui summè activus, & fermentativus est, nova in tenellis vegetantibus particulis excitat fermentatio, seu intestinus motus; itâ ut appellens nutritivus succus, & in transversalibus recollectus utriculis peregrinâ aurâ inspiratus fermentari incipiat, & turgere; ut frequenter in nobis, & sanguineis quibusdam perfectis animalibus, ex apum inflicto vulnere, & subinde infuso ichore experimur.

MARCELLO MALPIGHI, 1686[1]

GALL INSECTS are purely vegetarian, but they induce such profound reactions on the part of the plants they affect, that it is appropriate to treat them separately from the phytophagous forms dealt with in the previous chapter. It has been shown that the feeding of insects generally evokes scant changes in their food plants, aside from the actual damage done to foliage, roots, stems or flowers. In contrast to these destructive activities the gall insects present a very different picture. They cause their hosts to produce special structures within which they live, feed and grow to maturity. Moreover, these galls are very often highly specific, definitely organized entities, with peculiar morphological characters, reappearing from generation to generation, each after its own kind.

Galls were familiar to the ancients and are often called cecidia, a word of Greek origin, variously compounded as phytocecidium,

[1] "Thus, the very fertile family of flies emerging from the galls follow the dictates of Nature to wound and perforate the soft parts of plants with the file-like ovipositor. Then in accordance with their diverse nature and relations, tumors of similarly varied kinds arise. The liquid arising from the ovipositor is highly active and fermentative, and when injected, this excites a new fermentation or internal motion within the delicate, growing particles of the plant. The nutritive sap accumulated in the transverse vessels begins to ferment and swell when acted upon by the moving air, just as we ourselves and certain of the perfected sanguineous animals react to the wound inflicted by bees immediately after their secretion has been introduced."

zoöcecidium, mycetocecidium, entomocecidium, acarocecidium, helminthocecidium, and still others, to indicate that they are caused by plants, animals, fungi, insects, mites, worms or the like. The same root appears in the name *Cecidomyia*, literally, gall-fly, applied to the typical genus of gall midges. The laws of zoological nomenclature require the use of the earlier name, *Itonida*, and convenience is served by the word gall in place of long-winded substitutes. Gall undoubtedly refers to the bitter taste of certain oak galls, long useful in the making of ink on account of the large amount of bitter and styptic tannin that they contain.

Galls may be defined as abnormal growths developing in the tissues of plants, due ordinarily to the introduction of foreign substances. The latter are always derived from the secretions of living organisms, although, of course, the presence of most organisms does not call forth any such violent reaction. The gall develops and reaches its definitive form and structure as a result of the continued stimulus supplied by the gall maker, but it is entirely a product of the plant.

Although it is evident that the development of the gall is brought about by these foreign substances, it is not yet clear just what the substances are, nor how they act. It has frequently been assumed that they are complex materials, most likely similar to proteins, on account of their highly specific, determinate action. However, this interpretation is not necessarily correct as will appear later.

Aside from insects, the most diverse living organisms are known to cause galls, but the most elaborate and perfect, specific and uniform ones are produced by insects. A few of the insect galls are far less regular and these grade off into others where the response of the plant is less and less pronounced, finally reaching a point where they are scarcely definite structures.

One very common malformation, occurring in a variety of plants, is crown-gall. The cause of this was first shown many years ago by Erwin F. Smith to be a bacterium, now known as *Phytomonas tumefaciens*. The irregular hypertrophied growths that result have been referred to loosely as a sort of plant cancer. There is really some validity in this terminology since the tissues composing many insect galls frequently show nuclear abnormalities resembling those of malignant animal tumors. The root nodules on leguminous plants such as beans and clover are also of bacterial origin and appear to be the only class of galls that represent a mutualistic symbiosis. The ability of the gall makers (*Rhizobium*) to utilize gaseous nitrogen in their metabolism furnishes their host plants with this needed element. In return the plant furnishes the secluded underground retreat essential to the growth and reproduction of the bacterium.

Many galls result from invasion by fungi, although most parasitic fungi do not call forth any such conspicuously abnormal growths in their hosts.

Among animals, a very few of the numerous species of nematode worms produce galls, usually on roots. These are, at best, crude affairs from a domiciliary standpoint and such squalid quarters are quite in keeping with the lowly organization of their inhabitants, whose standard of living antedates the democracies by a sizable margin.

Certain minute Arachnida of the order Acarina are also gall makers. Members of one family, Eriophyidae, live entirely in galls, principally on the leaves of various trees. The gall mites are all extremely minute, but often appear in great abundance.

Insects are, however, the most conspicuous cause of plant galls. Most of these belong to two large families, the Itonididae or Cecido-

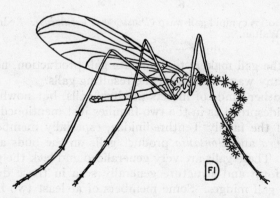

FIG. 19. A Gall-midge, *Hormosomyia* (Family Itonididae). Redrawn from Cole.

myiidae of the order Diptera and the Cynipidae of the order Hymenoptera. The Itonididae are rather primitive, structurally degenerate flies, of very small or minute size, known as gall midges (Fig. 19). Most members of the family produce galls, but many of the more generalized ones live on fungi or in decaying plant materials, and a few are carnivorous. The Cynipidae, or gall wasps (Fig. 20), are also small insects, all members of the family living in galls. Many gall insects are commonly attacked by entomophagous parasites, mainly small chalcid wasps, and these bear no relation to the formation of the galls. Their presence does not interfere with the

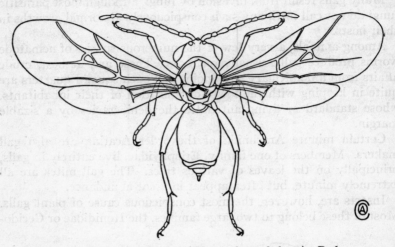

FIG. 20. A cynipid gall-wasp (*Diastrophus nebulosus*). Redrawn from Walton.

ability of the gall makers to stimulate gall production, nor does it modify in any way the form of the resulting galls.

Several other kinds of insects produce galls, but nowhere is the habit so widespread as in the two families just mentioned. Certain saw-flies of the family Tenthredinidae, especially members of the genera *Euura* and *Pontania* produce galls on the buds and leaves of willows. These galls are very generalized and lack the great constancy of form and structure generally seen in those due to gall wasps and gall midges. Some members of at least two families of chalcid wasps produce galls. These are particularly interesting as they represent a reversion to the vegetarian habits of primitive ancestors by representatives of a superfamily that is now almost exclusively parasitic. Reference has already been made to some vegetarian species (Family Callimomidae) that develop in seeds. Several genera of the family Eurytomidae have phytophagous habits, and the well known genus *Harmolita* contains a number of species that live in the culms of grasses. These cause swellings or galls of a simple type, as does also *Eurytoma orchidearum* in pseudobulbs of the showy Cattleya orchids. Another allied family, Peri-lampidae, includes a few tropical gall makers that depart from the usual parasitic habits of that family. There are some ants that produce galls and others inhabit preformed cavities in certain tropical plants. There is some question as to whether some of these should be considered to be true galls or simply the result of mechanical injury. In some cases the ants feed directly on the nutritive tissue that develops while in others they foster coccids in these habitations, which have been termed myrmecodomatia. In the latter type the formation of the gall tissues must be attributed to the coccids. In most of the "ant plants" it is known that the structures which house these insects are inherited and not due to the activities of the ants, except possibly in the development of callus tissue for the repair of traumatic modifications.

In addition to the gall midges, the Diptera include some other gall making flies, notably members of the Trypetidae and Agromyzidae. These are highly specialized families, far removed from the gall midges, all of whose members feed internally in plant tissues. The Trypetidae usually develop in fruits and the Agromyzidae in stems or leaves, and only a very small number produce galls.

A large series of galls are due to various Homoptera, almost exclusively plant lice or aphids, now classified into several related families. The jumping plant lice of the family Psyllidae include species that produce galls on various plants, especially in tropical countries. Notable among these is *Pachypsylla*, restricted to hackberry (*Celtis*), on which the several species produce stem and leaf galls. The Aphididae include a number of genera that cause galls on many diverse plants. Often a single species of aphid may affect more than one species of host plant during the year, just as the free living aphids exhibit an alternation of food plants during the course of the succeeding generations of the annual life cycle. The galls due to plant lice are formed by an overgrowth of the leaf or other tissues which encloses the prolific mother and her family of aphids within a cavity that always retains an opening to the outside world. This serves for the emergence of migrating individuals when they leave the parental home. At other times it bears no traffic as the mother and her pregnant spinster offspring sit continuously at table with their mouthparts imbedded in the sappy tissues of the gall.

A few other scattering examples of gall insects are known. In the enormous order Coleoptera, a very few members of several families (Buprestidae, Cerambycidae, Chrysomelidae, Ipidae, Curculionidae)

are known to cause galls of very simple type. The same is true of
a very few lepidopterous caterpillars in several families: one of these,
Gnorimoschema of the family Gelechiidae, is the maker of one of our
very abundant stem galls on goldenrod (*Solidago*). Some minute
insects of the order Thysanoptera, known as thrips, produce galls,
although most species are free living.

The galls produced by the minute eriophyid mites (Fig. 21) are
similar to insect galls and studies made on them by Kendall and

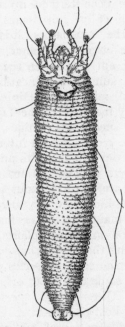

FIG. 21. A gall mite (*Eriophyes celtis*) which produces galls in the
buds of the hackberry (*Celtis occidentalis*). After Kendall.

others throw considerable light on the phenomena of gall growth
and the cytological peculiarities associated with it. The mite galls
are numerous, most generally appearing as small pouch-like malformations on leaves, particularly those of various trees.

Practically all parts of plants are subject to attack by gall insects,
both the aërial and subterranean portions, but most generally galls
originate in the buds or more actively growing regions where differentiation of tissues is still under way. Galls of the North American
Cynipidae, for example, are distributed as follows: on leaves, 54%;

twigs and woody parts, 14%; buds, 9%; roots and flowers, each 8%; fruits, 7% (Felt, '40). The site of origin is determined by the deposition of the egg or eggs of the gall insect, or in the case of aphids, by the selection of the spot where feeding will take place.

We know that the deposition of the egg in the tissues of the plant does not commonly initiate the changes that lead to the growth of the gall, as in many cases the eggs lie dormant for periods of several months without inciting any reaction. Slight changes are sometimes observed which may be attributed to substances introduced with the egg, or possibly to injury or some secretion from the insect's ovipositor. Once the egg hatches and larval feeding begins, development proceeds rapidly and the gall attains its final form under the continued stimulus of the contained larva. The same stimulus follows the insertion of the mouthparts by a gall-making aphid as this marks the beginning of the feeding process. It has already been mentioned in the case of aphids which do not cause galls, that sheaths are formed about the mouthparts and microscopic cellular changes occur adjacent to the feeding punctures. The aphid gall is thus a greatly exaggerated and highly integrated expression of the same, or of a closely similar phenomenon.

Insect galls are restricted almost exclusively to the seed plants and their comparative frequency varies greatly among the several natural families. Unfortunately we know little about them in many parts of the world, but data satisfactory for examination are available for several widely separated regions. Extensive series of galls have been made known and systematically treated from Europe and North America, as well as from the Dutch East Indies and tropical America.

Naturally those of Europe and the Mediterranean Basin are best known. In this region insect galls other than those occurring on flowering plants are practically nonexistent although seven species of Diptera affect higher fungi of the families Polyporaceae and Agaricaceae, and several species of eriophyid mite galls have been found on lichens of the family Usneaceae. The few galls on algae and bryophytes are due to nematode worms and do not concern us here. Among the pteridophytes insect galls really first appear. They are by no means numerous, numbering only about a dozen, all but one of which affect the true ferns. Thus in spite of their environmental association and similarity to the flowering plants, the ferns are no more acceptable to the gall makers than they are to the more independent phytophagous insects discussed in the preceding chapter.

The same is true of the neotropical region, where only eight species are known to form galls on ferns.

Some families of plants are especially attractive, but there is little correlation in the number of gall insects that affect them in different regions. This will be very apparent from the accompanying tabulation which shows the number of galls known on a few common families of plants in Europe and North America, for which extensive data are available.

These are arranged in order of their susceptibility in Europe and it is seen that twelve families harbor more than 70% of the known insect and mite galls. The flora and insect fauna of North America are essentially similar and we should expect some uniformity in these two regions, but such is not the case.

	Europe	No. Amer.		Europe	No. Amer.
Fagaceae	901	805	Labiatae	217	16
Compositae	664	181	Gramineae	193	68
Salicaceae	573	114	Umbelliferae	181	3
Rosaceae	500	133	Rubiaceae	162	5
Leguminosae	481	32	Scrophulariaceae	139	8
Cruciferae	256	3	Caryophyllaceae	107	0

Fundamental differences should exist in tropical countries on the same basis for both the New and Old World. There both the fauna and flora differ widely from those of the northern continents and from one another. Some groups of gall insects show much greater abundance in particular regions. In Europe and North America there exist many hundreds of species of gall wasps. These affect principally oaks which are rare in tropical countries and this fact accounts at least in part for the difference. In Malaya, Psyllidae and thrips are much more abundant than elsewhere. With due consideration to such variables, it is still strange to find so little agreement among the several faunas and floras. Such disparity leads one to believe that the various gall insect faunas must be of comparatively recent origin, and that they are undergoing speciation at an unusually rapid rate. Studies by Kinsey on the Cynipidae of North America seem to bear out this conclusion.

Even more so than most other vegetarian insects, gall makers are very specific in their selection of food plants. Nearly all are either strictly monophagous, or oligophagous with a very restricted series of host plants, rarely including more than related members of a single genus. Even in the case of plants like the willows (*Salix*) where botanists distinguish species with difficulty, some are known to develop button galls following oviposition by specific gall midges while other closely related ones do not. The situation is here more complex than it is with free living insects. In the latter, the egg-laying instinct of the mother must coincide with the specialized appetite of her offspring, but the gall insect must have also the coöperation of the plant in forming the gall before it can complete its own development. This added factor renders any shift in food habits much more difficult as an appropriate response by the plant is an absolute necessity. Examples of a failure to produce galls although the insect is present have been described in willows by Barnes. Eggs laid on a species of willow not normally affected may develop into larvae, but no gall is produced as the plant appears immune to stimulation.

With respect to the evolution of plant galls, paleontology offers very meagre information. A few fossil galls have been described from Tertiary formations, but they cannot be associated with specific insects. These are all leaf galls, and others have been seen on leaves that are preserved in great quantity in certain Cretaceous deposits. Kinsey has described two fossil Cynipidae, preserved in Baltic Amber of lower Oligocene age, and very recently another in Cretaceous amber from Canada. All these are generalized forms, quite surely ones that did not make galls on oaks, and the Cretaceous fossil is the most primitive genus of its family yet to be discovered. This places the origin of the gall wasps at least in Cretaceous times and strongly suggests that their present predilection for oaks is of rather recent origin. This view is further supported by studies on the living fauna, also by Kinsey, which show that the Cynipidae are in an active state of evolution and differentiation. Little is known concerning the age of the gall making habit in other groups of insects.

Galls are commonly grouped into two categories, known as cataplasmas and prosoplasmas. The cataplasma is the simpler type where a malformation develops in the affected portion of the plant, and the normal differentiation of its tissues is inhibited. This means that differentiation does not attain the complexity of corresponding normal parts, and, moreover, is not constant, representing a sort of hyperplasia of essentially embryonic tissue. Likewise, such galls vary considerably in size and form. The prosoplasma, on the other hand, is a structure almost as constant as the flower, fruit, or plant leaf, not only in form and size, but also in the uniform and specific differentiation of the tissues that compose it. Furthermore, it is specialized anatomically in a way that is truly its own, and not directly comparable with the corresponding parts of unaffected plants (Pl. XI A). Thus, something new has been produced by the plant under the influence of the gall maker. This new structure has many constant characters, involving size, color, and seasonal time of development as well as minute cellular structure and a specific type of tissue differentiation. Naturally the rudiments of these growths must originate in that portion of the plant, known as the meristem, which possesses the ability to produce the several types of tissue that appear in the mature gall. Consequently, a great majority of galls begin as derangements of buds in which meristem occurs (Pl. XI B). Occasionally galls on one part of a plant comically reproduce some peculiarity of another part. Thus, Felt has called attention to the similarity of the splitting of one of the hickory galls at maturity to the dehiscence of the husk of the ripening hickory-nut. This and similar cases, where characteristics of one part of a plant appear in galls developed elsewhere, suggest the possibility that some growth hormone or organizing substance may be produced in the gall, or that the presence of the gall maker may cause its action to localize there rather than at the site normally selected. No proof of such an assumption has been adduced, and such similarities may, of course, be purely coincidental.

Cataplasmas and prosoplasmas are not sharply to be distinguished as they form an ascending evolutionary series and there are intergrading forms which are difficult to place. Such a variety must naturally be expected considering the polyphyletic origin of the gall making habit among insects and the diverse portions of the multifarious plants that are concerned. Numerically, more than half of the described galls are cataplasmas. Phylogenetically the prosoplasma represents the more recent type. This cannot be determined on the basis of paleontological evidence, but appears certain, since in their ontogenetic development at least some prosoplasmas pass through a cataplasma stage. This has been interestingly shown by Wells in a histological study of the development of certain galls produced on hackberry leaves by jumping plant lice of the genus *Pachypsylla*.

Commonly a single insect inhabitant is present in each individual gall, but frequently there may be several or occasionally a considerable number. Those with a single inmate in solitary confinement are known as monothalamous, and structurally they are oriented and patterned with reference to the developing larva contained therein. Polythalamous galls contain a number of similar chambers, each containing a single enclosed larva. They are essentially an agglutination of such units, sometimes, however, with indications of a specific pattern involving the gall as a whole, not merely a summation of its parts. The polythalamous condition arises from the simultaneous deposition of a number of eggs by the gall maker at the site of origin of the gall.

When considering the anatomical and histological characteristics of galls it must be borne in mind that, although the insect is actually a parasite, it enters into a relationship with the plant that is actually symbiotic. This differs from the mutualistic symbioses in that the gall maker gains a shelter and all of its food from nutritive tissues produced especially at its order by the plant, which in turn receives no advantage whatsoever. The plant loses the food materials consumed by the gall maker and also considerable parts of the host frequently die or it fails to produce seed as a result. The sole contribution of the gall maker is the material, probably very minute in quantity, which initiates and later controls the processes of growth that ensue.

There has been much speculation concerning the nature of this stimulating material. Even in the seventeenth century, that remarkable zoologist, Malpighi, believed that galls originated from substances introduced into healthy plants, and to-day we agree with him, but have learned really nothing more as to what these substances may be. They are obviously highly specific, since a plant may develop at the same site widely dissimilar galls in response to different insects. Moreover, practically all of the thousands of prosoplasmatic galls now known are distinguishable on the basis of good morphological characters; commonly those caused by related species are more widely distinct than the insects themselves. For example, in one genus of gall wasps (*Cynips*) Kinsey states that the insects are more distinctive in some 54 species, the galls in 24, and both are equally distinctive in 17 species. On this basis it might be postulated that the irritating substances are legion, and on account of their great variety, complex, rather than simple in chemical composition. The possibility that they may be toxic proteins or viruses cannot be excluded.

Since some galls are caused by microörganisms, there is also the possibility that insect galls may be due to specific bacteria or fungi introduced into the plants by the insects. This seems plausible since all galls generally harbor numerous microörganisms and many insects are regularly infected with intracellular symbiotic microörganisms. Such an assumption presents two alternatives. One of these implies that the organisms may simply be transmitted to the plants by the gall makers, just as many bacterial, protozoan and virus diseases of plants are carried by specific insect vectors.

The other possibility is that the stimulating material may be produced by the symbionts (symbiotic microörganisms). There is a hint that such might be sometimes the case in the fact, for example, that certain mealy-bugs which feed on pineapple leaves are known to exist as two races with different symbionts, each causing a distinctive type of leaf spotting. Such microörganisms are transmitted in various ways from the mother insects to their offspring (see Chapter V), either within or on the surface of their eggs. In some gall insects it has been noted that the egg when laid is apparently infected with bacteria as these quickly develop about it after deposition. No such symbionts have, however, yet been demonstrated in gall insects, except in aphids and related groups, and although such an hypothesis is very enticing, it is unlikely that it apples generally to any series of insect galls. Some years ago, Dr. Rudolf W. Glaser and the writer isolated numerous bacteria and yeasts from some of our common insect galls and attempted to reproduce the galls by inoculating cultures of some of these into the appropriate plants. With a single exception, these experiments yielded entirely negative results and were never published. The exceptional case was the development of a typical gall of *Eurosta solidaginis* which appeared on a goldenrod plant close to the spot where it had received an injection. We recognized the gall and allowed it to ripen to full maturity. Then, on opening it, a full-blown Eurosta larva was disclosed, proving that a gall fly had discovered the plant and, unbeknown to us, laid her egg in it! A few years later, Lutz and Brown described a bacterium thought to be the possible cause of an insect gall produced by an aphid. Various bacteria, yeasts and simple fungi have frequently been reported from the internal tissues of numerous insect galls, but these have generally been regarded as coincidental invaders. Nevertheless, the presence of microörganisms in the actively proliferating inner layers of such galls may be a contributing factor in their development, and they cannot be excluded, even though it has not been proven that gall insects introduce specific organisms into their host plants.

This last statement must be qualified, however, in the case of certain plant deformities known as ambrosia galls. They are in a sense fungus galls, as the fungi are inoculated into the plant by the ovipositor of the insect when she lays her egg. The insects concerned are mainly species of gall midges (*Lasioptera* and *Asphondylia*) and their larvae feed on the ambrosia fungus which develops in the central cavity of the gall. This arrangement is clearly a symbiotic association of fungus and insect and either might be considered as the ringleader of the pair. Inasmuch as the ambrosia galls harbor larvae of cecidomyiid midges, a family notorious as gall makers, there is no valid reason to consider the fungi as more than constant invaders which have become a necessary food for the midge larvae. Ambrosia galls are known from Europe, North America and the Malayan tropics.

The very frequent transmission of virus diseases of plants by sucking insects suggests the possibility that gall insects may introduce viruses into the plants and that these substances may be the immediate cause of the gall's growth. No typical gall thus produced has so far been demonstrated, although various deformations of leaves, similar to the cataplasmatic galls of aphids, are due to viruses transmitted by aphids and leafhoppers. We may well believe that some galls inhabited by aphids are due to virus infection carried by the aphids.

Very recently Martin has produced stem galls on sugar cane following the artificial inoculation of extracts of the leafhopper and mealy-bug which normally cause these galls. They developed after the injection of material that had been sterilized in an autoclave, giving strong support to the belief that the active substance is a simple chemical rather than a biological product, as the latter are generally denatured by heat. These results are similar to those obtained by several earlier students (Kendall, Annand, Levine) who induced the development of tumors and gall-like malformations in plants injected with simple chemicals and carcinogenic agents.

The recent discovery of growth hormones in plants opens up another avenue of approach for investigating the etiology of galls. As has been pointed out by Went and Thimann, some of the growth phenomena associated with the auxins present striking similarities to those occurring in galls. Thus, it is possible that some kinds of galls may be due to the action of one or another of these substances. Such seems much more plausible with reference to the indeterminate cataplasmas rather than among the highly complex prosoplasmas.

The latter appear in such innumerable variety and specificity that they cannot reasonably be attributed to the action of a limited number of hormonal substances widely distributed in the plant kingdom. In the case of crown-gall developing in tomato plants, Henry, Riker and Duggar ('43) have shown recently that there is an increased amount of vitamin B_1 present in these bacterial galls during their early development, but they could trace no causal relation between this and the growth of the gall.

All of this conflicting evidence shows beyond a doubt that we have no clear understanding of the nature of the stimulus that determines the formation of insect galls, other than that they are initiated by the action of some substance introduced by the gall makers. Gall development appears rather to be directly comparable to embryonic growth and ontogenetic differentiation where the postulated organizing substances are supplied by the invading gall makers. Such an interpretation seems to be the only one which fits the observed processes at all closely. Furthermore, it must not be forgotten that galls are really pathological manifestations, most clearly so in the indefinite cataplasmas and apparently less so in the higher prosoplasmas where the plant produces a finished structure always true to plan. We can hardly escape the conclusion that both represent the same type of specific response by differentiating tissues in the presence of activating agents. This is really no more than the phenomenon of normal ontogeny which is generally believed to proceed under the guidance of organizing substances. The existence of numerous organizers of this sort in the production of galls would be a necessary corollary, as different insects developing in the same tissues of identical plants cause the formation of galls that differ widely in form, structure and color (Fig. 22). In gall wasps Kinsey has shown that the gall-producing stimuli act selectively on the several types of plant tissues. Thus, the galls produced by species of *Cynips* show morphological characters correlated with the several subgenera in which the species are classified. This shows very clearly that the physiological action of the stimulating materials is hereditary, remains constant over long periods, but is nevertheless undergoing gradual evolutionary change. Cases are known also in this genus of identical galls produced by two or more closely related species, indicating that at times speciation has proceeded without modification of the stimulating substances.

The cynipid gall wasps include a number of species in several genera whose life cycle is much modified by the occurrence of an alternation of generations. In such forms the larvae of the spring generation develop into individuals usually representing both sexes, males and females. They grow very rapidly and are mature within a month or less, giving forth the adult wasps at the time the leaves are unfolding on the oak trees. The succeeding generation is completed within the remaining portion of the summer, but the larvae develop into individuals which are all of one sex, females which do not emerge from the galls till winter, when they deposit their eggs

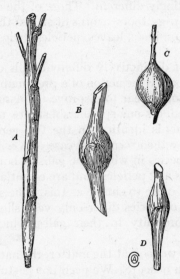

FIG. 22. Four goldenrod stem-galls, showing the widely different structures produced by diverse insects affecting the same plant. *A, Lasioptera cylindrigallae* and *B, Lasioptera solidaginis,* both gall-midges; *C, Eurosta solidaginis,* a trypetid fly; *D, Gnorimoschema gallaesolidaginis,* a small moth. Redrawn from Felt, 1917.

in the unopened buds. From these eggs the early spring brood develops. Thus, there is a bisexual generation, followed by an agamic one which lays eggs that develop parthenogenetically, giving rise in turn to the bisexual generation. Moreover, the females of the two generations are morphologically quite different and those of the agamic generation are frequently flightless, possessing only vestiges of wings. The conditions just described are not unique among insects as they parallel quite closely those that prevail among the aphids, already referred to in Chapter II, and earlier in the present chapter. They are of special interest here on account of the fact that these gall wasps produce complex, prosoplasmatic galls and the precise reactions of the host plants may be studied in detail.

Both generations produce galls, but only rarely on the same part of the plant. Of *Cynips* and several other genera, there are in the United States about a dozen species whose double life has been fully exposed. All of these inhabit galls on various species of oaks. Like the wasps reared in them, the galls of the alternating generations are almost always strikingly different. Those of the agamic generation may be on leaves, twigs, roots, acorns or stems; the bisexual generation produces galls on buds, leaves, petioles, twigs, stems, or on the flowering aments.

The formation of distinctively different galls on separate structures of the host tree, by the action of a presumably single stimulating material does not appear to disprove the assumption that each species of wasp produces one specific substance to which the plant reacts, and that this is identical in the two generations. This is further borne out by the exceptional case of *Neuroterus batatus* and *N. noxiosus*, two species in which the galls of both generations are formed on the twigs and petioles, and are essentially similar. However, it may not be quite so simple as this. The sexual and agamic females of these two species differ only very slightly in structure, and from this, uniformity in their gall-producing potentialities might be expected.

In whatever way we look at the matter, the nature of the specific stimuli must remain obscure. We need not postulate a great number of chemical substances, either on the basis of analogy with the ontogeny of normal structures, or as a result of experimental work done in stimulating the development of gall-like deformities by the use of chemical agents. On the other hand, the specificity of action exhibited by the gall wasps does not warrant the belief that the structural peculiarities of all galls represent unlike reactions to like stimuli. In cynipid galls on oak, for example, some species affect different hosts to cause similar galls. Such cases must be like reactions to like stimuli by different species of plants belonging to a genus that has shown itself capable of supplying almost any sort of fancy gall when given the required countersign. As the result of his prolonged researches on Cynipidae, Kinsey goes so far as to state that there must be nearly as many different kinds of gall-producing enzymes as there are species of gall wasps. Such a clear-cut conclusion could not be reached from a study of other gall insects as none of them confine their attacks to a restricted group of plants like the many hundreds of species of cynipids that occur only on oaks.

It is probable also that traumatism through the feeding process engaged in by the gall-making larva actually plays some part in determining the form of the gall. Thus the formation of callus tissue is common in many galls, most conspicuous in cataplasmas of the simpler type. Callus also develops from the feeding of boring, or even leaf-mining insects which do not cause galls.

In their histological structure all galls present certain features in common. They are usually poorly supplied with vascular bundles, and their bulk is composed almost entirely of parenchymatous elements. Frequently there is a very pronounced decrease in chlorophyll content, which causes the galls to become a paler green in color, and sometimes the green pigment may be entirely lacking. Again, other colors may be developed, frequently a rosy or red anthocyanin pigment is prominently displayed.

Histological differentiation in galls produced by insects falls into two types. The tissues of the cataplasmas consist generally of quite uniform parenchyma in which the cell elements are commonly much enlarged. Compared with the normal tissue of corresponding parts, that formed in the gall is less highly differentiated and resembles callus or wound-tissue. This type of structure is seen in practically all phytocecidia and in galls due to animals other than insects.

The prosoplasmas, which are due almost exclusively to insects, are produced mainly by the cynipid gall wasps and the gall midges. Numerically, they include about half of the insect galls. We have already defined as prosoplasmas those galls that introduce something new in the plant that develops them. The internal anatomy and histology of the prosoplasmas are equally specific and constant, and likewise they are histologically specialized, forming thoroughly differentiated associations of diverse tissues. Their architecture includes primarily a definite central cavity which houses the developing insect. This cavity, or several cavities in the case of polythalamous galls, is closed, completely shutting off its occupant from the outside world. With this cavity as a center, the surrounding tissues are concentrically arranged; sometimes in the more complicated type such as those due to certain gall wasps, there are also radiating, fibrous structures within or external to the body of the gall. The concentric layers form primarily an inner, or "nutritive" layer and a "protective" or supporting one, as these two functions dominate the entire structure of the gall, aside from various embellishments in the way of texture, color, adornment, and occasionally special glands which arise from the epidermis.

The secretion of a sweetish, honey-like substance by certain galls is noteworthy. This accumulates on the surface as a sticky coating, known as honeydew which is very attractive to many insects, especially ants. Best known in this country are several species of *Disholcaspis*, particularly *D. perniciosus* which furnishes the main food of the common honey-ant of Colorado (*Myrmecocystus*). Honeydew is produced almost exclusively by cynipid galls and is apparently a by-product of the metabolism of the gall rather than a secretion for some special purpose. In some cases even the punctures made by the gall wasps when inserting their eggs in twigs calls forth the exudation of droplets of a similarly sweet liquid, indicating a reaction to some material secreted by the insects. The apparently utter uselessness of the honeydew to the plants has called forth much controversy as to its origin through natural selection or otherwise. This matter has been interestingly reviewed by Bequaert.

One striking peculiarity of many, quite diverse galls is the presence of abnormally large, frequently multinucleate cells, particularly in the tissues which lie adjacent to the feeding insect inhabitant. Although not restricted to gall tissues, such hypertrophies and cellular maladjustments are highly characteristic features and they have attracted the attention of several investigators. In the galls produced by some species of cynipids, Kostoff and Kendall have shown that cytological changes begin soon after the eggs are laid, due probably to toxic materials introduced at the time of oviposition, possibly augmented by bacterial contamination on the surface of the eggs. After the egg hatches and as larval feeding progresses, differentiation of tissues within the gall proceeds rapidly. The cells in close proximity to the larva fail to divide and increase in size. Their nuclei enlarge and the nucleoli become increasingly prominent until finally the nuclei break down. These cells form the nutritive zone, upon which the larva feeds. This layer is best developed at an early stage and is progressively destroyed by the growing larva. Outside this lies a sclerenchyma zone and here the cells often contain two or three nuclei, a condition which extends also into the adjacent zone of parenchyma. The nuclei arise through mitotic division, but this is often irregular and the cells fail to separate completely. These phenomena in conjunction with evidently increased disturbances in proximity to the larva, indicate very clearly the more powerful stimulation by the irritating secretions next to their source and their diminishing effect on the plant cells that are further removed. The source of the secretions has not been demonstrated, but it is generally presumed, and with good reason, that they are products of the salivary glands as other sources seem to be precluded. No dermal glands are of general occurrence in gall insects, and in some at least (Cynipidae), the midgut ends blindly so that no excrements are voided by the larva until final transformation of the adult. In some other forms, like the common trypetid goldenrod gall-fly, even the nitrogenous wastes in the excretory malpighian tubules of the larvae are retained as large, crystalline "urate pearls," thus avoiding any danger of soiling the neat little cell with any materials from the digestive tract.

We might draw some interesting parallels between the cytological irregularities of galls and those which arise through the action of alkaloids like colchicine, sanguinarine and possibly others when acting on living cells. As is well known, the introduction of these into plants causes a derangement of the orderly process of mitosis in cell division, often resulting in the formation of nuclei containing a double or otherwise polyploid number of chromosomes. The frequent persistence of this abnormality without deleterious effects, results in the tetraploid plants which have been produced by treatment with these drugs. Their effect on animal cells is similar, but not so regular, resulting in the arrest of mitosis in the metaphase stage, after which abnormalities like those seen in plant galls frequently appear. On account of these similarities, it seems quite probable that the underlying causes may be related.

A still closer analogy between plant galls and malignant animal tumors is indicated by the occurrence of irregular or asymmetrical mitoses in the cells of the nutritive layer. Particularly similar also, is the enlargement of the nucleoli noticed in the disintegrating cells of certain cynipid galls. This has been described and figured by Kostoff and Kendall ('29) in *Andricus palustris*.

Since growth in plant galls and animal tumors, particularly malignant ones, is comparable in some respects, it is to be hoped that the field of gall cytology may be more thoroughly explored. It must be borne in mind, however, that there is a striking difference in the reactions of animal tissues in neoplasms and of plant tissues in galls. The gall continues to develop only under the sustained action of the stimulating agent, while the growth of the animal neoplasm is not thus limited.

No account of insect galls could be complete without some men-

tion of their usefulness to mankind. Insects in general are such an
unmitigated nuisance that it is nice once in a while to find one that
does not rank as an outlawed enemy of civilized man. The praises
of the honeybee, ant, silkworm, shellac-insect and a few others
that can be counted on the fingers, have been sung with, and without, reason for many centuries. Gall insects have generally been
forgotten in drawing up the roll of honor. To entomologists, and to
all biologists, their value as materials for the study of evolution,
speciation, growth and development is evident from the main contents of the present chapter. They furnish beautiful examples of
one-sided symbiosis in which diverse insects have exploited plants
to an extent paralleled only by certain intraspecific symbioses perfected by the human species itself. They propound a number of
interesting questions concerning the action of chemical substances
(toxins, enzymes, hormones, organizers or what not) on the processes
of orderly development and tissue differentiation. These are of wide
general interest and they offer a promising field for research.

To mankind in general, especially to the literary, legal and edu-
cational professions, in fact to all those who use ink, or misuse it to
preserve the written word, the gall insects have unwittingly made
their greatest contribution. For more than a thousand years certain
galls have formed a major ingredient of the best and most permanent writing inks. The most important of these is the Aleppo gall,
(*Cynips gallae-tinctoriae*) produced on several species of oaks in
southeastern Europe and Asia Minor. Its value for this purpose lies
in its high content of tannic acid, which amounts to 65% of the entire gall. Many other cynipid galls developing on oaks contain
tannin, but to a much lesser degree. One Chinese gall produced on
an Asiatic sumach (*Rhus semialata*), even richer in tannin than the
Aleppo gall, is also used for the same purpose. These galls were
formerly important as mordants in the dyeing of fabrics, but their
usefulness for this purpose has faded in recent years, although it appears that they are still valuable in the dyeing of furs. Whether the
flood of newly patented vinx, jinx and kinx will replace inks from
insect galls to serve as a preservative and pickle for future manu-
scripts is not a question upon which a mere biologist may pass judge-
ment. Some help is urgently needed to augment the ink supply and
support the orgy of documentation that envelops us. Obviously the
fecundity of the gall wasp is not equal to the task.

REFERENCES

Adler, H.

1877 Lege-Apparat und Eierlegen der Gallwespen. Deutsch. Entom. Zeits., vol. 21, pp. 305–332.

1881 Ueber den Generationswechsel der Eichengallwespen. Zeits. wiss. Zool., vol. 35, pp. 151–246.

Adler, H., and C. R. Lichtenstein

1881 La génération alternante chez les Cynipides. xvi + 141 pp. C. Coulet, Paris.

Anderson, J. A. T.

1936 Gall-midges (Cecidomyidae) Whose Larvae Attack Fungi. Journ. Southeast. Agric. Coll., Wye, Kent, No. 38, pp. 95–107.

Annand, P. N.

1927 Tumors in Kale. Science, vol. 65, pp. 553–554.

Ashmead, W. H.

1903 Classification of the Gall-wasps and the Parasitic Cynipids, or the Super-family Cynipoidea. Psyche, vol. 10, pp. 7–13; 59–73; 140–155; 210–216.

Baccarini, P.

1909 Sui micozoöcecidii od "Ambrosiagallen." Bull. Soc. Bot. Italiana, vol. 18, pp. 137–145.

Bailey, I. W.

1922 Notes on Neotropical Ant Plants, I. *Cecropia angulata.* Bot. Gaz., vol. 74, pp. 369–391, 1 pl., 8 figs.

1922 The Anatomy of Certain Plants from the Belgian Congo, with Special Reference to Myrmecophytism. Bull. American Mus. Nat. Hist., vol. 45, pp. 585–621.

1923 Notes on Neotropical Ant Plants, II. *Tachigalia paniculata.* Bot. Gaz., vol. 75, pp. 27–41.

1924 Notes on Neotropical Ant Plants, III. *Cordia nodosa. Ibid.,* vol. 77, pp. 32–49, 2 pls., 5 figs.

Balch, R. E.

1932 *Dreyfusia piceae* and its Relation to "Gout Disease" in Balsam Fir. 62nd. Ann. Rept. Entom. Soc. Ontario, pp. 61–65.

Barnes, H. F.

1930 On the Resistance of Basket Willows to Button Gall Formation. Ann. App. Biol., vol. 17, pp. 638–640.

Beauverie, J.

1928 La dégénérescence des plastes et les cas zoöcécidies et d'altération pathologique. C. R. Soc. Biol., Paris, vol. 99, pp. 1991–1993.

Beauvisage, G. E. C.

1883 Les galles utiles. 101 pp. Octave Doin, Paris.

Beccari, O.

1886 Piante ospitatrici ossia formicarie della Malesia e della Papuasia. Malesia, vol. 2, pp. 1–340.

Bequaert, J.

1913   Sur quelques cécidies observées en Algérie.   Rev. Zool. Afrique, vol. 3, pp. 245–259.

1922   Ants in their Diverse Relations to the Plant World.   Bull. American Mus. Nat. Hist., vol. 45, pp. 333–583.

1924   Galls that Secrete Honeydew; A Contribution to the Problem as to whether Galls are Altruistic Adaptations.   Bull. Brooklyn Entom. Soc., vol. 19, pp. 101–124.

Beutenmueller, W.

1904   The Insect Galls of the vicinity of New York City.   American Mus. Nat. Hist., Guide Leaflet, No. 16, pp. 1–38.

Beyerinck, M. W.

1883   Beobachtungen über die ersten Entwicklungsphasen einiger Cynipidengallen. Verh. K. Akad. Amsterdam, vol. 22, pp. 1–198, 6 pls.

Bignell, G. C.

1898   Oak Galls.   Entom. Monthly Mag., vol. 34, pp. 99–100.

Bloch, R.

1941   Wound Healing in Higher Plants.   Bot. Rev., vol. 7, pp. 110–146.

Brandes, G.

1893   Die Blattläuse und der Honigthau.   Zeitschr. Ges. Naturwiss., vol. 66, pp. 98–103.

Bullrich, O.

1913   Beiträge zur Kenntnis der Cynipidenlarven.   54 pp., H. Blanke, Berlin.

Büsgen, M.

1891   Der Honigtau.   Biologische Studien an Pflanzen und Pflanzenläusen.   Zeitschr. Naturwiss., vol. 25, pp. 339–428, 2 pls.   Biol. Centralbl., vol. 11, pp. 193–200.

Cameron, P.

1883   On the Origin of the Forms of Galls.   Trans. Nat. Hist. Soc., Glasgow, n.s., vol. 1, pp. 28–37; appendix, pp. 50–95.

1893   A Monograph of the British Phytophagous Hymenoptera.   Vol. 4, 248 pp., London.

Carter, W.

1939   Injuries to Plants Caused by Insect Toxins.   Bot. Rev., vol. 5, pp. 273–326.

Chevalier, A.

1923   Les galles de Chine et leur origine.   Rev. Bot. Appl. and Agric. Colon., vol. 3, pp. 513–522.

Cockerell, T. D. A.

1890   The Evolution of Insect Galls.   Entomologist, vol. 23, pp. 73–76.

1921   Oligocene Hymenoptera from the Isle of Wight.   Ann. Mag. Nat. Hist., vol. (9) 7, pp. 1–25.

Connold, E. T.

1902   British Vegetable Galls.   xii + 312 pp.   New York, E. P. Dutton.

1908   British Oak Galls.   xviii + 169 pp.   London, Adlard and Son.

Cook, M. T.

1902 Galls and Insects Producing Them. Pts. I and II. Ohio Nat., vol. 2, pp. 263–278.

1923 The Origin and Nature of Plant Galls. Science, vol. 57, pp. 6–14.

Cosens, A.

1912 A Contribution to the Morphology and Biology of Insect Galls. Trans. Canadian Inst., vol. 9, pp. 297–387.

1913 Insect Galls. Canadian Entom., vol. 45, pp. 380–384.

Crawford, D. L.

1914 A Monograph of the Jumping Plant Lice or Psyllidae of the New World. Bull. U. S. Nat. Mus., No. 85, 186 pp., 30 pls.

Darboux, G., and C. Houard

1901 Catalogue systématique des zoöcécidies de l'Europe et du Bassin Méditerranéen. xi + 544 pp. Paris, Lab. Êtres Organisés.

Decaux, F.

1896 La mouche des Orchidées (*Isosoma orchidearum*). Journ. Soc. Nat. Hort. France, (3), vol. 18, pp. 837–842.

Dermen, H., and N. A. Brown

1940 Cytological Basis of Killing Plant Tumors by Colchicine. Journ. Heredity, vol. 31, pp. 197–199.

Dieuzeide, R.

1929 Contribution a l'étude des néoplasmes végétaux. Actes Soc. Linn., Bordeaux, vol. 81, pp. 5–245, 64 figs.

Dittrich, R.

1924 Die Tenthredinocecidien, durch Blattwespen verursachte Pflanzengallen und ihre Erzeuger. Zoologica, vol. 24, pp. 583–635, 9 pls., 6 figs.

Docters Van Leeuwen-Reijnvaan, J., and W. M.

1926 The Zoöcecidia of the Netherlands East Indies. 601 pp., 1088 figs. 'Slands Plantentuin, Batavia, Java.

1928 Ueber ein von *Gynaikothrips devriesii* aus einer Gallmücken-Galle gebildetes Thysanoptero-Cecidium. Rec. Trav. Bot. Néer., vol. 25a, pp. 99–114.

Docters Van Leeuwen, W. M.

1938 Zoocecidia. In Verdoorn's Manual of Pteridology, pp. 192–195. The Hague, M. Nijhoff.

Fagan, M. M.

1918 The Uses of Insect Galls. American Naturalist, vol. 52, pp. 155–176.

Felt, E. P.

1909 Gall Midges of the Goldenrod. Ottawa Natural., vol. 22, pp. 244–249.

1911 Summary of Food Habits of American Gall Midges. Ann. Entom. Soc. America, vol. 4, pp. 55–62.

1911 A Generic Synopsis of the Itonidae. Journ. New York Entom. Soc., vol. 19, pp. 31–62.

1911 Hosts and Galls of American Gall Midges. Journ. Econ. Entom., vol. 4, pp. 451–475.

1912– A Study of Gall Midges. Pts. 1–7. Annual Repts., New York State Ento-
18   mologist, 1911–1917.
1917   Key to American Gall Insects. Bull. New York State Mus., No. 200, 310 pp.,
       250 figs., 16 pls.
1918   Gall Insects and Their Relations to Plants. Scientific Monthly, vol. 6, pp.
       509–525.
1925   Key to Gall Midges. Bull. New York State Mus., No. 257, 239 pp., 8 pls.,
       57 figs.
1936   The Relations of Insects and Plants in Gall Production. Ann. Entom. Soc.
       America, vol. 29, pp. 694–700.
1940   Plant Galls and Gall Makers. ix + 356 pp., 700 figs. Ithaca, N. Y., Comstock
       Pub. Co.

Fryer, J. C. F.

1916   Capsid Bugs. Journ. Board Agric., London, vol. 22, pp. 950–958.

Fullaway, D. T.

1911   Monograph of the Gall-making Cynipidae (Cynipinae) of California. Ann.
       Entom. Soc. America, vol. 4, pp. 311–380.

Giard, A.

1893   Sur l'organe appelé spatula sternalis et sur les tubes de Malpighi des larves de
       Cecidomyies. Bull. Soc. Entom. France, vol. 63, pp. lxxx–lxxxiv.

Guttenberg, H. von

1909   Cytologische Studien an Synchytrium-Gallen. Jahrb. f. Bot., vol. 46, pp.
       453–477, 2 pls.

Hassan, A. S.

1928   The Biology of the Eriophyidae, with special Reference to *Eriophyes tristriatus*.
       Univ. California Publ. Entom., vol. 4, pp. 341–394, 6 pls., 14 figs.

Hedicke, H.

1929   Insectengallen. Die Rohstoffe des Tierreichs, vol. 2, pp. 324–341.

Henry, B. W., A. J. Riker, and B. M. Duggar

1943   Thiamine in Crown Gall as Measured with the Phycomyces Assay. Journ.
       Agric. Res., vol. 67, pp. 89–110.

Hoffman, A.

1932   Miocene Insect Gall Impressions. Bot. Gaz., vol. 93, pp. 341–342.

Houard, C.

1908⎫ Les Zoocécidies des Plantes d'Europe et du Bassin de la Méditerranée. 2
1909⎬ vols., 1247 pp., 2 pls., 1365 figs. Paris.
1913⎭
1912   Les zoocécidies du nord de l'Afrique. Ann. Soc. Entom. France, 1912, pp. 1–
       236.
1921   Galles d'Europe. Deuxième Série. Marcellia, vol. 17, pp. 93–113.
1922– Les zoocécidies des plantes d'Afrique, d'Asie et d'Océanie. 2 vols, 1058 pp.,
23   1909 figs. Paris.
1933   Les zoocécidies des plantes de l'Amérique de Sud et de l'Amérique Centrale.
       519 pp. Libr. Sci., Hermann et Cie., Paris.

James, H. C.

1927   The Life History and Bionomics of a British Phytophagous Chalcidoid of
       the genus *Harmolita*. Ann. Appl. Biol., vol. 14, pp. 132–149, 12 figs.

Johannis, J. de

1922 Revision critique des espèces de Lépidoptères cécidogènes d'Europe et du bassin Méditerranéen. Am. Soc. Entom. France, vol. 91, pp. 73-155.

Karny, H., and J. Docters Van Leeuwen-Reijnvaan

1913 Beiträge zur Kenntnis der Gallen auf Java. Ueber die javanischen Thysanopterencecidien und deren Bewohner. Bull. Jard. Bot. Buitenzorg, sér. 2, No. 10.

Kendall, J.

1930 The Structure and Development of Certain Eriophyid Galls. Zeits. f. Parasitenkunde, vol. 2, pp., 477-501.

1930 Histological and Cytological Studies of Stems of Plants Injected with Certain Chemicals. Inaugural Dissertation, Sofia University. 40 pp., 3 pls. of 21 figs.

Kieffer, J. J.

1897- Les Cynipides. André, Spécies des Hyménoptères d'Europe et d'Algérie.
1903 Vol. 7 (1), pp. 1-687; (2) pp. 1-323, 507-631, 652-748. Paris, Froment-Dubosclard.

1902 Synopsis des Zoocécidies d'Europe. Ann. Soc. Entom. France, vol. 70, pp. 223-579.

1914 Die Gallwespen (Cynipiden) Mitteleuropas, insbesondere Deutschlands. Schröder, Insekten Mitteleuropas, vol. 3, pp. 1-94.

Kinsey, A. C.

1919 Fossil Cynipidae. Psyche, vol. 26, pp. 44-49.

1920 Life Histories of American Cynipidae. Bull. American Mus. Nat. Hist., vol. 42, pp. 319-357.

1920 Phylogeny of Cynipid Genera and Biological Characteristics. Bull. American Mus. Nat. Hist., vol. 42, pp. 357-402.

1923 The gall wasp genus *Neuroterus* (Hymenoptera). Indiana Univ. Studies, vol. 10, No. 58, pp. 1-150.

1930 The Gall Wasp Genus Cynips. A Study in the Origin of Species. Indiana Univ. Studies, vol. 16, Nos. 84-86, pp. 1-577, 429 figs.

Kinsey, A. C., and K. D. Ayres

1922 Varieties of a Rose Gall Wasp. Indiana Univ. Studies, vol. 9, No. 53, pp. 1-141.

Kolbe, H. J.

1924 Ueber den gallenbildenden Rüsselkäfer *Ceuthorhynchus ruebsaameni*. Entom. Mitt., vol. 13, p. 298.

Kostoff, D.

1933 Tumor Problem in the Light of Researches on Plant Tumors and Galls and its Relation to the Problem of Mutation. Protoplasma, vol. 20, pp. 440-456.

Kostoff, D., and J. Kendall

1929 Studies on the Structure and Development of Certain Cynipid Galls. Biol. Bull., vol. 56, pp. 402-458.

1929 Irregular Meiosis in *Lycium halimifolium* Produced by Gall Mites (*Eriophyes*). Journ. Genetics, vol. 21, pp. 113-115.

1930 Irregular Meiosis in *Datura ferox* caused by *Tetranychus telarius*. Genetica, vol. 12, pp. 140-144.

1933 Studies on Plant Tumors and Plant Polyploidy Produced by Bacteria and Other Agents. Arch. f. Mikrobiol., vol. 4, pp. 487-508.

Küster, E.

1900   Beiträge zur Kenntnis der Gallenanatomie. Flora, vol. 87, pp. 117–193.
1911   Die Gallen der Pflanzen. x + 437 pp. Leipzig, S. Hirzel. (Second Edition, 1927).
1911   Die Zoocecidien; allegemeiner Teil. Zoologica, vol. 24, pp. 105–165.
1916   Pathologische Pflanzenanatomie. Jena, G. Fischer.
1930   Anatomie der Gallen. Linsbauer's Handbuch der Pflanzenanatomie, Abth. 1, Teil 3, pp. 1–197.

Lacaze-Duthiers, H., and A. Riche

1854   Mémoire sur l'alimentation de quelques insectes gallicoles et sur la production de la graisse. Ann. Sci. Nat., (4), vol. 2, pp. 81–105.

La Rue, C. D.

1937   Cell Outgrowths from Wounded Surfaces of Plants in Damp Atmospheres. Pap. Michigan Acad. Sci. Arts and Lett., vol. 22, pp. 123–129.

Leiby, R. W.

1922   Biology of the Goldenrod Gall-maker *Gnorimoschema gallaesolidaginis*. Journ. New York Entom. Soc., vol. 30, pp. 81–94.

Levine, M.

1934   A Preliminary Report on Plants Treated with the Carcinogenic Agents of Animals. Bull. Torrey Bot. Club, vol. 61, pp. 103–118.

Lutz, F. E., and F. M. Brown

1928   A New Species of Bacteria and the Gall of an Aphid. American Museum Novitates, No. 305, 4 pp.

McLachlan, R.

1881   Dr. Adler's Second Memoire on Dimorphism in the Cynipidae which Produce Oak-galls. Entom. Month. Mag., vol. 17, pp. 258–259.

Malpighi, M.

1686   De gallis. Opera Omnia, vol. 2, pp. 17–38, 15 pls. Londini.

Martin, J. P.

1938   Stem Galls of Sugar Cane Induced with an Insect Extract. Hawaiian Planters' Rec., vol. 42, pp. 129–134.
1942   Stem Galls of Sugar Cane Induced with Insect Extracts. Science, vol. 96, p. 39.

Mayr, G.

1905   Ueber Perilampiden. Verh. zool.- bot. Gesellsch. Wien, vol. 55, pp. 549–571, 3 figs.

Meess, A.

1923   Die Zoocecidien, durch Tiere erzeugte Pflanzengallen Deutschlands und ihre Bewohner. Lief. III. Die Cecidogenen cecidocolen Lepidopteren, gallenerzeugende und gallenbewohnende Schmetterlinge und ihre Cecidien. Zoologica, vol. 24, pp. 499–584, 3 pls., 77 figs.

Meyer, F.

1912   Beiträge zur Kenntnis der anatomischen Verhältnisse der Eichen-Cynipidengallen mit Berücksichtigung der Lage der Gallen. 58 pp. Dieterichschen Univ. — Buchdrückerein, Göttingen.

Miller, D.

1921 The Gall Chalcid of Blue-gum. New Zealand Journ. Agric., vol. 23, p. 282, 4 figs.

Molliard, M.

1900 Sur quelques caractères histologiques des cécidies produites par l'*Heterodera radicicola*. Rev. Génér. Botanique, vol. 12, p. 157.

1912 Comparaison des galles et des fruits au point de vue physiologique. Bull. Soc. Botan. France, vol. 59, pp. 201–214.

1913 Recherches physiologiques sur les galles. Rev. génér. Botanique, vol. 25, pp. 225–252; 285–307; 341–370.

Nalepa, A.

1898 Eriophyidae (Phytoptidae). Das Tierreich, Lief. 4, pp. 1–74.

1917 Die Systematik der Eriophyiden, ihre Aufgabe und Arbeitsmethode. Verh. zool.– bot. Ges. Wien., vol. 67, pp. 12–38.

1922 Zur Kenntnis der Milbengallen einiger Ahornarten und ihrer Erzeuger. Marcellia, vol. 19, pp. 3–33.

1928 Neuer Katalog der bisher beschriebenen Gallmilben. *Ibid.*, vol. 25, pp. 67–183.

Parr, T. J.

1939 *Matsucoccus*, a Scale Injurious to Certain Conifers in the Northeast. Journ. Econ. Entom., vol. 32, pp. 624–630.

Parshley, H. M.

1930 Gall Wasps and the Species Problem. Entom. News, vol. 41, pp. 191–195.

Patch, E. M.

1920 The Life-cycle of Aphids and Coccids. Ann. Entom. Soc. America, vol. 13, pp. 156–167.

Phillips, W. J.

1920 Life History and Habits of the Jointworms, *Harmolita*. Bull. U. S. Dept. Agric., No. 808, 27 pp.

Phillips, W. J., and W. T. Emery

1919 Revision of the North American Species of *Harmolita*. Proc. U. S. Nat. Mus., vol. 55, pp. 433–471.

Rabaud, É.

1918 Ethologie et comportement de diverses larves endophytes. III. *Pontania proxima* dans les galles des feuilles de Saules. Bull. Biol. France Belgique, vol. 52, pp. 303–322, 4 figs.

Rathey, E.

1891 Ueber myrmecophile Eichengallen. Verh. zool.– bot. Ges. Wien, vol. 41, SB., pp. 83–93.

Rettig, E.

1904 Ameisenpflanzen-Pflanzenameisen. Bot. Centralbl., vol. 17, Beih., pp. 89–122.

Riley, C. V.

1880 Honey-producing Oak-gall. American Entom., n.s., vol. 1, p. 298.

Ross, H., and H. Hedicke

1927 Die Pflanzengallen (Cecidien) Mittel- und Nordeuropas, ihre Erreger und Biologie und Bestimmungstabellen. vii + 348 pp. Jena, Gustav Fischer.

Ross, H.

1932 Prakticum der Gallenkunde. x + 312 pp., 161 figs. J. Springer, Berlin.

Rössig, H.

1904 Von welchen Organen der Gallwespen-Larven geht der Reiz zur Bildung der Pflanzengalle aus? Zool. Jahrb., Abth. f. Syst., vol. 20, pp. 19–90, 4 pls.

Rübsaamen, E. H.

1899 Ueber die Lebensweise der Cecidomyiden. Biol. Centralbl., vol. 19, pp. 529–549; 561–570; 593–607.

1916 Die Zoöcecidien, durch Tiere erzeugte Pflanzengallen Deutschlands und ihre Bewohner. Zoologica, vol. 24, 498 pp.

Safford, W. E.

1923 Ant Acacias and Acacia Ants of Mexico and Central America. Rept. Smithsonian Inst. for 1921, pp. 381–394.

Schenck, H.

1914 Die myrmecophilen Acacia-Arten. Engler's Bot. Jahrb., vol. 50, pp. 449–487.

Schulze, P.

1923 Eriophyina. Gallmilben. Biol. Tiere Deutschlands, Teil 21, pp. 52–60.

Sjöstedt, Y.

1908 Akaziengallen und Ameisen auf den ostafrikanischen Steppen. Exped. Kilimandjaro Meru, vol. 2, pp. 97–118.

Smith, E. F.

1917 Mechanism of Tumor Growth in Crowngall. Journ. Agric. Res., vol. 8, pp. 165–186, 62 pls.

Smith, E. F., N. E. Brown, and L. McCulloch

1912 The Structure and Development of Crown Gall: a Plant Cancer. Bull. Bur. Plant Ind., U. S. Dept. Agric., No. 255, 60 pp.

Stebbing, E. P.

1905 On the Cecidomyid Forming the Galls or Pseudocones on *Pinus longifolia*. Indian Forester, vol. 31, pp. 429–434, 1 pl.

Stebbins, F. A.

1910 Insect galls of Springfield, Massachusetts, and Vicinity. Bull. Springfield Mus. Nat. Hist., vol. 2, pp. 1–64.

Thomas, F.

1897 Mimicry bei Eichenblatt-Gallen. S.B. Ges. naturf. Fr., Berlin, 1897, pp. 45–47.

Thompson, M. T.

1915 An Illustrated Catalogue of American Insect Galls. 116 pp., Nassau, Rensselaer Co., N. Y. (E. P. Felt, Editor.)

Trelease, W.

1924 The American Oaks. Mem. Nat. Acad. Sci., vol. 20, v + 255 pp., 420 pls.

Trotter, A.

1900   I micromiceti delle galle. Atti Ist. Veneto, Sci., vol. 59, pp. 715–736.

1905   Nuove ricerche sui micromiceti delle galle e sulla natura dei loro rapporti ecologici. Ann. Mycologici, vol. 3, pp. 521–547.

Ule, E.

1906   Ameisenpflanzen. Engler's Bot. Jahrb., vol. 37, pp. 335–352.

Verrier, M. S.

1928   Sur les particuliarités de l'appareil mitachondrial de quelques cécidies. C. R. Acad. Sci., Paris, vol. 187, pp. 611–613.

Vorms, Mme.

Caractères anatomiques résultant de l'arrêt du développement chez les galles. *Ibid.*, vol. 196, pp. 558–560.

Walsh, B. D.

1864   On Dimorphism in the Hymenopterous Genus *Cynips*. Proc. Entom. Soc. Philadelphia, vol. 2, pp. 443–500.

1864– On the Insects, Coleopterous, Hymenopterous and Dipterous Inhabiting the
67   Galls of Certain Species of Willow. *Ibid.*, vol. 3, pp. 543–641; vol. 6, pp. 223–288.

Weld, L. H.

1921   American Gallflies of the Family Cynipidae Producing Subterranean Galls on Oak. Proc. U. S. Nat. Mus., vol. 59, pp. 187–246, 10 pls.

1922   Notes on American Gallflies of the Family Cynipidae Producing Galls on Acorns. *Ibid.*, vol. 61, art. 19, 32 pp., 5 pls.

Wells, B. W.

1915   A Survey of the Zoocecidia on Species of Hicoria, by Parasites belonging to the Eriophyidae and Itonididae. Ohio Journ. Sci., vol. 16, pp. 37–57.

1916   The Comparative Morphology of the Zoocecidia of *Celtis occidentalis*. Ohio Journ. Sci., vol. 16, pp. 249–290.

1918   The Zoocecidia of the Northeastern United States and Canada. Botan. Gaz., vol. 65, pp. 535–542.

1920   Early Stages in the Development of Certain *Pachypsylla* Galls on *Celtis*. American Journ. Bot., vol. 7, pp. 275–284.

1921   Evolution of Zoocecidia. Bot. Gaz., vol. 71, pp. 358–377.

1923   Fundamental Classification of Galls. Science, vol. 57, pp. 469–470.

Went, F. W.

1940   Local and General Defense Reactions in Plants and Animals. American Naturalist, vol. 74, pp. 107–116.

1943   The Regulation of Plant Growth. American Scientist, vol. 31, pp. 189–210.

Went, F. W., and K. V. Thimann

1937   Phytohormones. xi + 294 pp. New York, Macmillan.

Wheeler, W. M.

1908   Honey Ants, with a Revision of the American Myrmecocysti. Bull. American Mus. Nat. Hist., vol. 24, pp. 345–397.

1913   Observations of the Central American Acacia Ants. Trans. 2nd. Entom. Congr., Oxford, 1912, vol. 2, pp. 109–139.

1921   A Study of Some Social Beetles in British Guiana and Their Relations to the Ant-plant, *Tachigalia*. Zoologica, New York, vol. 3, pp. 35–126.

Williams, F. X.

1909   The Monterey Pine Resin Midge—*Cecidomyia resinicoloides* n. sp. Entom. News, vol. 20, pp. 1–8.

1910   Anatomy of the Larva of *Cecidomyia resinicoloides*. Ann. Entom. Soc. America, vol. 3, pp. 45–57.

Zwiegelt, F.

1917   Blattlausgallen unter besonderer Berücksichtigung der Anatomie und Aetiologie. Zentralbl. f. Bakt., Abth. II, vol. 47, pp. 408–535.

1929   Gallenbildung und Specialization. Verh. Deutsch. Ges. angew. Entom., 7ten. Versamml., 1928, pp. 62–76.

1931   Blattausgallen. Histogenetische und Biologische Studien an Tetraneura- und Schizoneura-Gallen. Die Blattlausgallen im Dienste principeller Gallenforschung. Zeits. angew. Entom. Beiheft, vol. 17, pp. 684, 155 figs., 5 pls.

CHAPTER V

FUNGI AND MICROBES AS FOOD;
SYMBIOSIS WITH MICROÖRGANISMS

> The mushrooms have two strange properties; the one that
> they yeeld so delicious a meat; the other, that they come up so
> hastily, as in a night, and yet they are unsown.
>
> FRANCIS BACON, in *Naturall Historie*, 1627

IT IS A common observation that our abundant edible mushrooms
and other fleshy fungi are very frequently infested by the larvae of
insects whose food burrows permeate their substance and hasten
its decay. Due to the ephemeral life of these fruiting bodies the insects that feed on them are necessarily species that undergo a rapid
development, usually flies of the order Diptera. The tougher, more
persistent polyporoid shelf-fungi are commonly attacked by the
larvae of a variety of beetles and other insects undergoing more prolonged periods of development. Mycetophagous insects like these
are abundant and the members of numerous families and genera
are addicted to this sort of food.

It has long been recognized also that many insects develop in
decaying wood, beneath the bark of fallen logs or tree-stumps and
in other substances that have suffered invasion by the mycelia of
fungi or by bacteria. Here it is evident that such secondary organisms are necessarily eaten with the substratum on which they are
growing and form a part of the insect's diet. Among some insects
the utilization of fungi as food has progressed far beyond this point;
the fungi have become the main source of nutriment and clearly
symbiotic relations have been developed. We shall return to this
matter in the second part of the present chapter, to consider it at
some length.

Only about a quarter of a century has elapsed however, since it
was clearly recognized that various microörganisms, such as yeasts
and bacteria are essential to the growth of many of the insects that
develop in various fermenting and decaying substances such as
fruits, carrion, and excrementitious material. In some cases of this

sort it appears probable that such microörganisms form their sole sources of food.

The term saprophagous has long been applied to these microphagous forms, although it was not understood until quite recently that the microörganisms were the essential factors and that the nature of the substratum was frequently of little or no direct importance in the economy of the insects.

The pomace fly, *Drosophila*, so successfully used by geneticists to elucidate the processes of inheritance, has likewise served as an experimental animal to demonstrate some of the food relations of microphagous insects. In an attempt to rear the larvae of *Drosophila* on sterile media Delcourt and Guyénot ('10) discovered that sterile larvae placed on sterile media could not survive, but that the addition of living or killed yeast enabled them to develop in the normal way. Later work by Guyénot ('13), by Loeb and Northrop ('16; '17) and by Baumberger independently ('17; '19) clarified these conclusions and showed that so far as *Drosophila* is concerned, yeast supplies the materials essential for normal growth and reproduction. Consequently, *Drosophila* is now grown in genetical laboratories by feeding it a culture of yeast maintained on a suitable substratum. Baumberger ('19) extended his studies to include other kinds of flies and later workers have investigated many more insects (Glaser, Wollman, and others, some of whose papers are cited in Chapter II). Sometimes an unusual habitat will furnish a clue to the true nature of the food. For example, a moth of the genus *Mineola*, that usually lives in fruits has been known (Pack and Dowdle '30) to feed in the fungus galls on cherry twigs. It has been shown also that yeast forms an adequate diet even for certain mosquito larvae, cockroaches, and beetles if their instincts permit them to accept it.

It is beyond the scope of the present book to include any extensive discussion of the physiology of nutrition as it relates to these microorganisms and there is no close agreement among workers on many of the matters involved. It is clear, however, that saprophagous insects of many kinds depend upon varied fungi, yeasts and bacteria for accessory growth factors and in some cases at least for their entire food supply. As the insects included in this category are extremely diverse, it will be of interest to consider some of them individually.

Among the Diptera, the small flies of the family Mycetophilidae, commonly known as fungus gnats, are almost entirely mycetophagous with the exception of a very few aberrant forms that have become secondarily predatory (Fig. 23). Their larvae occur abundantly in many fleshy fungi, also beneath decaying bark and in rotten wood. Those infesting mushrooms are frequently present in enormous numbers due to the abundant food supply, while those dependent on looser mycelial growths in wood or humus enjoy a

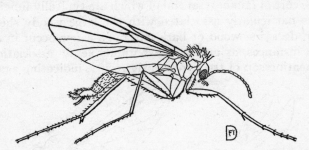

FIG. 23. A fungus-gnat, *Boletina* (Family Mycetophilidae), whose larvae feed on fungi. Redrawn from Cole.

less concentrated distribution. The adult fungus gnats form a considerable part of the forest insect fauna, occasionally appearing in swarms in the autumn when mushrooms are abundant.

Another primitive family of Diptera, the gall-midges of the family Itonididae are, as we have seen, primarily gall-producing insects affecting various flowering plants although a few of them are predatory. Another small scattering series are mycetophagous, commonly feeding on the fruiting bodies of smaller fungi, among the spores of smuts and rusts, in decaying wood, or beneath bark. Some of the latter, like the remarkable paedogenetic *Miastor*, often develop in dead, fermenting, sappy bark and in this case they are undoubtedly associated with yeasts. In another family, the Phoridae, with exceedingly varied habits, a few species occur in fleshy fungi and one or two of these are economically important as pests of cultivated mushrooms (Fig. 24). A similar fondness for fungi is evinced by several members of other families of Diptera (Syrphidae, Borboridae, Helomyzidae) and many such forms feed also in decaying animal and plant materials, dung, etc., indicating that their diet includes bacteria also. One small family of flies, the Platypezidae which are characteristic feeders in the pileus of agaric fungi have the hind legs of the male very strikingly modified and ornamented, a condition which prevails curiously enough, in several of the mushroom-eating Phoridae and in all Borboridae of both sexes. It frequently happens that the food of larval insects changes with age. An interesting instance of this occurs in the anthomyiid fly, *Egle spreta* where Trägårdh ('14) found that the newly-hatched larva feeds on a fungus occurring on grasses. Later it transfers its attention to the grass on which development is completed.

The Coleoptera include many mycetophagous beetles distributed among a score of families, several of which are typically fungivorous. They are particularly associated with the large woody shelf-fungi and with decaying wood or bark, although a few occur in agarics. In some instances, as among the Nitidulidae, an association with the fermenting sap of trees frequently recurs, indicating again that

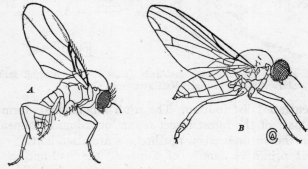

FIG. 24. Two small flies whose larvae feed in mushrooms. *A*, *Megaselia agarici* (Family Phoridae); *B*, *Calotarsa insignis*. *A*, original, *B*, redrawn from Cole.

yeasts form a part of their diet. It is by no means unusual to find the larvae of predatory beetles in fungi where they satisfy their carnivorous cravings at the expense of their vegetarian cohabitants. Such a glut of food naturally attracts many carrion beetles like *Necrophorus* or *Hister* and has led to erroneous assumptions concerning nature of food which appears to be strictly confined to living insects, mainly dipterous larvae.

If one might imagine a classical scholar wasting his precious moments scanning through a catalogue of Coleoptera, he could easily pick out a miscellaneous assortment of mycetophagous beetles. Many genera bear tell-tale names such as *Boletobius*, *Mycetophagus*, *Boletotherus*, *Lycoperdina*, *Mycotretus*, *Cryptophagus*, *Mycetoporus*, *Mycetina*, *Mycetaea*, etc. These names are fully as descriptive as tippler, guzzler, bacchante, milksop, boozer and the like as applied to the human species to indicate beverage addicts. Also, it must be remembered that the intimate association of man with alcohol was developed as a result of the extensive synthesis of this material by yeasts. *Drosophila* is again a prime factor in this process. Although the production of wine by the fermentation of grapes is a very ancient art, it has only quite recently been demonstrated that *Drosophila* and other flies of similar habits are the agents that serve generally to disseminate yeasts where they will do the most good, or harm, according to one's personal viewpoint. Bred in a medium heavily populated by yeasts, the flies bear these minute plants or their spores abundantly on their bodies. When they visit grapes, for example, in order to deposit their eggs, they infect the fruit with the yeasts which will later form the food for their larvae. There is thus a constant relationship between the yeasts and the development of the eggs. Moreover, molds do not commonly infest grapes, probably because these are eaten up by the fly-larvae. This phenomenon of mutualism between the flies and yeast and a corresponding antagonism against molds has been discussed by Sergent and Rougebief ('26–'27) in several interesting papers. As the yeast grows and ferments the grape sugar, a plentiful supply of food is developed for the fly larvae and incidentally the crop remains heavily infected with yeast. It is therefore ready for fermentation without further preparation and the discovery of this fact by the human species made possible the ancient art of wine making and brewing. To this day grapes are the fruits most generally used for making wine as they require no artificial treatment to insure proper fermentation. *Drosophila* is thus, in the light of our present knowledge, the god or at least the father of wine-making, not Bacchus, *alias* Dionysus for whom this honor has long been claimed.

In certain ants the head contains a diverticulum opening into the pharynx, known as the infrabuccal pocket (Fig. 25), in which small particles of food and other material accumulate. The pellets thus formed are fed to the ant larvae where they are retained in a special modification of the mouth, termed the trophorinium, which serves to triturate them while bathed in a digestive salivary secretion. Many fungus spores and bits of fungal mycelium are contained in these pellets, undoubtedly serving as an important part of the diet of many species.

It has been noted by various observers that the cavities of plants inhabited by ants, especially those of the true myrmecophytes where the plants are highly modified to this end, commonly contain fungi. Bathed in a damp, still atmosphere these grow profusely on

the moist walls. Such mycelial growths are regularly cropped by
the ants and must form a part of their diet. Bailey has shown that
such conditions undoubtedly exist in the case of many African
myrmecophytes, although he doubts whether the ants are able to do
more than express the juices from bits of hyphae and spores that are
taken into the infrabuccal pocket of ants which collect such material
more or less indiscriminately. It is also generally believed that certain seed-storing ants of dry regions, such as *Messor*, supplement
their granivorous diet with fungus spores. Furthermore, other ants

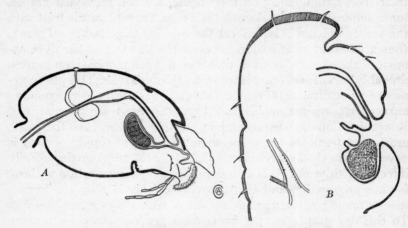

FIG. 25. *A*, Infrabuccal jacket and contained food pellet in an
adult worker ant (*Camponotus*); *B*, anterior end of an ant larva
(*Viticicola*) showing food pellet contained in the trophorinium.
Redrawn from Wheeler and Bailey.

such as *Crematogaster* which commonly build carton shelters or more
elaborate arboreal nests in tropical countries, make use of the fungi
to some extent as food. In such nests, containing a large proportion
of woody particles cemented with salivary secretion, fungi grow
readily. These are cleaned off by the ants and probably serve to
supplement their diet. The European garden ant, *Lasius fuliginosus*,
also builds similar carton nests in which an apparently specific fungus regularly develops and is believed to be an essential item in the
diet of the ants (Donisthorpe '15; Elliott '15).

The fact that bees make use of fungus spores on occasion in place
of the pollen which normally forms the protein portion of the bee-
bread on which their larvae are nourished, was noted years ago by
Long ('01). Even some of the smaller lady-bird beetles sometimes

PLATE XI

A. Mossy rose gall, produced on the stems of roses by the cypnid gall-wasp, *Diplolepis rosae.* Original.

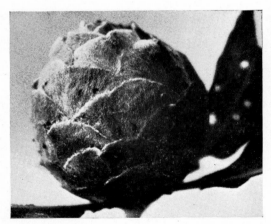

B. The "pine-cone" gall of willows, produced by the gall-midge, *Rhabdophaga strobiloides.* Original.

PLATE XII

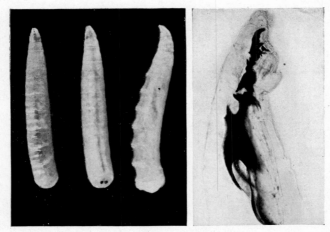

Three larvae of the common housefly (*Musca*), with the cephalopharyngeal skeleton bearing the mouth hooklets shown at the right and more highly magnified. Original photographs.

modify a highly carnivorous diet by feeding on the spores of fungi.

The caterpillars of a few Lepidoptera develop in fungi. These are mainly those of small moths, several of which are close relatives of our common household clothes moth.

Another type of plants that furnish food for various lepidopterous caterpillars are the lichens. In this case it is difficult to say whether we should regard the food plants as green plants or fungi, for the lichens are symbiotic organisms in which a fungus and alga are so intimately combined that their tissues are not separable, at least from the standpoint of an insect which consumes them for food. It is a notable fact that lichens are fed on by certain moths, again forms related to the clothes moth, and by a considerable series of moths of the family Lithosiidae. The former are evidently derived from saprophagous ancestors and the latter from insects that fed on green plants, so that the lichen fauna is, to some extent at least, composed of these two elements of independent derivation.

Large numbers of bacteria are regularly present in many of the substances eaten by insects. This is particularly true of decaying animal materials, excrement, etc. These substances constitute the food of a great variety of insects and such forms regularly ingest the bacteria that are present. Moreover, they are capable of digesting them as is shown by their disappearance to a large extent during passage through the alimentary canal. Experiments by Bogdanow ('o6; 'o8), Wollman ('19) and numerous others have shown that many insects of this type, particularly blow-flies such as *Calliphora* may develop normally under aseptic conditions just as we have seen to be the case with insects like *Drosophila*, where yeasts are an essential part of their diet. However, supplementary materials are equally necessary if the normal bacterial flora is lacking. A few years ago when sterile maggots were commonly employed in the treatment of infected wounds and osteomyelitis the technique of rearing sterile blowfly larvae was brought to a high degree of perfection (*cf.* Robinson '33).

Many extensive investigations of these blowflies have been conducted in order to elucidate their food-habits, not only in connection with their bactericidal activities, but also in relation to their importance in causing myiasis in animals, including man. This term is applied in a very general way to include the invasion of wounds, uninjured tissues, or cavities in the body by the larvae of flies. Some of these are true parasites, others are always associated with previous bacterial invasion and some appear to be scavengers that may be accidentally or secondarily associated with living animals. With such varied and frequently inconstant habits, it would be futile to discuss them without going to too great length. Briefly it may be noted that those occurring in wounds, generally known as screwworms, include members of several genera. These have been studied by Knipling and Rainwater ('37) in our country. They found that most of the cases were due to *Cochliomyia americana*, which is regularly a primary invader, but occasionally other species of related genera, including *Lucilia*, *Cynomyia*, *Phormia* and *Sarcophaga*, are found associated with it. The loss to livestock in the southern states due to screwworms is enormous. Other blowflies are a great pest in Australia where they develop from eggs laid in the soiled wool of sheep. The larvae frequently penetrate the skin and the ensuing bacterial invasion supplies them with food similar to that of the screwworm.

There is no great difference in habits between these flies and some of the bot-flies like *Oestrus*, but the latter are truly parasitic as their development is not associated with bacterial infection. Screwworm and blowfly maggots may be reared readily in the absence of bacteria on media containing meat extracts and blood serum.

Numerous flies and beetles develop in animal excrement; in fact one of the worst nuisances of civilized life, the housefly, depends principally upon horse manure for its larval food. This material always supports a rich growth of fungi and bacteria which we know are digested by the larva and essential to its development. Other related muscid and calliphorid flies develop in other kinds of excrementitious substances, decaying meat, fermenting vegetable substance, grass, straw and the like. There is a wide difference in the hydrogen-ion concentration of such materials; in other words, the range of acidity or alkalinity varies according to their origin and composition. This is undoubtedly one of the main factors which determine the type of organisms present and secondarily the kind of fly larvae that will be found therein. The reactions of the adult flies in laying their eggs are also correlated closely with such conditions. Thus the housefly (Pl. XII) which normally breeds in horse manure may be attracted to other substances by the addition of an ammoniacal odor which the flies associate with the normal food of their larvae. Blowflies of the genera *Lucilia* and *Calliphora* and the stable-fly, *Stomoxys*, on the other hand breed in materials having an acid reaction such as meat, rotting straw or lawn clippings.

There are many diverse coprophagous or dung inhabiting insects and we can be quite positive that most, if not all, would be unable to exist were it not for the ever present fungi, molds, yeasts and bacteria that develop in such substances. These simpler organisms are capable of synthesizing the proteins and accessory food materials from the otherwise totally indigestible detritus, thus making them available to the insects. Aside from Diptera, numerous beetles are commonly encountered in dung, or even in owl pellets (*cf.* Howard '00; Hubbell '39; Kolbe '04; Davis '09; etc.).

An interesting series of saprophagous Diptera belong to the family Sarcophagidae, commonly known as "flesh-flies." Many members of the family are actually entomophagous parasites, but some feed on decaying fish, snails or insects and others on animal excrement. Such diversity in habits is generally believed to indicate that the parasitic habit is only now developing from previously more generalized food relations, especially as some species may be either saprophagous or parasitic on different occasions. Another group which show similarly variable habits are the tiny flies of the family Phoridae. Some of these develop in the bodies of dead caterpillars and the like, while others are internal parasites of living insects. Much yet remains to be learned concerning such insects and the same is true of many others which we must, for the present, class simply as scavengers as we know little about them.

There are, however, various insects whose food is restricted to very unusual substances where they enjoy a monopoly on materials unsuitable for other animals. Some of these would be difficult to classify in any general scheme, but most of them would be casually regarded as scavengers. One of these is a small caterpillar, the waxmoth (*Galleria mellonella*), that occurs commonly in hives of the honeybee where it feeds on the waxen comb. Inasmuch as beeswax contains no nitrogen it is obvious that this is not the whole story. Investigation by several workers, notably Metalnikow, Sieber, Dickman, Borchet and Wollman, involving much controversy, has made it clear that these caterpillars are actually able to digest some of the wax eaten, but that they depend on accumulations of nitrogenous material present on the comb and perhaps also on bacterial organisms to round out an apparently most extraordinary and impossible diet. More recently Haydak ('36) has reared them on food devoid of wax, and noted that under conditions in the hive, they consume pollen in quantity. Other strange habits like the feeding of *Dermestes* beetles in bath sponges undoubtedly depends upon the presence of accessory food materials. These beetles normally infest the drying carcasses of animals already partially consumed by blow-fly larvae, a diet rich in proteins and fats. Even the strange fly-larva, *Psilopa petrolei* that develops in petroleum seeps or pools most likely depends entirely upon plant or insect remains that are trapped and accumulate in the oil.

In the light of the information presented, many such cases are easily interpreted, although in themselves quite incomprehensible. Thus the ptinid beetle, *Trichogenius*, breeds in argol, the crude tartar consisting mainly of acid potassium tartrate that forms a sediment in wine casks (Scott '21). This in itself is, of course, no food for an insect larva, but it contains much sedimented yeast and here lies the perfectly obvious explanation. The same holds true probably for the boring of a caterpillar (*Oenophila*) in wine corks (Manor '31) and of another cork-eating species of *Tinea* (Stellwaag '24).

That some molds like *Mucor* or *Penicillium* furnish nutritional advantages similar to those supplied by yeasts has been demonstrated by incorporating them in artificial media for rearing certain insects, for example, the tortricid moth, *Argyroploce leucotreta*, destructive to orange fruits (Ripley, *et al.* '39). Penicillin and other similarly potent bacteriostatic substances that are elaborated in the mycelia of some species of such fungi do not appear to be noticeably toxic to insects. Penicillin certainly is no more so to insects than to the human organism.

The clothes moths and some of their relatives are typical scavengers. Their food is of animal origin and contains a considerable quantity of keratin, a material not ordinarily entering into the diet of insects. Keratin is present in hair, fur, wool, feathers and horn, so when the moths start to work on your best suit, you may be sure that some of these materials are included in its make-up. They will eat silk too, an entirely different substance composed of amino acids with varying amounts of other materials. They balk at cotton, and artificial silk is impossible, as this, like cotton, is cellulose, which insects cannot ordinarily digest. Careful studies have been made of the food-habits of the common clothes moth, *Tinea biselliella*, especially by Titschak, which indicate that the keratin is digested by the caterpillars. This strange habit is also seen in a number of other moths which occur in the fur of living animals, like the South American sloths, or in the base of the horns of African antelopes. The South American forms are members of another family, Pyralidae (*cf.* Brues '36).

Such an account might be drawn out to an interminable length and any reader who wishes to pursue further details may be referred to the literature cited at the end of the chapter.

From what has been said it is clear that sarcophagous, saprophagous, mycetophagous and microphagous insects, all commonly known as scavengers, are by no means uniform in their food-habits. Moreover, casual observation leads one astray on account of the relations existing between many of these insects and microörganisms that are present in the materials which form their ostensible diet.

Their relations to fungi are more readily apparent as we may trace an easy sequence from the large fleshy fungi, through smaller forms to the diffuse mycelia that penetrate decaying material of many kinds, both plant and animal, and even into the organic components of the soil. The frequent association of specific insects and fungi is not surprising as it repeats the similarly restricted diets of many insects that feed upon the flowering plants.

The way in which mycetophagous and microphagous insects have found their own special niches in the environment is nicely illustrated by a study of the sequence of insect life in the living, moribund, dead and gradually disintegrating trunks of trees in a forest. This process has been described at length by Blackman and Stage ('24). The trees were a grove of hickories that were attacked by a bark-beetle (*Eccoptogaster quadrispinosa*) which killed practically all the whole stand during the course of four seasons. This insect, like other bark-beetles develops under the bark, consuming the cambium and adjacent tissues. After death, the sapwood decays rapidly and is much changed after a year. The next season another series of beetles appears, including mainly longicorns and buprestid borers together with other bark-beetles and ambrosia beetles. The following and succeeding years the fauna changes with the disappearance of some species and the addition of others. Meanwhile predators and parasites appear, and finally in the well rotted wood, burrowing bees make their nests together with numerous insects characteristic of such material. In such a sequence we can readily trace the ascending importance of fungal and bacterial agents which disintegrate the wood and come to serve as food for the changing fauna of insect invaders. This is a commonplace progression, so much so that we are apt to overlook its true significance.

Many other similar sequences might be cited, for instance, the early work of Mégnin ('85) who followed the somewhat gruesome progression of insects during the disintegration of corpses. Others have later added to his findings.

Other evidence on the intimate association of insects with microorganisms is forthcoming from observations on the reactions and behavior of gravid females in choosing sites to deposit their eggs. Even the actual hatching of eggs is sometimes influenced by the presence of yeasts and bacteria as was first noticed by Atkin and Bacot ('17) in the case of the yellow-fever mosquito (*Aëdes aegypti*). They found that hatching was inhibited by the absence of such organisms in the water, but their addition to the medium furnished the necessary stimulus for prompt hatching. Later, however, Barber ('28) was unable to confirm this observation. It must be remembered in this case that the eggs of *Aëdes* mosquitoes commonly remain for considerable periods without hatching when kept dry, not giving forth the brood until immersed in water, an anticipation of the fact that this medium is an absolute essential for their subsequent larval development. The normal food of most mosquito larvae appears to be a mixture of bacteria, infusoria and algae. At least Barber ('27; '28) found that growth usually progressed most satisfactorily on such mixtures although some variation exists among different species. Yeast alone is a very satisfactory food, and even dried milk combined with liver extract suffices. However such discoveries are easily interpreted on the basis of the chemical composition of the ingredients and innumerable others might be devised from the products of the modern corner grocery. Insects themselves are, of course, quick to discover opportunities of this sort as is evident from the numerous kinds that ruin stored food products held for human consumption or higher prices.

Several entomologists (Crumb, Lyon, Richardson) have studied the attractiveness of various substances to the common housefly and the reactions of the flies in ovipositing on substances thus perfumed. Richardson ('16) found that the emission of ammonia attracts the flies and is the main factor in inducing oviposition, but that eggs are laid more abundantly when butyric and valerianic acids are present in addition. However, the texture of the preferred food, horse manure, also influences the flies. Consequently we may assume that odor is the prime factor, tempered by tactile stimuli. The instincts concerned are therefore complex, but they are obviously related to the fermentation processes which depend in turn on the microörganisms present. In the case of the beetle, *Carpophilus*, that develops in dried fruits, inoculation of stewed fruit with any of several molds or yeast serves as a powerful attractant according to Wildman ('33). The same in general is true of many microphagous insects affecting fruits, sap, etc., where the alcohol and acetic acid are without doubt the indicators of suitable media for oviposition, recognizable to the insects and eliciting the appropriate response. In this way the instincts of the microphagous insects have attained the high correlation with their food supply necessary for successful and abundant multiplication.

Although many of the insects considered in the preceding pages are filthy creatures when measured by human standards they have not attained this condition through purely degenerative changes. Instinctively and physiologically they are highly adapted to a variety of food-habits, some of which indicate unusual specialization. Structurally they are ordinarily quite drab.

2

Un grand nombre d'êtres vivants appartenant aux groupes les plus divers sont normalement parasités par des microörganismes dont la plus grande partie appartiennent au groupe des Bactéries. La présence obligatoire de ces microörganismes dans les tissus a fait supposer qu'ils jouaient un rôle utile dans le métabolisme de l'hôte et c'est pour cette raison qu'on les a considérés comme des microörganismes symbiotiques.

A. PAILLOT, in *L'infection chez les insectes*

So FAR, the relations between insects and microörganisms to which we have referred deal with the use of such organisms solely as food by the insects. It will be readily seen, however, that certain more intimate relationships follow as a matter of course. The contamination of the hands, mouth or other parts of our own anatomy may serve to distribute pathogenic organisms where they may cause disease in other people. The same situation applies to certain insects, like the housefly which is notoriously obnoxious in this respect. Here the insect becomes a public menace as it serves to infect food with the bacteria that cause certain enteric diseases. Insects feeding on wood infested with fungi, or on leaves similarly affected, may readily carry the spores to other susceptible plant materials that they visit. Thus, the fungus that causes the chestnut blight is occasionally transferred from infected to healthy trees by wood-boring insects that become contaminated and then carry the fungus on their bodies when they seek out another tree for oviposition. In another similar case the relationship is much more definite. The

Dutch elm disease is due to a fungus (*Ceratostomella ulmi*) that develops beneath the bark. In the parts of Eastern North America where this disease has been accidentally introduced, the sickly or moribund elm trees very commonly harbor a bark-beetle *Scolytus multistriatus*, whose larvae develop in burrows, likewise beneath the bark. When the beetles transform into adults they bore their way out and leave the parental tree, bearing on their bodies an inoculum of the fungus. When they then dig their way into the bark of another host tree they carry the fungus with them and the disease is transferred unerringly to a fresh host. Thus a perfectly casual or accidental association becomes very constant, and it is believed that the Dutch elm disease spreads only very rarely in any other way. Nevertheless, there is no direct biological association of this beetle with the fungus further than their coincident occurrence. This, coupled with the habitual contamination of the beetles by the spores serves to spread the disease with great regularity although the fungus is in no way essential to the welfare of the beetles. Wright ('35) has described a fungus similarly associated with another species of *Scolytus* where it seems that the fungus is beneficial to the development of the beetle larvae. At this point we may recall the fact already referred to in the previous section that the mycetophagous pomace-fly, *Drosophila* regularly contaminates grapes with yeasts which are later necessary for the development of its larvae.

It is an easy step from this condition to a state of actual symbiosis which exists in other bark-beetles of the same family, Scolytidae, where we find each species associated with a specific fungus which it carries along from one host tree to another and passes continually to succeeding generations. Furthermore, these fungi serve as the sole food for the developing larvae of the beetles and the relationship is truly symbiotic since the fungi are dependent upon the insects for their dissemination and propagation.

On account of the nature of their food these insects are known as ambrosia beetles. They belong mainly to several groups of Scolytidae, together with a smaller related family, the Platypodidae. Instead of excavating their galleries just beneath the bark, the ambrosia beetles tunnel more deeply into the wood, selecting freshly dead or dying trees, and their brood chambers appear to be located with reference to the optimum moisture appropriate for the growth of the symbiotic fungi. The biology of these beetles has been carefully studied by a number of entomologists. Of these Hubbard ('96; '06) was the first to make extensive observations on several American species and he was followed by a number of Europeans, particularly Neger ('08; '11) and Schneider-Orelli ('11; '12; '13). The fungi are specific for the particular beetles (Fig. 26), developing in the burrows and being constantly eaten by the larvae which have thus transferred their diet from the substance of the tree to the growing fungus. The spores of the fungus are carried as a contamination by the emerging beetles and when these enter another tree to excavate burrows in which to lay their eggs and rear their larvae, the fungus is "planted." Thus we may quite properly refer to the ambrosia beetles as mushroom growers although we may not envy them their invariable diet, however delectable it may be on occasion. One type of provision for insuring the transport of

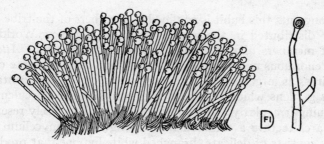

FIG. 26. Ambrosia fungus which is cultivated in the burrows of the beetle, *Xyleborus xylographus*. Redrawn from Hubbard.

spores by the beetles is a brush of hairs on the front of the head, but many of them actually carry infective material in the stomach. Some of the related Platypodidae have most elaborate cavities and bunches of bristles on the head of the female, serving to carry the fungus. The beetle larvae live either in pocket-like excavations along the side of the burrow or in chambers where they are commonly fed by the adult beetles. Pure cultures of the fungi, as maintained in the burrows of each species, vary greatly in form and color, undoubtedly representing a considerable number of different types, but mycologists are not agreed as to their exact relationships.

A most unusual type of symbiosis between certain scale-insects (*Aspidiotus osborni*) and a parasitic fungus (*Septobasidium retiforme*) has been described by Couch ('31). This is really a not too complicated biocenosis in which some individual scale insects are infested with the fungus which lives at their expense. In turn the mycelium of the fungus covers the insects and protects those that escape infection. Young scales migrating to fresh bark on which they feed may disseminate the fungus, but a considerable proportion of the insects never acquire the disease. Consequently the association with this parasite actually benefits the host insect as well as its parasite.

A still more closely integrated method of transfer for the fungus is accomplished by a wood-boring lymexylonid beetle, *Hylecoetus*, in which the spores of the fungus are smeared upon the eggs at the time they are laid from a supply in special pouches associated with the genital apparatus.

The best known fungus-growing insects are certain ants and termites that cultivate "mushroom gardens" in their nests upon a specially prepared substratum which they provide specially for this purpose.

Among ants this habit is confined to members of the tribe Attini, widely distributed in the warmer parts of the New World. The largest members of this group, belonging to the genus *Atta*, construct enormous nests deep in the soil where they excavate cavities sometimes as large as good-sized waste baskets. These contain the fungus gardens which consist of a brownish white flocculent mass containing irregularly anastomosing cavities and closely resembling in its architecture a common bath sponge. The mycelium of the fungus consists of delicate, branched white hyphae that produce on stalks innumerable minute globular bodies (Fig. 27) known as bromatia, or more jocularly, "kohlrabi globules." The latter serve as the food for the entire colony. The smaller worker ants keep the garden weeded and maintain a pure culture of the mycelium, although they prevent it from attaining the fruiting stage by their continual cropping of the bromatia. The substratum consists of innumerable bits of leaves cut from trees or other vegetation in the vicinity and carried into the nests by the larger worker ants who incorporate them into the compost. These ants like the bark-beetles transfer the fungus through the female who fills a cavity, the infrabuccal pouch, in her throat with a mass of mycelium when she leaves the parental nest to establish one of her own. Settling down at a chosen site she nurses the little packet of precious material until her first small brood of workers has developed. They then take up the task of securing leaf-material for the garden and barring unforeseen accidents in the long and tedious process the new colony prospers to grow and persist, probably over a long series of years. Numerous related genera have essentially similar habits, although their colonies are far smaller and less populous. Some of the fungi grown by the attine ants are true mushrooms and have been found by various observers (Moeller '93; Spegazzini, '21; Weber, '38) fruiting on the ground above deserted Atta nests (Pl. XIII). Some, however, appear to belong to other groups of fungi,

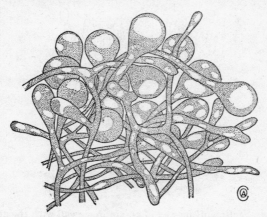

FIG. 27. A small bit of mycelium in the fungus-garden of the South American ant, *Moellerius*, showing the swollen bromatia or "Kohlrabi" bodies that serve the ants as food. Redrawn from Bruch.

and the one grown in the nests of the Texan *Cyphomyrmex rimosus* grows as a mass of yeast-like cells which may represent a vegetative form of one of the fungi imperfecti.

We have already referred in the previous chapter to the ambrosia galls, certain insect galls where specific fungi (Fig. 28) are constantly present.

Certain termites, all confined to Africa and Indomalaya, raise fungi in their nests much after the manner of the ants and some of these fungi are likewise true mushrooms so far as is known, although others obviously are not. From the standpoint of their symbiotic relations they appear to present conditions essentially similar to those described above.

A still more intimate association with microörganisms is encountered among many termites and Hemiptera, and in a very few other insects, which constantly harbor in the alimentary canal large numbers of specific Protozoa or bacteria. In many, if not all such cases, the microörganisms play a definite role in the digestive processes of the insects. They are consistently present in all individuals of the species concerned and there is always some mechanism which provides for them to be passed on to later generations. We can thus refer to them as hereditary, although the actual process of transfer is not in the nature of true inheritance since it does not depend upon the constitution of the germ cells.

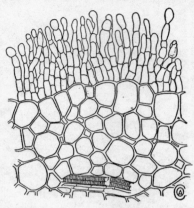

FIG. 28. Section of the wall of an ambrosia gall with the growing fungus. Redrawn from Ross.

Best known among these symbionts are the Protozoa that occur in termites. These are peculiarly modified flagellates, of very complex structure and frequently bearing long, conspicuous cilia (Fig. 29). They are present in large masses in the gut of the termites, often making up a very considerable part of the bulk of the insect. Although the food eaten by these termites consists entirely of wood, the insects themselves do not supply the enzymes capable of digesting cellulose and must rely upon their protozoan symbionts to produce these essential elements before they can secure any benefit from the ingested wood. Many species of such Protozoa are known, all closely restricted in their association with specific termites, although frequently a single species of termite may harbor several kinds of Protozoa. The termites are, of course, social insects, living in colonies and their unsanitary habits of feeding one another with regurgitated food and fecal excretions provide a ready means for distributing the Protozoa among all members of the colony. Although the probable importance of termite Protozoa in digestion was suspected by earlier writers, actual proof was not forthcoming until 1923 when Cleveland showed by cleverly devised experiments that termites could be deprived of their symbionts, or "defaunated" by the application of heat or excess oxygen, whereupon they were wholly incapable of digesting wood and starved miserably even though supplied with an abundance of their normal diet. The same investigator showed later ('34) that the wood-eating cockroach, *Cryptocercus*, supports a rich fauna of closely similar Protozoa enjoying the same symbiotic relationship. They have not been found in any other cockroaches, however, and their presence in this species cannot readily be explained, especially since the Protozoa are very

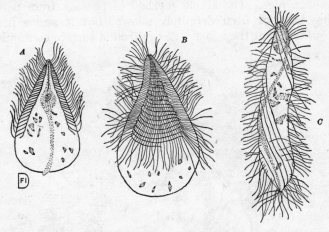

FIG. 29. Symbiotic Protozoa from the intestinal tract of termites. *A* and *B, Spirotrichonympha polygyra* from a species of *Neotermes; C, Dinenympha fimbriata* from *Reticulitermes hesperus. A* and *B* after Cupp, *C* after Kirby.

closely related to those found in termites, several of them being actually members of the same genus, *Trichonympha*, represented by numerous species in termites.

Various true bugs of the order *Hemiptera* contain large masses of bacteria in the alimentary tract. These, like the Protozoa just described, live in a presumably symbiotic association with the bugs although actual proof of this is still lacking. These bacteria are confined to large crypts or coeca that arise from the posterior part of the mid-intestine, occurring regularly in the members of several families like the Pentatomidae and in certain genera distributed in numerous other families. Their occurrence was first accurately described by the very astute, pioneer entomologist, Stephen A. Forbes in 1882 when bacteriological science was still in its infancy. Although they had been noticed much earlier by Leydig in 1857, it was at that time naturally impossible to recognize their significance.

Forbes found that the alimentary canal of the chinch bug (*Blissus leucopterus*) (Fig. 30) contained immense numbers of bacteria, and later ('92) examined a series of other Hemiptera where he recognized them to be of constant occurrence. Later Glasgow ('14) found that specific types were characteristic of each species examined. He was able to grow them on artificial media, and to prove the identity of his cultivated bacteria by immune serum reactions with guinea pigs. The actual method of passage from parent to offspring has not been definitely shown, but it seems probable (Kuskop, '24) that they enter the egg from a surface contamination

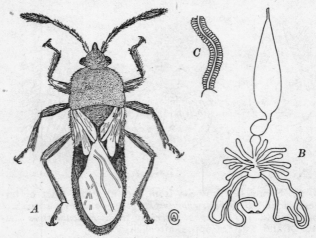

FIG. 30. Gastric coeca containing coecal bacteria in Hemiptera. *A*, Chinch bug (*Blissus leucopterus*), *B*, alimentary canal showing five pairs of coeca; *C*, the numerous short coeca of *Oedancala dorsalis*, a related member of the same family, Lygaeidae. *A*, redrawn from Luginbill; *B*, redrawn from Glasgow.

at the time of oviposition, whence they reach the alimentary canal before hatching, later to accumulate in the special coeca to which they are generally confined.

Other cases of similar nature involving bacteria in gastric coeca are known in the olive fly (*Dacus oleae*) (Petri '05) certain beetles, and other insects. Here also the eggs are contaminated and the symbionts are swallowed by the young larva, sometimes as a result of eating the freshly vacated egg-shell. Also, it is probable that certain other types of Protozoa, for example intestinal gregarines, are in a sense symbiotic in their relations in that their presence hastens growth and lessens mortality (Sumner '36) among the infected insects. Beetles of the family Lagriidae studied by Stammer ('29) harbor bacteroids that occupy a pair of intersegmental pouches which discharge into the genital aperture and thus contaminate the eggs. These pouches appear to be of a glandular nature and are elaborately branched in some species of *Lagria*. In the larva the symbionts are contained in three closed median pouches which lie within the thorax above the alimentary tract. Infection occurs as the result of eating the egg-shell. Similar pouches occurring in some longicorn beetles that eat wood have been described by Heitz ('27). In one of these, *Rhagium*, the yeast-like symbionts of the larva are in the cells lining coeca that open into the alimentary tract (Ekblom '32).

A series of primitive Hymenoptera known as wood-wasps (Siricoidea) also have glands at the base of the ovipositor containing fungi which are introduced into the wood on which the larva feeds. The infection of the wood occurs at the time of oviposition (*v.* Müller '34; Clark '35; Cartright '38).

It will be noted here that we encounter a relationship quite similar to that seen in *Hyloecetus* to which we referred on a previous page. The origin in the two instances can hardly have been the same, however, as the fungi carried by the wood-wasps do not appear to be other than types that regularly occur in decaying wood. Nevertheless they are carried over from the larval stage into the adult as they are present in the pupa. Müller has successfully cultivated them as well as those from some of the longicorn beetles.

It would appear that the symbiotic Protozoa and bacteria discussed above may have originated as true parasites which have become innocuous to their hosts after long association and finally come to aid them in their digestive processes. There are, however, many very obvious gaps in such a sequence which must be bridged before we may safely assume that this has been the course of events.

The most highly evolved associations with microörganisms are those in which the symbionts occur actually within the cells of the insects, at least during the greater part of the life-cycle. In contrast to the types previously described this phenomenon is generally known as intracellular symbiosis. The symbionts very commonly penetrate into the ovaries and thence into the eggs while the latter are in the process of formation, consequently their passage to the offspring is truly hereditary process since they are contained in the cytoplasm of the egg cell before fertilization. For the same reason they are, of course, derived from the female parent alone.

There would seem to be a clear gap between the extracellular and intracellular symbionts, but there are cases known in which there is a transition from one to the other condition during the ontogeny of certain insects. Thus in a chrysomelid beetle, *Donacia*, Stammer ('35) found that the larvae harbor a large mass of bacteroid symbionts in each of four blind sacs which open near the anterior end of the mid-intestine. Later, at the time of pupation the symbionts enter the lumen of the malpighian tubules in large numbers, multiply enormously and penetrate the cells which become greatly hypertrophied. They later pass in abundance into the rectum where they are exuded to form a mass at the anterior end of the egg when oviposition occurs. Thence they enter the egg probably through the micropyle which serves to admit the sperm to the egg at the time of fertilization. Finally they become enclosed in the alimentary canal during the development of the embryo.

There is very great variety in the position occupied by the intracellular symbionts in the bodies of the insects where they occur. Frequently they are distributed generally throughout the fat-body or in certain specialized portions of it. In the latter case the gross appearance of the fatty tissue is frequently modified and is sometimes recognizable by a distinctive color. The cells have been termed mycetocytes or bacteriocytes with reference to the probable fungous or bacterial relationship of the included symbionts and when these form discrete masses the latter are known as mycetomes or bacteriotomes. Intracellular symbionts are by no means restricted to the fat-body, but may occur in various other organs or tissues, particularly the lining of certain parts of the alimentary canal.

Ever since intracellular organisms have been known to occur in a viable condition and to multiply in the cells of the insect body there has been a divergence of opinion as to whether they are truly symbiotic. It has been thought by some biologists that they are inquilines or commensals, deriving nourishment and the advantages of seclusion in the tissues of their host and in turn causing it negligible harm or inconvenience. Others regard them as parasites to which their hosts are practically immune so that they suffer only slight damage. These possibilities grade into one another more or less completely and our interpretation depends in considerable part upon what we believe to have been the original nature of the association before it grew to perfection through a long evolutionary process. We will be better able to answer this question intelligently after inquiring more closely into the occurrence, behavior and, so far as may be, the physiology of the intracellular symbionts.

One of the better known, and apparently more primitive types are the bacteria or "bacteroids" that are present generally in the cells of the fat-body of cockroaches (Fig. 31). These have been extensively studied by several competent investigators (cf. Bode '36;

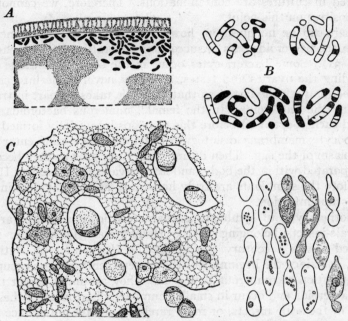

FIG. 31. Bacteroid and yeast-like intracellular symbiotic microorganisms in insects. *A*, bacteroids in a mature egg of a cockroach: *B*, morphology of the same bacteroids: *C*, yeast-like symbionts in the fat body of a coccid (*Pulvinaria*); *D*, morphology of yeast-like symbionts from the lac insects (*Lakshadia*). *A* and *B* redrawn after Geier, *C* from Brues and Glaser, *D* redrawn after Mahdihassan.

Fraenkel '21; Gier '36; Glaser '30a, '30b; Hertig '21; Hollande '31; Milovidov '28; Neukomm '27). All are agreed that the bacteroids are invariably present in every individual cockroach, massed within the cytoplasm of the fat cells in enormous numbers. The mycetocytes are in the central part of the strands of fatty tissue usually surrounded by a layer of fat cells that remain devoid of bacteroids. The individual bacteroids are short, slightly curved rods of variable length, but averaging 3.5μ in length and about 1μ in thickness, with rounded tips. They are thus like bacteria in appearance and multiply in the same way by transverse fission, but their staining reactions are slightly different from those of most bacteria. Several workers (Mercier, Glaser, Bode) have apparently been successful in growing them in artificial media, but others (Hertig and Gier) have failed after repeated trials and are convinced that the organisms secured in cultures are contaminations. Therefore, we cannot be too positive on this point.

There can be no question, however, concerning the manner in which the bacteroids pass to succeeding generations of cockroaches (Fig. 31). Some bacteriocytes occur in the connective tissue surrounding the ovaries and testes, but they never come into closer connection with the latter, so that the male takes no part in transmitting the bacteroids. In the female, numerous bacteroids surround each oöcyte and before this develops into a fully formed egg the oöcyte membrane disintegrates and the bacteroids enter the cytoplasm of the egg. Then during embryonic growth they become incorporated within the body, and later invade the fat cells. Here, therefore the symbionts are truly hereditary as no individual insect can escape infection.

Another type of symbiont is characteristic of a great variety of scale-insects belonging to the Homoptera. These were first noticed nearly a century ago by Leydig ('54), later by Putnam ('79), and since the beginning of the present century have occupied the attention of numerous biologists. They are much larger than the bacteroids and occur in smaller numbers, distributed in the fat-body of the scale insects, or more rarely in greater abundance and forming large, discrete mycetomes. In some cases at least (Schrader '23) the mycetocytes are derived from giant cells formed by the fusion of two very early embryonic cells which later enter into association with the symbionts and are incorporated in the fat body. These are generally believed to be yeast-like organisms as they multiply by forming an apical bud which is later constricted off from the parent cell. However, they have been grown on artificial media in several instances and, sometimes at least, produce mycelia under such conditions. This indicates that they are fungi of a higher type whose development within the body of the scale insect is so greatly inhibited that it does not progress beyond a yeast-like stage. Such a modification is scarcely surprising when we recall that the fungus-growing ants are able to prevent the fruiting of the fungi in their gardens simply by keeping them cropped like a well-mowed lawn. In the case of the symbionts of the soft-scale, *Pulvinaria innumerabilis* (Fig. 31), cultivation on artificial media (Brues and Glaser '21) induces the development of isolated, budding cells at first, but later a branched mycelium appears and in old cultures free conidia are present. The production of protease, lipase and diastase by this fungus has been demonstrated.

The symbionts of several other scale insects have likewise been grown on artificial media, although a mycelial stage has not always been secured (Schwartz '24; Benedek and Specht '33).

Yeast-like symbionts are known in many other Homoptera. They are particularly conspicuous in aphids where they form a pair of large mycetomes in the abdomen. The symbionts pass into the eggs and become incorporated into the body during the development of the embryo. They were first noticed by Leydig in 1850 who thought that the mass which invades the egg was a sort of yolk which he termed the pseudovitellus. Later investigators regarded them as nutritive material until 1889 when Krassilstchik suspected their symbiotic nature after finding the organisms in the eggs, although he failed to associate them with the large mycetome. Since then various authors (Pierantoni, Sulc, Webster and Phillips, Buchner, Uichanco and others) have worked out the details of the mycetome in various aphids and demonstrated the manner in which the symbionts are transmitted to the offspring. So far, however, all attempts to cultivate them on artificial media have failed.

There is a very fascinating possibility that the luminescence of insects such as fire-flies, click-beetles (*Pyrophorus*) and certain other light-producing forms like the mycetophilid fly known as the New Zealand glow-worm (Fig. 32) is due to the presence of luminescent symbionts in the body. Luminescent bacteria and fungi are well known and at least some of the luminous organs of other invertebrate animals, *e.g.*, certain cephalopod mollusks, are known to be due to the regular presence of such luminescent bacteria massed in special glandular structures associated with the ink-sac. These were first elucidated by Pierantoni ('18), who has also vigorously upheld the view that the production of light by insects is due to the luminescence of symbiotic bacteria contained in the bacteriocytes that form the luminous organs. It has long been known that the luminous organs of insects as well as those of all other animals invariably contain cells filled with granules or rods that resemble bacteria. Most investigators (*cf.* Buchner '30; Harvey '40) believe these to be cellular constituents or products. Nevertheless, the granules appear to be transmitted hereditarily like the symbionts previously described. They are present in the fire-fly egg which shows a faint, diffuse luminescence and are retained in the larva in a pair of discrete organs. After metamorphosis the adult beetles develop a very elaborate organ composed of modified fat-cells filled with granules and emitting the very brilliant light characteristic of the flashing fire-fly, although diffuse luminescence is maintained during the pupal stage. The life cycle of the beetle with its granules thus

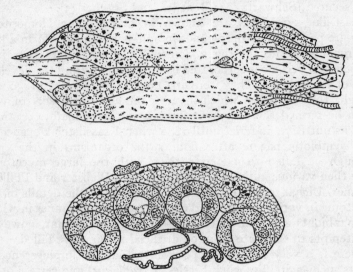

FIG. 32. Luminous organ of the New Zealand glow-worm, *Arachnocampa luminosa*. Swollen malpighian tubules with cells containing the luminous granules, and cross-section of tubules below. After Wheeler and Williams.

repeats very closely the general sequence that prevails among other insects with intracellular symbionts (Williams '16). Harvey does not accept the symbiotic theory of insect luminescence and has actually succeeded (Harvey and Hall '29) in surgically extirpating the luminous glands of the larval fire-fly, after which the adult developed a perfectly normal organ. From this he believes that the luminescence of the adult structure bears no relation to that of the larval one. However, it must be remembered that the removal of the larval organ would not necessarily eliminate bacteria from the body, nor prevent the independently developing adult organ from acquiring them. Hence, it will be seen that a definitely final decision cannot be reached on the basis of our present knowledge. Certainly the matter merits further attention, particularly with reference to the possible extracellular migration of granules and the exact nature of their occurrence during the several stages in the life cycle of the insects. Williams ('16) and Hess ('22) have already made a fine start in this direction, but further cytological studies will be required. Many attempts have been made to cultivate the "granules" on artificial media, but these have failed completely.

There is great variety among insects in the position that the symbionts occupy in the body and in the manner by which they pass to the offspring. Moreover, there are occasional cases where the symbionts are extracellular at one stage and intracellular at another and other instances where the larval and adult mycetocytes or mycetomes develop independently in different parts of the body, although the symbionts regularly pass from one to the other in the orderly process of development.

In many instances it seems at the first blush that the microörganisms must bear some relation to the digestive processes of the insects, especially where they occur in the lumen of the alimentary canal or in pouches opening into it. The demonstrated importance of the intestinal Protozoa of termites strengthens this conclusion. Their presence in the cells lining the gut suggests the same possibility and they are frequently in this location. Their frequent occurrence in the fat-body would seem to remove them from any direct connection with alimentation and it has been suggested that in such cases they bear a relation to the utilization of fat for the production of eggs. This seems especially plausible in the scale insects where almost the whole contents of the body are transformed into a mass of eggs by the mature female who gradually shrivels away during the process.

In some of the cases where intracellular symbionts have been grown in culture their ability to produce specific enzymes has been investigated. It has been found that there are marked differences just as would happen with any miscellaneous series of microörganisms taken at random. Sometimes however, correlations exist which may indicate that the symbionts are of value to the insects. This is especially noticeable in the case of wood-eating insects in which it appears that the symbionts produce a cellulose digesting enzyme (cellulase). Some other forms occurring in the fat cells produce lipase, capable of digesting fat. However these facts may be interpreted equally well if we assume that these aid the symbionts in their own digestion without reference to the insects. It is evident that all insects which subsist on wood must either produce a cellulose-digesting enzyme or depend upon symbiotic organisms, unless they utilize sap or materials stored in the wood, such as starches. Actually decayed or rotten wood, is, of course, a different matter as this is permeated with fungi and bacteria and becomes a source of food as the result of such invasion. It has been shown by Ripper ('30) and others that a few insects devoid of symbionts actually secrete cellulase, but most of them do not, at least in useful amounts, and those lacking symbionts generally confine their feeding to wood that actually contains much material other than cellulose. The main valid argument which supports the idea that the symbionts are actually symbiotic and not parasitic is their constant occurrence in the specific insects, their obviously innocuous nature and the development of special glands or other structures by the insects to house the microörganisms and to provide for their regular passage not only "unto the third and fourth generation," but forever.[1] This does not mean necessarily that the symbionts may not have originated in the dim past as low-grade parasites. Whether they existed for example in the Palaeozoic cockroaches we do not know, but there is reason to believe that they did since the most primitive living termite, *Mastotermes* of Australia, harbors bacteroids in the fat-body essentially similar to those of present day cockroaches (Koch '38), and the termites were probably derived from cockroach-like insects before the dawn of Mesozoic times. The acquisition by the higher termites of the utterly unrelated symbiotic Protozoa upon which they are now dependent is clearly a later development.

---

[1] Since this chapter was written, further light has been cast on the relationship of the symbionts to their hosts in two diverse insects. Fraenckel and Blewett (Proc. R. Soc. London, B, vol. 132, pp. 212–221 [1944]) have shown in the case of two beetles (*Sitodrepa* and *Lasioderma*) where the microörganisms contaminate the egg-shell, which is eaten by the newly hatched larva, that eggs washed with a bactericidal solution give rise to larvae devoid of symbionts. Such larvae are not viable unless supplied with vitamins of the B group. The evidence is clear that the symbionts supply these substances. Also, it has been shown by Brues and Dunn (Science, vol. 101, pp. 336–337 [1945]) that cockroaches treated with penicillin suffer a great reduction in the numbers of bacteroids in the fat-body which results in their premature death. It appears, therefore, that in some instances, at least, the symbionts are necessary to the normal metabolism of their hosts.

REFERENCES

PART 1

NOTE: Some other relevant literature is cited at the end of Chapter II.

Abbott, C. E.

1937 The Necrophilous Habit in Coleoptera. Bull. Brooklyn Entom. Soc., vol. 32, pp. 202–204.

Abderhalden, E.

1925 Beitrag zur Kenntnis der synthetischen Leistungen des tierischen Organismus. Zeits. physiol. Chemie, vol. 142, pp. 189–190.

Adolph, E. F.

1920 Egg-laying Reactions in the Pomace Fly, *Drosophila*. Journ. Exper. Zool., vol. 31, pp. 327–341.

Aldrich, J. M.

1915 The Economic Relations of the Sarcophagidae. Journ. Econ. Entom., vol. 8, pp. 242–247.

Alessandrini, G.

1909 Studi ed esperienzi sulle larve della *Piophila casei*. Arch. Parasitol., vol. 13, pp. 337–382, 32 figs.

Alfieri, A.

1931 Les Insectes de la tombe de Toutankhamon. Bull. Soc. R. Entom. Egypte, 1931, pp. 188–189.

Anderson, J. A. T.

1936 Gall-midges whose Larvae Attack Fungi. Journ. S. E. Agric. Coll., Wye, Kent, No. 38, pp. 95–107.

Arndt, W.

1929 Schäden an lagernden und an in Gebrauch befindlichen Badeschwämmen. S. B. Gesellsch. naturf. Freunde, Berlin, 1928, pp. 213–214.

1931 Zur Frage der Verdaubarkeit der Badenschwammgerustsubstanz. Zool. Anz., vol. 93, pp. 199–207, 7 figs.

Arnold, G.

1913 The Food of Ants. Proc. Rhodesia Sci. Assoc., vol. 12, pp. 11–24.

Atkin, E. E., and A. Bacot

1917 The Relation between the Hatching of the Eggs and the Development of the Larvae of *Stegomyia fasciata* and the Presence of Bacteria and Yeasts. Parasitology, vol. 9, pp. 482–536.

Awati, P. R.

1920 Bionomics of House-flies. I. Outdoor Feeding Habits of House-flies. Indian Journ. Med. Res., vol. 7, pp. 548–552.

Awati, P. R., and C. S. Swaminath

1920 Bionomics of House-flies. III. A Preliminary Note on Attraction of House-flies to Certain Fermenting and Putrefying Substances. Indian Journ. Med. Res., vol. 7, pp. 560–567.

Bailey, I. W.

1920 Some Relations Between Ants and Fungi. Ecology, vol. 1, pp. 174–189, 3 pls.

Barber, M. A.

1927 The Food of Anopheline Larvae. Food Organisms in Pure Culture. Public Health Repts., vol. 42, pp. 1494–1510.

1928 The Food of Culicine Larvae. U. S. Pub. Health Repts., vol. 43, pp. 11–17.

Barbey, Aug.

1920 Contribution à l'étude des Diptères xylophages (*Ctenophora atrata*). Bull. Soc. vaud. Sci., vol. 53, Proc.-Verb., pp. 27–28; 259–262.

Baumberger, J. P.

1917 The Food of *Drosophila melanogaster*. Proc. Nat. Acad. Sci., vol. 3, pp. 122–126.

1919 A Nutritional Study of Insects, with Special Reference to Microörganisms and their Substrata. Jour. Exper. Zool., vol. 28, pp. 1–81, 18 figs.

Baumberger, J. P., and R. W. Glaser

1917 The Rearing of *Drosophila ampelophila* on Solid Media. Science, vol. 45, pp. 21–22.

Bequaert, J.

1922 Ants in their Diverse Relations to the Plant World. Bull. American Mus. Nat. Hist., vol. 45, pp. 333–583, 100 figs., 4 pls.

Bessell, T. L.

1935 The Screw Worm. Bull. Georgia State Expt. Sta., No. 189, 11 pp., 6 figs.

Bishopp, F. C.

1915 Flies which Cause Myiasis in Man and Animals — Some Aspects of the Problem. Journ. Econ. Entom., vol. 8, pp. 317–329.

Bishopp, F. C., and E. W. Laake

1915 A Preliminary Statement Regarding Wool Maggots of Sheep in the United States. Journ. Econ. Entom., vol. 8, pp. 466–474.

Blackman, M. W., and H. H. Stage

1924 On the Succession of Insects Living in the Bark and Wood of Dying, Dead and Decaying Hickory. New York State Coll. Forestry, Syracuse, Tech. Publ., No. 17, 240 pp., 13 pls.

Bogdanow, A. E.

1906 Ueber das Züchten der Larven der Fleischfliege (*Calliphora*) in sterilisierten Nährmitteln. Arch. f. d. ges. Physiol., vol. 113, pp. 97–105.

1908 Ueber das Abhängigkeit des Wachstums der Fliegenlarven von Bakterien und Fermenten und über Väriabilität und Vererbung bei den Fliegenlarven. Arch. f. Anat. Physiol., 1908, suppl., pp. 173–200.

Bonnamour, S.

1926 Élévages et nouvelle liste de Diptères fongicoles. Ann. Soc. Linn., Lyon, n.s., vol. 72, pp. 85–93.

Börchert, A.

1935 Ueber den Frasschäden und die Ernährung der Larven der grossen Wachsmotte. Zool. Jahrb., Abth. f. Syst., vol. 66, pp. 380–400.

Brues, C. T.

1936   Aberrant Feeding Behavior Among Insects and its Bearing on the Development of Specialized Food Habits. Quart. Rev. Biol., vol. 11, pp. 305–319, 1 fig.

Busck, A.

1910   Notes on a Horn-feeding Lepidopterous Larva from Africa. Smithsonian Misc. Coll., vol. 56, No. 8, 2 pp., 2 pls.

Campbell, W. G., and S. A. Bryant

1940   A Chemical Study of the Bearing of Decay by *Phellinus cryptarum* and other Fungi on the Destruction of Wood by the Death Watch Beetle. Biochem. Journ., vol. 34, pp. 1404–1414.

Clark, C. U.

1895   On the Food Habits of Certain Dung and Carrion Beetles. Journ. New York Entom. Soc., vol. 3, p. 61.

Colman, W.

1932   Effect of Yeast on Clothes Moth Larvae. Journ. Econ. Entom., vol. 25, p. 1242.

Cousin, G.

1929   La vie larvaire de *Lucilia sericata*. C. R. Soc. Biol., Paris, vol. 101, pp. 788–790.

Crawford, D. L.

1912   The Petroleum Fly in California, *Psilopa petrolei*. Pomona Journ. Entom., vol. 4, pp. 686–697, 1 pl.

Creager, D. B., and F. J. Spruit

1935   The Relation of Certain Fungi to Larval Development in *Eumerus tuberculatus*. Ann. Entom. Soc. America, vol. 28, pp. 425–436.

Crumb, S. E., and S. C. Lyon

1917   The Effect of Certain Chemicals upon Oviposition in the House-fly (*Musca domestica*). Journ. Econ. Entom., vol. 10, pp. 532–536, 1 fig.

1921   Further Observations on the Effect of Certain Chemicals upon Oviposition in the House-fly (*Musca domestica*). Journ. Econ. Entom., vol. 14, pp. 461–465.

Davidson, W. M.

1924   Observations on *Psyllobora taedata*, a Coccinellid Attacking Mildews. Entom. News, vol. 32, pp. 83–89.

Davis, W. T.

1909   Owl Pellets and Insects. Journ. New York Entom. Soc., vol. 17, pp. 49–51.

Delcourt, A., and E. Guyénot

1910   De la possibilité d'étudier certains Diptères en milieu défini. C. R. Acad. Sci., Paris, vol. 151, pp. 255–257.

Dickman, A.

1933   Studies on the Wax Moth, *Galleria mellonella*, with Reference to the Digestion of Wax by the Larvae. Journ. Cellul. Comp. Physiol., vol. 3, pp. 223–246.

Dingler, M.

1928 *Cartodere filiformis* und *C. filum* als Schädlinge in Hefepräparaten. Zeits. Angew. Entom., vol. 14, pp. 189–224, 24 figs., 1 pl.

Donisthorpe, H.

1915 British Ants, their Life-history and Classification. 373 pp. Plymouth, England.

Dyar, H. G.

1908 A Pyralid Inhabiting the Fur of the Living Sloth. Proc. Entom. Soc. Washington, vol. 9, pp. 142–144. Also, *ibid.*, vol. 14, pp. 169–170.

Elliott, J. S. B.

1914 Fungi in the Nests of Ants. Trans. British Mycol. Soc., vol. 5, pp. 138–142, 1 pl.

Farquharson, C. O., W. Lamborn, and R. C. L. Perkins

1914 The Growth of Fungi on the Shelters Built over Coccidae by Cremastogaster-ants. Trans. Entom. Soc. London, Proc., pp. 42–50.

Felt, E. P.

1911a *Miastor americana*, an Account of Paedogenesis. Rept. New York State Entom., 1910, pp. 82–104, 14 pls.

1911b Hosts and Galls of American Gall Midges. Journ. Econ. Entom., vol. 4, pp. 451–475.

Fluke, C. L., Jr., and T. C. Allen

1931 The Role of Yeast in Life History Studies of the Apple Maggot, *Rhagoletis pomonella*. Journ. Econ. Entom., vol. 25, pp. 77–80.

Forbes, W. T. M.

1933 Two Wasp-guests from Puerto Rico (Microlepidoptera). Psyche, vol. 40, pp. 89–93, 1 pl.

Fricklinger, H. W.

1920 Die Kleidermotte (*Tineola biselliella*) als Schädling in Zoologischen Sammlungen. Zeits. f. angew. Entom., vol. 6, pp. 400–404, 5 figs.

Friedrichs, K.

1912 Beobachtungen über *Phosphuga atrata*, ihre Nahrung und die einiger anderen Silphini. Zeits. wiss. Insektenbiol., vol. 8, pp. 348–352.

Frost, M., W. B. Herms, and W. M. Hoskins

1936 The Nutritional Requirements of the Larvae of the Mosquito, *Theobaldia incidens*. Journ. Exper. Zool., vol. 73, pp. 461–480.

Fuller, M. E.

1934 The Insect Inhabitants of Carrion. Bull. No. 82, Australian Council Sci. and Ind. Res., 62 pp., 1 pl., 2 figs.

Galtsoff, P. S., *et al.*

1927 Culture Methods for Invertebrate Animals. Insects, pp. 259–518. Ithaca, N. Y., Comstock Pub. Co.

Gay, F. J.

1938 A Nutritional Study of the Larvae of *Dermestes vulpinus*. Journ. Exper. Zool., vol. 79, pp. 93–107.

Giglioli, Italo

1897 Insects and Yeasts. Nature, vol. 56, pp. 575–577, 4 figs.

Glaser, R. W.

1923 The Effect of Food on Longevity and Reproduction in Flies. Journ. Exper. Zool., vol. 38, pp. 383–412.

1924 The Relation of Microörganisms to the Development and Longevity of Flies. American Journ. Trop. Med., vol. 4, pp. 85–107.

1924 Rearing Flies for Experimental Purposes, with Biological Notes. Journ. Econ. Entom., vol. 17, pp. 486–496, 2 pls.

1938 A Method for the Sterile Culture of House-flies. Journ. Parasitol., vol. 24, pp. 177–179.

Guyénot, E.

1906 Sur le mode de nutrition de quelques larves de mouches. C. R. Soc. Biol., Paris, vol. 61, pp. 634–635.

1913 Role de levures dans l'alimentation. *Ibid.*, vol. 74, pt. 1, pp. 178–180.

1913 Nutrition des larves et fécondité. *Ibid.*, vol. 74, pt. 1, pp. 270–272.

1917 Recherches expérimentales sur la vie aseptique et le développement d'une organisme (*Drosophila ampelophila*) en fonction du milieu. Bull. Biol. France et Belgique, vol. 51, pp. 1–135, 4 pls.; pp. 137–330, 10 figs.

Hafez, M.

1939 Some Ecological Observations on the Insect-fauna of Dung. Bul. Soc. Fouad I. Entom., vol. 23, pp. 241–287, 8 figs.

Handlirsch, A.

1926 Die Nahrung. Schroeder's Handbuch der Entomologie, vol. 7, pp. 35–52.

Harris, R. G.

1923 Sur la culture des larves de Cécidomyies paedogénétiques (*Miastor*) en milieu artificiel. C. R. Soc. Biol., Paris, vol. 88, pp. 256–258.

Haub, J. G., and D. F. Miller

1932 Food Requirements of Blowfly Cultures Used in the Treatment of Osteomyelitis. Journ. Exper. Zool., vol. 64, pp. 51–56.

Haydak, M. H.

1936 Is Wax a Necessary Constituent of the Diet of Wax Moth Larvae? Ann. Entom. Soc. America, vol. 29, pp. 581–588.

Herfs, A.

1933 Untersuchungen zur Oekologie und Physiologie von *Anthrenus fasciatus.* Congr. Int. Entom. Paris, 1932, vol. 5, pp. 295–302.

Hewitt, C. G.

1914 The House-fly, *Musca domestica*. Its Structure, Habits, Development, Relation to Disease and Control. 382 pp., Cambridge, England.

Heymons, R., and H. von Lengerken

1929 Biologische Untersuchungen an Coprophagen Lamellicorniern. Zeits. Morph. Oekol. Tier., vol. 14, pp. 531–613, 29 figs.

Holland, W. J.

1922 *Eois ptelearia* (Geometridae) Detected in the Herbarium of the Carnegie Museum. Journ. Econ. Entom., vol. 15, pp. 433–434.

Howard, L. O.

1900 The Insect Fauna of Human Excrement, with Special Reference to the Spread of Typhoid. Proc. Washington Acad. Sci., vol. 2, pp. 541–604.

Hubbell, T. H., and C. C. Goff

1939 Florida Pocket-Gopher Burrows and Their Arthropod Inhabitants. Proc. Florida Acad. Sci., vol. 4, pp. 127–166.

Ihering, R. von

1914 As traças que vivem sobre a pregiuça. *Bradypophila garbei.* Rev. Mus. Paulista, vol. 9, pp. 123–127.

Ingles, L. G.

1933 The Succession of Insects in Tree Trunks as Shown by the Collections from the Various Stages of Decay. Journ. Entom. & Zool., Pomona College, vol. 25, pp. 57–59.

Jablonowski, J.

1925 Ist der Getreideschmalkäfer, *Silvanus surinamensis* ein Getreideschädling? Zeits. angew. Entom., vol. 11, pp. 77–112, 3 figs.

Johannsen, O. A.

1909– The Fungus Gnats of North America. Bulls. Maine Agric. Expt. Sta., Nos.
12   172, 180, 196 and 200, 315 pp., 18 pls.

Keilin, D.

1914 Sur la biologie d'un Psychodidae à larve xylophage, *Trichomyia urbica.* C. R. Soc. Biol., Paris, vol. 76, pp. 434–437.

Knipling, E. F., and H. T. Rainwater

1937 Species and Incidence of Dipterous Larvae Concerned in Wound Myiasis. Journ. Parasitol., vol. 23, pp. 451–455.

Kolbe, H. J.

1904 Ueber die Lebensweise und die geographische Verbreitung der coprophagen Lamellicornier. Zool. Jahrb. Suppl., pp. 475–594, 3 pls.

Lennox, F. G.

1939 Development of a Synthetic Medium for Aseptic Cultivation of Larvae of *Lucilia cuprina.* Pamph. No. 90, Australian Council Sci. and Indus. Res., 24 pp., 1 pl., 8 figs.

Lichtenstein, J. L.

1917 Observations sur les Coccinellides Mycophages. Bull. Soc. Entom. France, 1917, pp. 298–302.

Linderstrøm-Lang, K., and F. Duspiva

1935 Die Verdauung von Keratin durch die Larven der Kleidermotte. Zeits. physiol. Chemie, vol. 237, pp. 131–158.

Lindquist, A. W.

1935 Notes on the Habits of Certain Coprophagous Beetles and Methods of Rearing them. Circ. U. S. Dept. Agric., No. 351, 9 pp.

Loeb, J., and J. H. Northrop

1916　Nutrition and Evolution (Second note).　Journ. Biol. Chem., vol. 27, pp. 309–312.

1917　On the Influence of Food and Temperature upon the Duration of Life. Journ. Biol. Chem., vol. 32, pp. 103–121.

Long, W. H.

1901　Fungus Spores as Bee-bread.　Plant World, vol. 4, pp. 49–51.

McConnell, W. R.

1915　A Unique Type of Insect Injury.　Journ. Econ. Entom., vol. 7, pp. 261–267.

McIndoo, N. E.

1933　Olfactory Responses of Blowflies, with and without Antennae.　Journ. Agric. Res., vol. 46, pp. 607–625, 4 figs.

Manon

1931　Au sujet des parasites des bouchons des bouteilles à vin, en particulier d'*Oenophila v-flavum*.　Rev. Zool. Agric., vol. 29, pp. 184–188.

Manunta, C.

1933　Sul metabolismo dei grassi nella tignuola degli alveari (*Galleria mellonella*). Rend. Accad. Lincei, Cl. Sci. Fis. Math. Nat. (6), vol. 17, pp. 309–312.

Martelli, G.

1910　Sullo micofagia del Coccinellide, *Thea 22-punctata*.　Boll. Lab. Zool. Portici, vol. 4, pp. 292–294, 1 fig.

1915　Notizie su due Coccinellidi micofagi.　Boll. Lab. Zool. Portici, vol. 9, pp. 151–160.

Mason, F. A.

1924　The Khapra Beetle in British Maltings.　Bull. Bur. Bio-Technol., Leeds, vol. 2, pp. 118–123.

Mégnin, P.

1885　La faune des cadavres.　214 pp.　Paris, Gauthier Villars and Son.

Melvin, R., and R. C. Bushland

1940　Nutritional Requirements of the Screw Worm Larvae.　Journ. Econ. Entom., vol. 33, pp. 850–852.

Metalnikov, S.

1908　Recherches expérimentales sur les chenilles de *Galleria mellonella*.　Arch. Zool. expér. (4), vol. 8, pp. 489–588.

Michelbacher, A. E., W. M. Hoskins, *et al.*

1932　The Nutrition of Flesh Fly Larvae.　Journ. Exper. Zool., vol. 64, pp. 109–128.

Motter, M. G.

1898　A Contribution to the Study of the Fauna of the Grave.　Journ. New York Entom. Soc., vol. 6, pp. 201–231.

Mrak, E. M., and L. S. McClung

1940　Yeasts Occurring on Grapes and Grape Products in California.　Journ. Bact., vol. 40, pp. 395–407, 1 fig.

Nadson, G. A.

1927   How Do Yeasts Get on Grapes? [In Russian.] Vestnik Vinodel Ukrain., vol. 28, pp. 323–328.

Northrop, J. H.

1917   The Role of Yeast in the Nutrition of an Insect (*Drosophila*). Journ. Biol. Chem., vol. 30, pp. 181–187.

Pack, H. J., and V. Dowdle

1930   A Wild Host of *Mineola scitulella*. Journ. Econ. Entom., vol. 23, p. 321.

Petch, T.

1907   Insects and Fungi. Science Progress, vol. 2, pp. 229–238.

Popenoe, C. H.

1912   Insects Injurious to Mushrooms. United States Dept. Agric., Bur. Entom., Circ. No. 155, 10 pp., 7 figs.

Portier, P.

1919   Developpement complet des larves de *Tenebrio molitor* obtenu au moyen d'une nourriture stérilisée à haute température (130°). C. R. Soc. Biol., Paris, vol. 82, pp. 59–60.

Pratt, F. C.

1912   Insects Bred from Cow Manure. Canadian Entom., vol. 44, pp. 180–184.

Richardson, C. H.

1916   The Response of the Housefly to Ammonia and Other Substances. Bull. New Jersey Expt. Sta., No. 292, 19 pp.

1917   The Response of the Housefly to Certain Foods and their Fermentation Products. Journ. Econ. Entom., vol. 10, pp. 102–109.

1925   The Oviposition Response of Insects. Bull. United States Dept. Agric., No. 1324, 17 pp.

Ripley, L. B., G. A. Hepburn, and J. Dick

1939   Mass Breeding of False Codling Moth in Artificial Media. Sci. Bull. Dept. Agric. South Africa, No. 207, 18 pp.

Ritzema, B. J.

1917   Mestkevers van het geslacht Aphodius Ill., als Uijanden van de Champignon-Kultur. Tijdschr. Plantenziekten, vol. 23, pp. 31–32.

Röber, H.

1933–   Sind die Wachsmotten Schädlinge? Entom. Rundsch., vol. 49, pp. 32–
34   53; 63; vol. 50, pp. 142–144.

Robinson, W.

1933   The Culture of Sterile Maggots for Use in the Treatment of Osteomyelitis and Other Suppurative Infections. United States Dept. Agric., Bur. Entom., 10 pp.

Ross, H.

1914   Ueber verpilzte Tiergallen. Ber. Deutsch. Bot. Ges., vol. 32, pp. 574–597.

Roubaud, E.

1922   Recherches sur la fécondité et la longévité de la mouche domestique. Ann. Inst. Pasteur, vol. 36, pp. 765–783.

Roy, D. N.

1936 The Nutrition of Larvae of the Bee Wax Moth. Zeits. f. vergl. Physiol., vol. 24, pp. 638–643, 1 fig.

Roy, D. N., and L. B. Siddons

1939 On the Life History and Bionomics of *Chrysomyia rufifacies*. Parasitology, vol. 31, pp. 442–447.

Rozeboom, L. E.

1935 Relation of Bacteria and Bacterial Filtrates to the Development of Mosquito Larvae. American Journ. Hyg., vol. 21, pp. 167–179.

Schulz, F. N.

1925 Die Verdauung der Raupe der Kleidermotte (*Tinea pellionella*). Biochem. Zeitschr., vol. 156, pp. 124–129.

Scott, H.

1921 The Ptinid Beetle, *Trigonogenius globulum*, Breeding in Argol. Bull. Entom. Res., vol. 12, pp. 133–134.

Sergent, E., and H. Rougebief

1926a Du mutualisme entre les Drosophiles et les levures de vin. Verh. 3ten Internat. Entom. Congr., Zürich, vol. 2, pp. 94–99.

1926b De l'antagonisme entre les Drosophiles et les moissures. C. R. Acad. Sci., France, vol. 182, pp. 1238–1239.

1926c Des rapports entre les moucherous du genre Drosophile et les microbes du raisin. I. Mutualisme a l'égard des levures. II. Antagonisme à l'égard des moissures. Ann. Inst. Pasteur, vol. 40, pp. 901–921.

Sieber, N., and S. Metalnikov

1904 Ueber Ernährung und Verdauung der Bienenmotte (*Galleria mellonella*). Arch. ges. Physiol., vol. 102, pp. 169–286.

Simmons, S. W.

1933 A Bactericidal Principle in Excretions of Surgical Maggots. Journ. Bacteriol., vol. 30, pp. 253–267.

Sitowski, L.

1905 Biologische Beobachtungen über Motten. Bull. Internat. Acad. Sci. Cracovie, 1905, pp. 534–548, 1 pl.

Spuler, A.

1906 Ueber einen parasitisch lebenden Schmetterling, *Bradypodicola hahneli* n. sp. Biol. Centralbl., vol. 26, pp. 690–697, 7 figs.

Stellwaag, F.

1924 *Tinea cloacella* und *Tinea granella*. Zeits. angew. Entom., vol. 10, pp. 181–188.

1924 Die Tierwelt tiefer Weinkeller. Wein u. Rebe, 1924, pp. 277–297.

Stephan, J.

1923 Moos- und Flechtenfresser unter den Raupen. Entom. Jahrb., vol. 32, pp. 95–100.

Stewart, M. A.

1934 The Role of *Lucilia sericata* Larvae in Osteomyelitis Wounds. Ann. Trop. Med. and Parasitol., vol. 28, pp. 445–460.

Strouhal, H.

1926 Pilzfressende Coccinelliden. (Tribus Psylloborini). Zeits. wiss. Insektenbiol., vol. 21, pp. 131–143.

Tatum, E. L.

1939 Nutritional Requirements of *Drosophila melanogaster*. Proc. Nat. Acad.
Sci., vol. 25, pp. 490–497.

Thomas, C. A.

1939 The Animals Associated with Edible Fungi. Journ. New York Entom. Soc.,
vol. 47, pp. 11–37.

Tillyard, R. J., and H. R. Seddon

1933 The Sheep Blowfly Problem in Australia. Sci. Bull. Dept. Agric. N. S. Wales,
No. 40, 136 pp., 6 pls., 15 figs.

Titschak, E.

1923 Beiträge zu einer Monographie der Kleidermotte. Zeits. tech. Biol., vol. 10,
pp. 1–168, 4 pls., 91 figs.

Trägårdh, I.

1914 En svampätande Anthomyidlarv. Ark. f. Zool., vol. 8, No. 5, 16 pp., 1 pl., 10
figs.

Trager, Wm.

1935 The Culture of Mosquito Larvae Free from Living Microörganisms. American Journ. Hyg., vol. 22, pp. 18–25.

1941 The Nutrition of Invertebrates. Physiol. Reviews, vol. 21, pp. 1–35.

Turner, A. J.

1923 A Lepidopterous Scavenger Living in Parrots' Nests. Trans. Entom. Soc.
London, pp. 170–175.

Twinn, C. R.

1934 The Dermestid, *Trogoderma versicolor*, a New Pest of Dried Milk Products.
Canadian Entom., vol. 66, pp. 49–51.

Uvarov, B. P.

1928 Insect Nutrition and Metabolism. Trans. Entom. Soc. London, 1928, pp.
255–343.

Van der Merwe, C. P.

1923 The Destruction of Vegetable Ivory Buttons. The Ravages of the Button
Beetle (*Coccotrypes dactyliperda*) and Suggestions for its Control. Dept.
Agric. Union South Africa, 4 pp.

Vaternahm, Th.

1924 Zur Ernährung und Verdauung unserer einheimischen Geotrupes-Arten.
Zeits. f. wiss. Insektenbiol., vol. 19, pp. 20–27.

Wasmund, Erich

1926 Biocœnose und Thanatocoenose. Arch. Hydrobiol., vol. 17, pp. 1–116, 4 pls.,
6 figs.

Plate XIII

A mushroom, *Lentinus atticolus*, which grew from an abandoned fungus garden on the periphery of a nest of *Atta cephalotes* in British Guiana. Photograph by N. A. Weber.

PLATE XIV

Three types of predatory insect larvae. *A*, larva of the worm lion (*Vermileo comstocki*) which traps its prey in pits dug in dusty soil; *B*, larvae of a coccinellid beetle (*Brachyacantha quadripunctata*), coated with wax derived from the coccids on which they feed; *C*, the imported Calosoma beetle (*Calosoma sycophanta*), having finished with one gipsy moth caterpillar, is attacking another. Photographs by the author.

Weiss, H. B.

1920 The Insect Enemies of Polyporoid Fungi. American Natural., vol. 54, pp. 443–447, 1 pl.

1921a Diptera and Fungi. Proc. Biol. Soc. Washington, vol. 34, pp. 85–88.

1921b A Bibliography on Fungous Insects and their Hosts. Entom. News, vol. 32, pp. 45–47.

1922 A Summary of the Food-habits of North American Coleoptera. American Natural., vol. 56, pp. 159–165.

1923 More Notes on Fungus Insects. Canadian Entom., vol. 55, pp. 199–202.

Weiss, H. B., and E. West

1920– Fungus Insects and their Hosts. Proc. Biol. Soc. Washington, vol. 33, pp. 1–
21   20. Additions, *ibid.*, vol. 34 (1921), pp. 59–62; 167–172.

Wellhouse, W. H.

1919 An Itonid Feeding on Rust Spores. Entom. News, vol. 30, pp. 144–145.

Wheeler, W. M.

1906 The Habits of the Tent-building Ant (*Cremastogaster lineolata*). Bull. American Mus. Nat. Hist., vol. 22, pp. 1–18, 6 pls.

Wheeler, W. M., and I. W. Bailey

1920 The Feeding Habits of Pseudomyrmine and Other Ants. Trans. American Philos. Soc., vol. 22, pp. 235–279, 5 pls.

Wildman, J. D.

1933 Note on the Use of Microörganisms for the Production of Odors Attractive to the Dried Fruit Beetle. Journ. Econ. Entom., vol. 26, pp. 516–517.

Wilson, G. F.

1926 Insect Visitors to Sap-exudations of Trees. Trans. Entom. Soc. London, vol. 74, pp. 243–253, 3 pls., 1 fig.

Wollman, E.

1911 Sur l'élévage des mouches stériles. Contribution à la connaissance du rôle des microbes dans les voies digestives. Ann. Inst. Pasteur, vol. 25, pp. 79–88.

1919 Élévage aseptique de larves de la mouche à viande (*Calliphora vomitoria*), sur milieu stérilisé à haute température. C. R. Soc. Biol., Paris, vol. 82, pp. 593–594.

1921 La méthode des élévages aseptiques en physiologie. Arch. Internat. Physiol., vol. 18, pp. 194–199.

1922 Biologie de la mouche domestique et des larves de mouches à viande, en élévages aseptiques. Ann. Inst. Pasteur, vol. 36, pp. 784–788.

1926 Observation sur une ligne aseptique de Blattes (*Blatella germanica*) datant de cinq ans. C. R. Soc. Biol., Paris, vol. 95, pp. 164–165.

Young, P. A.

1934 Wheat Bunt, a new Food for Grasshoppers. Journ. Econ. Entom., vol. 27, p. 548.

Zabinski, Jan

1926 Observations sur l'élévage des cafards nourris avec des aliments artificiels. C. R. Soc. Biol., Paris, vol. 94, pp. 545–548, 1 fig.

PART 2

Aschner, Manfred

1931 Die Bakterienflora der Pupiparen. Eine Symbiose Studie an blutsaugenden Insekten. Zeits. wiss. Biol. abth. A., Zeits. Morph. u. Oekol. Tiere, vol. 20, pp. 368–442, 5 pls., 5 figs.

1932 Experimentelle Untersuchungen ueber die Symbiose der Kleiderlaus. Naturwiss., vol. 20, pp. 501–505.

1934 Studies on the Symbiosis of the Body Louse. I. Elimination of the Symbionts by Centrifugalisation of the Eggs. Parasitology, vol. 26, pp. 309–314, 1 pl.

Aschner, M., and E. Ries

1933 Das Verhalten der Kleiderlaus bei Ausschaltung ihrer Symbionten. Zeits. Morph. u. Oekol. Tiere, vol. 26, pp. 529–590.

Assmuth, J.

1913 Wood-destroying White Ants of the Bombay Presidency. Journ. Bombay Nat. Hist., vol. 22, pp. 372–384, 4 pls.

Bathellier, J.

1927 Contribution a l'étude systématique biologique des termites de l'Indo-chine. Les cultures mycéliennes des termites de l'Indo-chine. Faune Colon. Franç., vol. 1, pp. 125–365, 13 pls., 113 figs.

Beeson, C. F. C.

1916 Ambrosia Beetles, or Pin-hole and Shot-hole Borers. Indian Forester, Allahabad, vol. 42, pp. 216–223, 1 pl.

Benedek, T., and G. Specht

1933 Mykologischbakteriologische Untersuchungen über Pilze und Bakterien als Symbioten in Kerbtieren. Zentralbl. Bakt. I. Abt. Orig., vol. 130, pp. 74–90, 7 figs.

Bequaert, J.

1921 On the Dispersal by Flies of the Spores of certain Mosses of the Family Splachnaceae. Bryologist, vol. 24, pp. 1–4.

Blackman, M. W.

1933 Insect Vectors of the Dutch Elm Disease. U. S. Dept. Agric., Bur. Entom., pp. 4, 2 pls.

Blochmann, F.

1888 Ueber das regelmässige Vorkommen von Bakteriennähnlichen Gebilden in den Geweben und Eiern verschiedener Insekten. Zeits. Biol., vol. 24, pp. 1–51.

Bode, H.

1936 Untersuchungen ueber die Symbiose von bakterienähnlichen Gebilden und das Verhalten von Blattiden bei aseptischer Aufzucht. Arch. Mikrobiol., vol. 7, pp. 391–403.

Breitsprecher, E.

1928 Beiträge zur Kenntnis der Anobiidensymbiose. Zeits. Morphol. Oekol. Tiere, vol. 11, pp. 495–538, 25 figs.

Brown, W. H.

1918   The Fungi Cultivated by Termites in the Vicinity of Manila and Los Baños. Philippine Journ. Sci., vol. 13, C, pp. 223–231, 2 pls.

Brues, C. T., and R. W. Glaser

1921   A Symbiotic Fungus Occurring in the Fat-body of *Pulvinaria innumerabilis*. Biol. Bull., vol. 40, pp. 299–324, 3 pls., 2 figs.

Buchner, P.

1919   Zur Kenntnis der Symbiose niederer pflanzlicher Organismen mit Pedikuliden. Biol. Zentralbl. Leipzig, vol. 39, pp. 535–540.

1921   Tier und Pflanze in intrazellularer Symbiose. xi + 462 pp. Berlin, Bornträger.

1921   Studien an intracellularen Symbionten. III. Die Symbiose der Anobiinen mit Hefepilzen. Arch. Protistenk., vol. 42, pp. 319–336.

1922   Rassen- und Bakteroidenbildung bei Hemipterensymbionten. Biol. Centralbl., vol. 42, pp. 38–46.

1923   Studien an intracellularen Symbionten. IV. Die Bakteriensymbiose der Bettwanze. Arch. Protistenk., vol. 46., pp. 225–263, 3 pls., 3 figs.

1924   System und Symbiose. Verh. deutsch. Zool. Ges., vol. 29, pp. 48–54.

1928   Holznahrung und Symbiose. 64 pp., 22 figs. Berlin, J. Springer.

1930   Tier und Pflanze in Symbiose. xx + 900 pp., 336 figs. Berlin, Borntraeger.

1932   Der gegenwärtige Stand der neuen Symbioseforschung. Der Biologe, vol. 2, pp. 1–7.

Bugnion, E.

1913   Les moeurs des termites champignonnistes. Bibliothèque, Univ. Lausanne, pp. 552–583, 1 pl.

Campbell, W. G.

1941   Relation between Nitrogen Metabolism and the Duration of the Larval Stage of the Death-watch Beetle, *Xestobium*, in wood Decayed by Fungi. Biochem. Journ., vol. 35, pp. 1200–1208, 3 figs.

Carter, W.

1935   The Symbionts of *Pseudococcus brevipes*. Ann. Ent. Soc. America, vol. 28, pp. 60–71, 4 pls.

1936   The Symbionts of *Pseudococcus brevipes* in Relation to a Phytotoxic Secretion of the Insect. Phytopathology, vol. 26, pp. 176–183.

Cartwright, K. St. G.

1938   A Further Note on Fungus Association in the Siricidae. Ann. Appl. Biol., vol. 25, pp. 430–432.

Cholodowsky, N.

1891   Die Embryonalentwicklung von *Phyllodromia germanica*. Mém. Acad. St. Pétersbourg (5), vol. 38, pp. 1–124.

Clark, A. F.

The Horntail Borer and its Fungal Association. New Zealand Journ. Sci. Techn., vol. 15, pp. 188–190.

Clark, C. U.

1895   On the Food Habits of Certain Dung and Carrion Beetles. Journ. New York Entom. Soc., vol. 3, p. 61.

Cleveland, L. R.

1923a Symbiosis between Termites and their Intestinal Protozoa. Proc. Nat. Acad. Sci., vol. 9, pp. 424–428.

1923b Correlation Between the Food and Morphology of Termites and the Presence of Intestinal Protozoa. American Journ. Hygiene, vol. 3, pp. 444–461.

1924 The Physiological and Symbiotic Relationships between the Intestinal Protozoa of Termites and their Host, with Special Reference to *Reticulitermes flavipes*. Biol. Bull., vol. 46, pp. 177–225.

1925a The Ability of Termites to Live Perhaps Indefinitely on a Diet of Pure Cellulose. *Ibid.*, vol. 48, pp. 289–293.

1925b The Feeding Habit of Termite Castes and its Relation to their Intestinal Flagellates. *Ibid.*, vol. 48, pp. 295–308, 1 pl.

1925c The Effects of Oxygenation and Starvation on the Symbiosis between the Termite *Termopsis* and its Intestinal Flagellates. *Ibid.*, vol. 48, pp. 309–326, 1 pl.

1926 Symbiosis Among Animals with Special Reference to Termites and their Intestinal Flagellates. Quart. Rev. Biol., vol. 1, pp. 51–60.

1928 Further Observations and Experiments on the Symbiosis Between Termites and their Intestinal Protozoa. Biol. Bull., vol. 54, pp. 231–237.

Cleveland, L. R., S. R. Hall, E. P. Sanders, and J. Collier

1934 The Wood-feeding Roach, *Cryptocercus*, its Protozoa, and the Symbiosis between Protozoa and Roach. Mem. American Acad. Arts Sci., vol. 17, pp. 185–342, 60 pls., 40 figs.

Cohen, W. E.

1933 An Analysis of Termite (*Eutermes exitiosus*) Mound Material. Journ. Council Sci. Industr. Res., Australia, vol. 6, pp. 166–169.

Cook, S. F.

1943 Nonsymbiotic Utilization of Carbohydrates by the Termite *Zoötermopsis angusticollis*. Physiol. Zool., vol. 16, pp. 123–128.

Cook, S. F., and A. E. Cook

1942 Metabolic Relations in the Termite-Protozoa Symbiosis. Journ. Cell. and Comp. Physiol., vol. 19, pp. 211–220.

Couch, J. N.

1931 The Biological Relationship between *Septobasidium retiforme* and *Aspidiotus osborni*. Quart. Journ. Micros. Sci., vol. 74, pp. 384–437, 5 pls., 60 figs.

Doflein, F.

1905 Die Pilzkulturen der Termiten. Verh. Deutsch. Zool. Ges., vol. 15, pp. 140–149.

Dubois, R.

1886 Contribution a l'étude de la production de la lumière par les êtres vivants. Les Elaterides lumineux. Bull. Soc. Zool. France, vol. 11, pp. 1–275.

Eidmann, H.

1932 Beiträge zur Kenntnis der Biologie, insbesondere des Nestbaues der Blattschneiderameise, *Atta sexdens*. Zeits. Morph. Oekol. d. Tiere, vol. 25, pp. 154–183.

Ekblom, T.

1931– Cytological and Biochemical Researches into the Intracellular Symbiosis in
32 the Intestinal Cells of *Rhagium inquisitor*. Skand. Arch. Physiol., vol. 6, pp. 35–48, 17 figs., 4 pls. (1931). Part 2, *ibid.*, vol. 64, pp. 279–298, 4 pls. (1932).

Escherich, K.

1900 Ueber das regelmässige Vorkommen von Sprosspilzen in dem Darmepithel eines Käfers. Biol. Centralbl., vol. 20, pp. 350–358.

1909 Die Termiten oder weissen Ameisen. xii + 198 pp., 51 figs., 1 pl. Leipzig.

1909 Die pilzzüchtenden Termiten. Biol. Centralbl., vol. 29, pp. 16–27.

Felt, E. P.

1934 Dutch Elm Disease Control and the Elm Bark Borer. Journ. Econ. Entom., vol. 27, pp. 315–319.

1940 European Elm Bark Beetle and Dutch Elm Disease Control. Ibid., vol. 33, pp. 556–558.

Forbes, S. A.

1895 Bacteria Normal to Digestive Organs of Hemiptera. Bull. Illinois State Lab. Nat. Hist., vol. 4, pp. 1–7.

Fraenkel, H.

1921 Die Symbionten der Blattiden im Fettgewebe und Ei insbesondere von Periplaneta orientalis. Zeits. wiss. Zoologie, vol. 119, pp. 53–66, 2 pls.

Gier, H. T.

1936 The Morphology and Behavior of the Intracellular Bacteroids of Roaches. Biol. Bull., vol. 71, pp. 433–452.

Glaser, R. W.

1920 Biological Studies on Intracellular Bacteria. Biol. Bull., vol. 39, pp. 133–145.

1930 On the Isolation, Cultivation and Classification of the So-called Intracellular "Symbiont" or Rickettsia of Periplaneta americana. Journ. Expt. Med., vol. 51, pp. 59–82; 903–907.

Glasgow, H.

1914 The Gastric Coeca and Coecal Bacteria of the Heteroptera. Biol. Bull., vol. 26, pp. 101–107, pls. 1–8.

Goetsch, W., and R. Stoppel

1940 Die Pilze der Blattschneiderameisen. Biol. Centralbl., vol. 60, pp. 393–398.

Grossman, H.

1930 Beiträge zur Kennitis der Lebensgemeinschaft Zwischen Borkenkäfern und Pilzen. Zeits. f. Parasitenk., vol. 3, pp. 56–102, 19 figs.

Guyot, R.

1928 De l'influence des insectes xylophages dans la propagation de l'armillaire. C. R. Assoc. Av. Sci. Franç., vol. 52, pp. 391–393.

Harris, J. A.

1907 The Fungi of Termite Nests. American Naturalist, vol. 41, pp. 536–539.

Harvey, E. N.

1940 Living Light. xv + 328 pp., 79 figs. Princeton, N. J., Princeton Univ. Press.

Harvey, E. N., and R. T. Hall

1929 Will the Adult Fire-fly Luminesce if its Larval Organs are Entirely Removed? Science, vol. 69, pp. 253–254.

236 INSECT DIETARY

Hecht, O.

1924 Embryonalentwicklung und Symbiose bei *Camponotus ligniperda*. Zeits. wiss. Zool., vol. 122, pp. 173–204, 28 figs.

Hedgcock, G. C.

1906 Studies upon some Chromogenic Fungi which Discolor Wood. Ann. Rept. Missouri Bot. Gard., No. 17, pp. 58–114, 10 pls.

Hegh, E.

1922 Les Termites. 756 pp., 460 figs. Imprim. Indust. et Financière, Brussels.

Heitz, E.

1927 Ueber intrazellulare Symbiose bei holzfressenden Käferlarven. Zeits. Morphol. Oekol. Tiere, vol. 7, pp. 279–305.

Hendee, E. C.

1933 The Association of the Termites, *Kalotermes minor*, *Reticulitermes hesperus* and *Zoötermopsis angusticollis* with Fungi. Pubs. Univ. California, Zool., vol. 39, pp. 111–113, 1 fig.

Hertig, Marshall

1921 Attempts to Cultivate the Bacteroids of the Blattidae. Biol. Bull., vol. 41, pp. 181–187.

Hess, W. N.

1922 Origin and Development of the Light-organs of *Photuris pennsylvanica*. Journ. Morph., vol. 36, pp. 245–266, 5 pls.

Hinman, E. H.

1932 The Presence of Bacteria within the Eggs of Mosquitoes. Science, vol. 76, pp. 106–107.

1933 The Role of Bacteria in the Nutrition of Mosquito Larvae, the Growth-stimulating Factor. American Journ. Hyg., vol. 18, pp. 224–236.

Hollande, A. C., and R. Favre

1931 La structure cytologique de *Blattabacterium cuenoti* symbiote du tissu adipeux des Blattides. C. R. Soc. Biol., vol. 107, pp. 752–754, 15 figs.

Hubbard, H. G.

1897 The Ambrosia Beetles of the United States. Bull. U. S. Dept. Agric. Div. Entom., No. 7, pp. 9–30, 34 figs.

1906 Ambrosia Beetles. Yearb. United States Dept. Agric., pp. 421–430, 7 figs.

Huber, J.

1905 Ueber die Koloniengründung bei *Atta sexdens*. Biol. Centralbl., vol. 25, pp. 606–619; 625–635, 26 figs.

1907 The Founding of Colonies by *Atta sexdens*. Ann. Rept. Smithsonian Inst., 1906, pp. 355–367, 5 pls.

Hudson, G. V.

1926 The New Zealand Glow-worm, *Boletophila* (*Arachnocampa*) *luminosa:* Summary of Observations. Ann. Mag. Nat. Hist., (9), vol. 17, pp. 228–235, 1 pl.

Hungate, R. E.

1936 Studies on the Nutrition of Zoötermopsis; the Role of Bacteria and Molds in Cellulose Decomposition. Zentralbl. f. Bakt., 2te Abt., vol. 94, pp. 240–249.

1938 The Relative Importance of the Termite and the Protozoa in Wood Digestion. Ecology, vol. 19, pp. 1–25.

1939 Experiments on the Nutrition of Zoötermopsis; the Anaerobic Carbohydrate Dissimilation by the Intestinal Protozoa. Ecology, vol. 20, pp. 230–245.

1941 Experiments on the Nitrogen Economy of Termites. Ann. Entom. Soc. America, vol. 34, pp. 467–489.

1943 Quantitative Analyses on the Cellulose Fermentation by Termite Protozoa. Ibid., vol. 36, pp. 730–739.

Jumelle, H., and H. Perrier de la Bathie

1910 Termites champignonnistes et champignons des termitières à Madagascar. Rev. Gén. Bot., Paris, vol. 22, pp. 30–64.

Karawaiew, W.

1899 Ueber Anatomie und Metamorphose des Darmkanals der Larve von *Anobium paniceum*. Biol. Centralbl., vol. 19, pp. 122–130; 161–171; 196–202.

Keilin, F.

1921 On the Life-history of *Dasyhelea obscura*, with some Remarks on the Parasites and Hereditary Bacterium Symbionts of the Midge. Ann. Mag. Nat. Hist., (9), vol. 8, pp. 576–590, 2 pls.

Kirby, H.

1926 On *Staurojoenina assimilis*, an Intestinal Flagellate from the Termite, *Kalotermes minor*. Univ. California Pub., Zool., vol. 29, pp. 25–102, 7 pls., 7 figs. (Contains a list of termite Protozoa.)

1937 Host-parasite Relations in the Distribution of Protozoa in Termites. Univ. California Publ., Zool., vol. 41, pp. 189–212.

Klevenhusen, F.

1927 Beiträge zur Kenntnis der Aphidensymbiose. Zeits. Morph. Oekol. d. Tiere, vol. 9, pp. 97–165, 2 pls., 41 figs.

Kligler, I. J., and M. Aschner

1931 Cultivation of Rickettsia-like Microörganisms from Certain Blood-sucking Pupipara. Journ. Bacter., vol. 22, pp. 103–118.

Koch, A.

1931 Die Symbiose von *Oryzaephilus surinamensis*. Zeits. Morph. Oekol. Tiere, vol. 23, pp. 389–424, 15 figs.

1936a Symbiosenstudien. I, II. Ibid., vol. 32, pp. 92–180.

1936b Bau und Entwicklungsgeschichte der Mycetome von *Lyctus linearis*. Verh. Deutsch. Zool. Ges., vol. 38, pp. 252–261.

1938 Symbiosen Studien, III. Die intrazellulare Bakteriensymbiose von *Mastotermes darwiniensis*. Zeits. Morph. Oekol. Tiere, vol. 34, pp. 584–609.

1941 Ueber den gegenwartigen Stand der experimentellen Symbiosenforschungen. Verh. VIIten Internat. Kongr. Entom. Berlin, vol. 2, pp. 760–771.

Kofoid, C. A., and R. F. MacLennan

1930– Ciliates from *Bos indicus*. California Univ. Publ., Zool., vol. 33, No. 22; vol.
33 37, No. 5; vol. 39, No. 1.

Krassilstchik, M. J.

1890 Ueber eine neue Kategorie von Bakterien (Biophyten) die im Inneren eines Organismus leben und ihm Nutzen bringen. Biol. Zentralbl., vol. 10, p. 421.

Kuskop, M.

1924   Bakteriensymbiosen bei Wanzen. Arch. Protistenk., vol. 47, pp. 350–385, 3 pls., 7 figs.

Leach, J. G.

1933   The Method of Survival of Bacteria in the Puparia of the Seed-corn Maggot (*Hylemyia cilicrura*). Zeits. angew. Entom., vol. 20, pp. 150–161, 9 figs.

1940   Observations on Two Ambrosia Beetles and their Associated Fungi. Phytopathology, vol. 30, pp. 227–236, 4 figs.

Leydig, F.

1854   Zur Anatomie von *Coccus hesperidum*. Zeits. wiss. Zool., vol. 5, pp. 1–12, 6 figs.

Lilienstern, M.

1932   Beiträge zur Bakteriensymbiose der Ameisen. Zeits. Morph. Oekol. Tiere, vol. 26, pp. 110–134.

Mahdihassan, S.

1923   Classification of Lac Insects from a Physiological Standpoint. Journ. Sci. Assoc. Maharajah's College Vizianagaram, vol. 1, pp. 47–49.

1929   Specific Symbiotes of a Few Indian Scale Insects. Zentralbl. f. Bakt. (2) vol. 78, pp. 254–259, 7 figs.

1932   The Symbionts of *Lakshadia communis*. Arch. Protistenkunde, vol. 78, pp. 514–521, 1 pl., 5 figs.

1933   Sur les différents symbiotes des Cochenilles productrices ou non productrices de cire. C. R. Acad. Sci., vol. 196, pp. 560–562.

Mansour, K.

1934   On the So-called Symbiotic Relationship between Coleopterous Insects and Intercellular Microörganisms. Quart. Journ. Micr. Sci., vol. 77, pp. 255–271, 2 pls.

Mansour, K., and J. J. Mansour-Bek

1934   On the Digestion of Wood by Insects. Journ. Expt. Biol., vol. 11, pp. 243–256, 1 fig.

Mansour-Bek, J. J.

1934   Digestion of Wood by Insects and the Supposed Role of Microörganisms. Biol. Rev. and Biol. Proc. Cambridge Philos. Soc., vol. 9, pp. 363–382.

Melampy, R. M., and G. F. MacLeod

1938   Bacteria Isolated from the Gut of the Larval *Agriotes mancus*. Journ. Econ. Entom., vol. 31, p. 320.

Mercier, L.

1907   Recherches sur les bactéroïdes des Blattides. Arch. Protistenkunde, vol. 9, pp. 346–358.

Milovidov, P. F.

1928   A propos des bactéroïdes des Blattes (*Blatella germanica*). C. R. Soc. Biol., Paris, vol. 99, pp. 127–128.

Moeller, A.

1893   Die Pilzgärten einiger südamerikanischer Ameisen. Schimper, Botanische Mittheilungen aus den Tropen, vol. 6, 127 pp., 7 pls.

Müller, H. J.

1941 Die intrazellulare Symbiose bei *Cixius nervosus* und *Fulgora europaea* als Beispiele polysymbionter Zyklen. Verh. VIIten Internat. Kongr. Entom., Berlin, vol. 2, pp. 877–894.

Müller, W.

1934 Ueber die Pilzsymbiose holzfressender Insekten. Arch. Mikrobiol., vol. 5, pp. 84–147, 31 figs.

Neger, F. W.

1908a Ambrosiapilze. Ber. Deutsch. Bot. Gesellsch., vol. 26, pp. 735–754 (also in vols. 27, 28, 29).

1908b Die Pilzzüchtenden Bostrychiden. Naturw. Zeits. f. Forst. u. Landwirtsch., vol. 6, pp. 274–280.

1908c Die Pilzkulturen der Nutzholzborkenkäfer. Centralbl. f. Bakt., vol. 20, pp. 279–282.

1911 Zur Uebertragung des Ambrosiapilzes von *Xyleborus dispar*. Naturw. Zeits. f. Forst. u. Landwirtsch., vol. 9, pp. 223–225.

Neukomm, A.

1927a Sur la structure des bactéroïdes des blattes (*Blattella germanica*). C. R. Soc. Biol., Paris, vol. 96, pp. 306–308.

1927b Action des rayons ultra-violets sur les bactéroïdes des Blattes (*Blatella germanica*). *Ibid.*, vol. 96, pp. 1155–1156.

Nolte, H. W.

1938 Die Legeapparat der Dorcatominen (Anobiidae) unter besonder Berücksichtigung der symbiontischen Einrichtungen. Zool. Anz. Suppl., vol. 11, pp. 147–154, 5 figs.

Paillot, A.

1929 Sur l'origine infectieuse des micro-organismes des Aphides. C. R. Soc. Acad. Sci. Paris, vol. 189, pp. 210–213, 2 figs.

1930 Sur la spécificité parasitaire des bactéries infectant normalemant les pucerous. *Ibid.*, vol. 103, pp. 89–90, 1 fig.

1931a Les Variations morphologiques du bacille symbiotique de *Macrosiphum tanaceti*. *Ibid.*, vol. 193, pp. 1222–1224.

1931b Parasitisme bactérien et symbiose chez *Aphis atriplis*. C. R. Acad. Sci. France, vol. 193, pp. 676–678.

1932 Les variations du parasitisme bactérien normal chez la *Chaitophorus lyropictus*. C. R. Acad. Sci. Paris, vol. 194, pp. 135–137.

1933 L'infection chez les insectes. Immunité et symbiose. 535 pp., 279 figs. Trévoux, France, G. Patissier.

Parkin, E. A.

1942 Symbiosis and Siricid Wood Wasps. Ann. Appl. Biol., vol. 29, pp. 268–274, 2 figs.

Petch, T.

1913 Termite fungi: a Résumé. Ann. R. Botan. Gard. Peradenyia, vol. 5, pp. 303–341.

Petri, L.

1910 Untersuchungen über die Darmbakterien der Olivenfliege. Centralbl. f. Bakt., vol. 26, pp. 357–367.

Pierantoni, U.

1914a Sulla luminosita e gli organi luminosi di *Lampyris noctiluca*. Boll. Soc. Nat. Napoli, vol. 27, pp. 83–88, 1 pl.

1914b La luce degli insetti luminosi e la simbiosi ereditaria. Rend. R. Accad. Sc. Fis. Math., Napoli (3) vol. 20, pp. 15–21.

1918 Gli organi simbiotici e la luminescenza batterica dei Cephalopodi. Publ. Staz. Zool. Napoli, vol. 2, pp. 104–146, 3 pls.

Portier, P.

1911 Digestion phagocytaire des chenilles xylophages des Lépidoptères. Exemple d'union symbiotique entre un insecte et un champignon. C. R. Soc. Biol., Paris, vol. 70, pp. 702–704.

Pospelov, V. P.

1929 Intracellular Symbiosis and its Relation to Insect Diseases [In Russian]. Ann. St. Inst. Expt. Agron., vol. 7, pp. 551–568, 9 figs.

Putnam, J. D.

1880 Biological and other Notes on Coccidae. Proc. Davenport Acad. Sci., vol. 2, pp. 293–347.

Ries, E.

1930 Die symbiontischen Einrichtungen der Läuse und Mallophagen. Jahrb. Schlesischen Ges. Vaterländ. Kultur, vol. 103, pp. 35–38.

1932 Experimentelle Symbiosenstudien. Zeits. Morph. Oekol. d. Tiere, vol. 25, pp. 184–234.

1935 Ueber den Sinn der erblichen Insektensymbiose. Naturwiss., vol. 23, pp. 744–749.

Ripper, W.

1930 Zur Frage des Cellulosabbaues bei der Holzverdauung xylophager Insektenlarven. Zeits. vergl. Physiol., vol. 13, pp. 314–333.

Romeo, Antonio

1935 Sugli "zoocecidii a fungaia" di *Coronilla emerus*. Ann. R. Ist. Sup. Agr. Portici, (3) vol. 7, pp. 81–120, 15 figs.

Roubaud, E.

1919 Les particularités de la nutrition et la vie symbiotique chez les mouches Tsétsés. Ann. Inst. Pasteur, vol. 33, pp. 489–536, 15 figs.

Scheinert, Willi

1933 Symbiose und Embryonalentwicklung bei Rüsselkäfern. Zeits. Morph. Oekol. Tiere, vol. 27, pp. 76–128, 34 figs.

Schneider-Orelli, F.

1911a Ueber die Symbiose eines einheimischen pilzzüchtenden Borkenkäfers (*Xyleborus dispar*) mit seinem Nährpilz. Verh. Schweiz. naturf. Gesell., vol. 94, pp. 279–280.

1911b Die Uebertragung und Keimung des Ambrosia-Pilzes von *Xyleborus dispar*. Naturw. Zeits. f. Forst- und Landwirtsch., vol. 9, pp. 186–192; 223–225.

1912 Der ungleiche Borkenkäfer (*Xyleborus dispar*) an Obstbäumen und sein Nährpilz. Naturwirtsch. Jahrb. Schweiz., 1912, pp. 326–334.

1913 Untersuchungen über den pilzzüchtenden Obstbaumborkenkäfer *Xyleborus* (*Anisandrus*) *dispar* und seinen Nährpilz. Centralbl. Bakter. Parasit., vol. 38, pp. 25–110.

1920 Beiträge zur Biologie des pilzzüchtenden Käfers, *Hylecoetus dermestoides*. Mitt. Schweiz. Entom. Ges., vol. 13, pp. 64–67.

Schomann, Hans

1937 Die Symbiose der Bockkäfer. Zeits. Morph. Oekol. d. Tiere, vol. 32, pp. 542–612, 31 figs.

Schrader, Franz

1923 The Origin of the Mycetocytes in *Pseudococcus*. Biol. Bull., vol. 45, pp. 279–302, 3 pls.

Schwartz, W.

1924 Untersuchungen über die Pilzsymbiose der Schildläuse. Biol. Zentralbl., vol. 44, pp. 487–527, 21 figs.

1932 Untersuchungen über die Symbiose von Tieren mit Pilzen und Bacterien. II Mitteilung: Neue Untersuchungen über die Pilzsymbiose der Schildläuse. Arch. Mikrobiol., vol. 3, pp. 453–472.

1941 Die physiologischen Grundlagen der Symbiosen von Tieren mit Pilzen und Bakterien. Verh. VIIten Internat. Kongr. Entom. Berlin, vol. 2, pp. 916–926.

Spegazzini, C.

1921 Descripción de Hongos Myrmecófilos. Dev. Mus. La Plata, vol. 26, pp. 166–173.

Stahel, G., and D. C. Geijskes

1941 Weitere Untersuchungen über Nestbau und Gartenpilz von *Atta cephalotes* und *A. sexdens*. Rev. Entom., vol. 12, pp. 243–268.

Stammer, H. J.

1929a Die Symbiose der Lagriiden. Zeits. Morph. Oekol. Tiere, vol. 15, pp. 1–34, 26 figs.

1929b Die Bakteriensymbiose der Trypetiden. *Ibid.*, vol. 15, pp. 481–523.

1935 Die Symbiose der Donaciinen. *Ibid.*, vol. 29, pp. 585–608, 16 figs.

Steinhaus, E. A.

1940a A Discussion of the Microbial Flora of Insects. Journ. Bact., 40, No. 1, pp. 161–162.

1940b The Microbiology of Insects. Bacteriol. Reviews, vol. 4, pp. 17–57.

1942 Catalogue of Bacteria associated extracellularly with Insects and Ticks. iii + 206 pp. Minneapolis, Burgess Pub. Co.

Strohmeyer, H.

1911 Die biologische Bedeutung secundärer Geschlechtscharaktere am Köpfe weiblicher Platypoden. Entom. Blätter, vol. 7, pp. 103–107, 2 pls.

Studhalter, R. A., and A. G. Ruggles

1915 Insects as Carriers of the Chestnut Blight Fungus. Bull. Pennsylvania Dept. Forestry, No. 12, 33 pp.

Šulc, K.

1910 "Pseudovitellus" und ähnliche Gewebe der Homopteren sind wohnstätten symbiotischer Saccharomyceten. S. B. Kön. Böhm. Ges. Wissensch., Math. Klasse, vol. 3, pp. 1–39.

Sumner, R.

1933 The Influence of Gregarines on Growth in the Mealworm. Science, vol. 78, p. 125.

1936 Relation of Gregarines to Growth and Longevity in the Mealworm, *Tenebrio molitor*. Ann. Entom. Soc. America, vol. 29, pp. 645–648.

Tarsia in Curia, Isabella

1933   Nuove osservazioni sull'organo simbiotico di *Calandra oryzae*. Arch. Zool. Italiano, vol. 18, pp. 247–264, 9 figs.

1935   La simbiosi ereditaria in *Trioza alacris. Ibid.*, vol. 20, pp. 215–235, 2 pls., 4 figs.

Toth, L.

1937   Entwicklungszyklus und Symbiose von *Pemphigus spirothecae*. Zeits. Morph. u. Oekol. d. Tiere, vol. 33, pp. 412–437, 16 figs.

Uichanco, L. B.

1924   Studies on the Embryogeny and Postnatal Development of the Aphididae with Special Reference to the History of the "Symbiotic Organ." Philippine Journ. Sci., vol. 24, pp. 143–247, 13 pls.

Verrall, A. F.

1943   Fungi Associated with Certain Ambrosia Beetles. Journ. Agric. Res., vol. 66, pp. 135–144.

Weber, N. A.

1938   The Sporophore of the Fungus Grown by *Atta cephalotes* and a Review of Reported Sporophores. Rev. Entom., vol. 8, pp. 265–272.

Webster, F. M., and W. J. Phillips

1912   The Spring Grain Aphis. Bull. United States Dept. Agric., Bur. Entom., No. 110.

Werner, E.

1926a   Die Ernährung von *Potosia cuprea*. Ein Beitrag zum Problem der Cellulose-versdauung bei Insektenlarven. Zeits. Morph. u. Oekol. Tiere, vol. 6, pp. 150–206.

1926b   Der Erreger der Zelluloseverdauung bei der Rosenkäferlarve (*Potosia cuprea*). Centralbl. f. Bakt. Parasitenk., 1 Abth., vol. 100, pp. 293–300.

Wheeler, W. M.

1907   The Fungus-growing Ants of North America. Bull. American Mus. Nat. Hist., vol. 23, pp. 669–807, 5 pls., 31 figs.

Wigglesworth, V. B.

1929   Digestion in the Tsetse-fly, a Study of Structure and Function. Parasitology, vol. 21, pp. 288–321.

1936   Symbiotic Bacteria in a Blood-sucking Insect, *Rhodnius prolixus*. Parasitology, vol. 28, pp. 284–289, 4 figs.

Williams, F. X.

1916   Photogenic Organs and Embryology of Lampyrids. Journ. Morphol., vol. 28, pp. 145–186, 10 pls.

Zimmermann, A.

1908   Ueber Ambrosiakäfer und ihre Beziehungen zur Gummibildung bei *Acacia decurrens*. Centralbl. Bakt. Parasitenk., vol. 10, pp. 716–724.

CHAPTER VI

PREDATORY INSECTS

Nature seems to have intended that her creatures feed upon
one another. At any rate she has so designed her cycles that
the only forms of life that are parasitic directly upon Mother
Earth herself are a proportion of the vegetable kingdom that
dig their roots into the sod for its nitrogenous juices and spread
their broad chlorophyllic leaves to the sun and air.

HANS ZINSSER, in *Rats, Lice and History*, 1935

PREDATISM is such a commonplace mode of sustenance among
animals, including man himself, that it is difficult to deal with its
widespread occurrence and significance among insects without
entering into great detail. Nevertheless, predatism cannot compare
in complexity with parasitism or some of the other similarly bizarre
series of adaptations that may be fished with equal ease from the
seething biological cauldron. Parasitic insects offer such an alluring
field that they have never failed to hold the attention of a goodly
number of admiring entomologists, and their almost endless structural and behavioristic diversity excites increasing wonder as its
details become known. It is most natural, therefore, that these
parasites have occupied the center of the entomological stage while
their humdrum vegetarian relatives have been generally cast aside
as of economic rather than of biological interest.

Predatory insects have been relegated to an intermediate position
in point of interest. The more typical ones, of ferocious mien and
insatiable appetite recall more primitive times when physical
strength and prowess were the necessary attributes of the successful
pirate or highway robber. Others depend upon stealth and super-
ficially respectable manners to conceal an equally vicious *Gestalt*.
Both modes of behavior may, furthermore, be combined in varying
proportions with the addition of special morphological structures
and even artifacts. With these several possibilities, the evolution
of predatism among insects has followed divergent paths, the end
results of which are varied in the extreme. Racketeers and gunmen
really have yet much to learn from a careful survey of insect be-

havior. Let us hope that not they, but more altruistic entomologists will in the future more thoroughly explore the realm of the insects that prey on their more helpless fellows. So far predatory insects have not received the share of attention that they deserve at the hands of the biologist.

Even though it may shock our sensibilities to observe a gorgeous Calosoma beetle, clad in shining metallic armor and armed to the teeth, sally forth to pounce upon one helpless flabby caterpillar after another, we are bound to admire the trim and powerful asilid fly that adroitly impales a large bee upon its sword-like beak with no less assurance than the gaudy matador pierces a floundering bull. Either act might furnish the plot for a melodrama, especially if we add that the burnished villain may die of lead arsenate extracted from the alimentary tract of her juicy victims.

These are examples taken from a vast range of predatory behavior and they give no hint as to the origin of predatism if in truth it has been derived from something else, nor do they in the least aid us in tracing its evolution to such final perfection or disgusting gluttony.

The great majority of predatory insects depend upon other, less active kinds of insects for food. Their prey includes therefore, mainly vegetarian or saprophagous species. Under such conditions it is difficult to believe that predatism is a primitive method of feeding among insects as it would appear impossible for a group of animals to appear and prosper if their living depended upon eating one another. That they might at first have fed upon other small animals seems equally improbable, for if such were true we should still find many existing relicts subsisting upon earthworms, snails, crustaceans, spiders and the like. Such do not exist in any abundance at least outside the more recent and highly specialized orders. Furthermore, if we turn to the evidence derived from palaeontology and rely to some extent upon analogies drawn from our knowledge of the behavior of modern insects we find additional, although not complete, confirmation of this premise. Among the most primitive types of insects we might mention the Protorthoptera, cockroaches and Protodonata, known from the Permian and Upper Carboniferous which is as far back as we may go safely if we are to consider food-habits with any degree of assurance. There can be no reasonable doubt that these locust-like creatures and cockroaches were vegetarians like their present-day descendants, but on the other hand it is probable, indeed quite surely true that the gigantic dragon-flies of that time were predatory, both as nymphs and imagines, although the occurrence of may-flies and stone-flies at this time shows the presence of other, most probably phytophagous types.

It would appear therefore that among the already quite diverse assemblage of insects extant in the late Carboniferous and early Permian times there were both phytophagous and predatory types, the vegetarian ones appearing first and the predatory ones developing soon afterwards when a supply of insect food had become available. This is a natural sequence and there seems to be no valid reasons for any other assumption. This of course does not entail the direct development of all the modern predators from primitive ones, for as we shall see in a moment there is indubitable evidence that predatism has developed independently many times in species, genera or larger groups in typically phytophagous orders like the Lepidoptera.

There is the possibility, also, that the first insects may have depended upon dead animal material for food. Such seems to have been the case with the once abundant trilobites which have been considered by some as the precursors of insects. Raymond ('20) gives good reasons for believing that the early trilobites so abundant in Cambrian times existed at a peaceful period in the earth's history when large numbers of animals died a natural death and supplied so much food of this sort that the trilobites thrived and multiplied. Such habits are at the present time characteristic of many crustaceans and may have been prevalent also among the Palaeodictyoptera and the nymphs of some other types of early amphibiotic insects. Specialized insects appeared, however, so much later than this "peaceful period" that the presence of necrophagous tendencies in their constitution must be duly discounted. That dead animal matter formed any great proportion of the diet of the Protoblattoids or Protorthopteroids does not seem at all probable, as it is difficult to conceive of the existence of any sufficiently abundant and suitable food-supply of this kind in the Carboniferous forests.

If, as seems quite certain, the early true insects were vegetarian, where and how did the predaceous forms arise? As has already been implied, it is quite evident that some of the first predaceous insects were dependent upon other insects for food, and it is equally clear that the development of predatism began soon after the evolution of the less aggressive insects was well under way. The ferocious mantid, with its enlarged prothorax and enormous spiny raptorial fore legs, is so well fitted for its purpose in life that its ancestry undoubtedly dates much farther back through a long series of less beautifully adapted creatures.

There is nevertheless little difficulty in understanding the development of predatory habits from vegetarian ones with little or no structural modification, for we see the same change occurring at the present time. The jaws of plant-eating insects are sharp and powerful and when opportunity presents itself some of them will occasionally bite at and actually feed upon other insects, even members of their own species (Fig. 33). Such behavior, especially if food be scarce, is not surprising. Numerous cases where such insects of the most diverse kinds have become temporarily predatory are well known and have been described by various entomologists. It is,

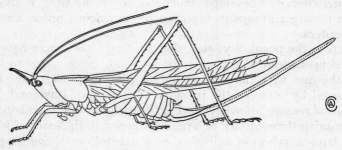

Fig. 33. A phytophagous katydid, *Conocephalus*, which will on occasion become predatory and devour other insects. Redrawn from Blatchley.

therefore, clear that when vegetarian insects actually become carnivorous, they find their masticatory and digestive apparatus quite well fitted for the new mode of life.

In nearly all groups where the predatory habit is well established and widespread we find, however, striking anatomical modifications which assist in the capture and handling of prey. As bipeds we ourselves are able to move more or less steadily on solid earth, and quadrupeds are even more sure-footed. The hexapodous insect, provided with claws, pulvilli, arolia, empodia, cupules, brushes, spurs and other anti-skid devices on its feet so far exceeds other animals in sticking to the substratum that it can easily spare one pair of legs for some other purpose. The front pair is frequently dispensed with for locomotion and may be put to some frivolous use. Thus the male dolichopodid fly whose fore feet sometimes bear a plume or paddle-shaped appendage finds them suitable for a sort of courtship semaphore or "wig-wag" signal apparatus. In the grind of everyday existence, however, the predatory insect frequently utilizes the front legs for the sordid act of seizing prey and for this purpose they have often become exquisitely adapted (*cf.* Reichert '06; Rabaud '23; Cuénot and Poisson '22; Corset '31).

In the mantids already referred to, the front legs are raptorial (see Pl. II B) and suited for grasping the prey, as well as for holding it while it is being devoured, by the pincers-like closure of the tibia against the femur. On account of their large size and consequently great food requirements the mantids are useful in destroying noxious insects (Didlake '26; Thierwolf '28; Weiss '13; Rollinat '26). Such a remarkable resemblance to *Mantis* occurs in a group of Neuroptera that the typical genus has been named *Mantispa*. It has the same long prothorax and pincers-like fore legs and, needless to say, is also predaceous.

In dragonflies the legs are spiny also, with a brush of stiff bristles. They are not greatly modified in form, and are used as a rake or sort of loose basket to retain the prey. As the thorax is less supple, the head has become so freely movable that it may be whirled about dizzily as the insect rushes through the air and catches in rapid succession the small winged insects that form its prey (*cf.* Campion '14; '21). Some large and powerful dragonflies may even capture and devour honey-bees (Goodacre '23), frequenting the vicinity of beehives for this purpose.

In the various predatory Hemiptera the fore-legs are quite generally and uniformly modified for grasping the prey. Here the thin inner edge of the tibia opposes a groove that extends along the femur (Fig. 34). This structure compensates for the absence of mandibles and prehensile mouthparts, for, although the suctorial beak of these insects is admirably suited for impaling the prey and feeding on body-juices, it must be supplemented by some device for holding the victim.

Many Diptera with predaceous habits have the legs spiny and thickened, usually the fore pair, for the same purpose, and these insects likewise lack prehensile mouthparts.

The mouthparts of predatory insects are usually modified in one way or another to facilitate feeding. In mandibulate types where the prey is of comparatively large size, the mandibles are generally well-developed and powerful, especially in those which disable their victims by crushing them in the jaws, or in those that malaxate or tear them into bits after capture. Many structural adaptations of this sort might be cited, as large numbers of insects exhibit them.

Much more interesting and instructive are certain other adaptations whereby equally blood-thirsty predators feed with much greater decorum. In several quite diverse groups of insects we find that the mandibles are sickle-shaped, and acutely pointed, and that they are widely separated where their bases are attached to the head capsule. With such an apparatus, supplied with powerful muscles for closing the jaws, the prey is pierced from opposite sides, the jaws sink in deeply and its juices may be extracted at leisure. When the victim

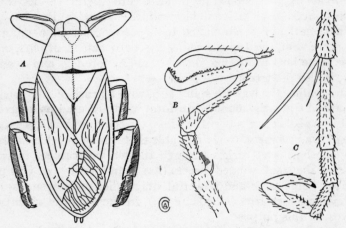

FIG. 34. Modifications of the front legs for grasping. *A*, a predatory bug, *Lethocerus americanus; B*, front tarsus of a parasitic hymenopteron, *Dryinus* ♀ ; *C*, tarsus of a predatory mecopteron, *Bittacus*. *A* after Garman, *B* and *C* original.

forcibly resents such treatment, he may be stilled by the injection of a powerful secretion with proteolytic and paralyzing properties. When the tissues begin to liquefy under the action of this material, a process of extra-intestinal digestion takes place and the partially digested food is withdrawn by the mandibles whence it passes into the mouth. With slight, but interesting variations this process occurs among a series of unrelated mandibulate insects, while a somewhat similar though simpler method is adopted by several haustellate types.

In the larva of the highly predatory waterbeetle, *Dytiscus*, studied by Meinert and later by others (Burgess '82; Rungius '11; Korschelt '24) the sickle-shaped jaws are furnished with a hollow tube near the inner margin which has an opening at the apex (Fig. 35). This tube is not continuous with the mouth-opening, but when the jaws are closed upon the prey the base of the mandibular tube is brought tightly against a lateral opening of the mouth cavity and by a dilatation of the pharynx the liquid food passes into the mouth cavity.

In several families of predatory Neuroptera, including the well known ant-lions (*Myrmeleon*) the larval method of feeding is ex-

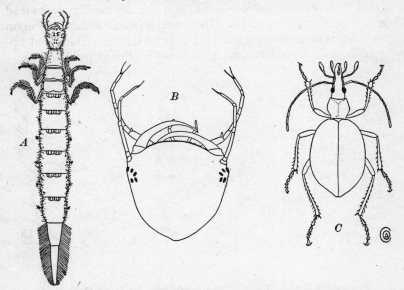

FIG. 35. Modifications of the mouthparts for predatory feeding. *A*, larva of an aquatic dytiscid beetle (*Cybister*); *B*, head of a closely related form (*Dytiscus*); *C*, adult terrestrial ground beetle (*Cychrus*) which has the head long and narrowed to facilitate feeding on snails. *A* and *B* redrawn from Wilson, *C* from Beutenmuller.

tremely similar, but here the mandibular tube is of an entirely different type (Fig. 36). Each mandible is deeply furrowed and into this groove the slender elongate maxilla, which is also grooved, fits tightly, thus forming a tubular substitute through which the fluids may be aspirated (Dewitz '81). The mandibles of *Myrmeleon* are provided in addition with several sharp basal teeth to aid in holding the prey, but in those of *Chrysopa*, *Hemerobius* and other Neuroptera with these habits (Meinert '79; Wildermuth '16; Blanchard '18; Withycombe '25; Pariser '19), they are usually simple and sickle-shaped as in *Dytiscus*. Strangely enough, when confronted by a lack of aphids, these confirmed predators may turn to vegetable food and Rabaud ('26) has actually seen them feeding on nectar which he believes may suffice to bring them to maturity.

More recently Miss Haddon ('15), Bugnion ('22) and Vogel ('15) described the mouthparts of the larval firefly, *Lampyris*, which are somewhat similar. Here the mandibles are perforated by a tube, but this has no direct connection with the mouth which is surrounded by hairs that serve to strain and also, by capillarity, to lead the fluids into the oesophagus. These beetle larvae which feed upon snails have glands opening near the base of the mandibles that appear to secrete materials similar to those produced by *Dytiscus*.

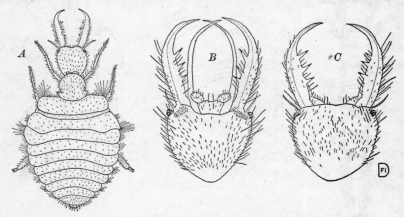

Fig. 36. *A*, larva of an ant-lion (*Myrmeleon*); *B* and *C*, head of the larva from beneath and above; seen from below at *B*, the maxillae lie at each side, separated from the groove on the toothed mandible in which they normally fit. *A* redrawn from Wheeler, *B* and *C* from Doflein.

Compared with such methods for neatly disposing of its meal, the actions of many insects in feeding are as inept as those of the visitor in the tropics when consuming his first mango. In the case of the carabid beetle, *Calosoma sycophanta*, so highly predatory and successful that it has been colonized in the United States as an enemy of the gipsy moth, the beetle makes an incision in the body wall of a caterpillar by means of its sharp jaws and then clumsily extracts the semi-liquid contents. The less active larva of the same beetle (Pl. XIV) crawls up the trunks of trees in search of caterpillars and feeds after much the same fashion burying its head and smaller jaws in the body of its victim (Burgess '10; '11; Burgess and Collins '12; '17).

Countless other predators with similar habits appear to have been equally successful in the struggle for existence, and one wonders why the larvae of *Myrmeleon*, *Dytiscus* and *Lampyris* have found it to their advantage to develop such highly modified mouthparts. This question does not seem to be easily answered and I am at a loss to offer any wholly satisfactory answer. In the case of the aquatic *Dytiscus* larva it seems evident that the watery medium would dissipate the food before it could be swallowed in the ordinary way and that the sand in which the ant-lion lives would absorb the liquids in the same manner. In the case of the *Chrysopa* feeding upon aphids on leaves, however, no such explanation will suffice and the development of the structure here must date back to forbears with different habits. In this connection it must be noted that *Myrmeleon* and *Chrysopa* are related to one another as well as to other Neuroptera with similar mouthparts.

A still more remarkable modification of the mouthparts for a predatory function is seen in the "mask" of the dragonfly nymph (Fig. 37). This instrument is entirely separate from the mandibles, although provided with a pair of pincer-like jaws which serve to seize the prey. It is formed entirely from the labium which may be shot forward in front of the head by the elongated mentum and submentum. Attached to the anterior angles of the mentum are the labial palpi which are blade-like and bear a strong inwardly curved movable spine at the tip. Drawn forcibly together in the horizontal plane these jaw-like palps and spines retain the prey in a relentless grip while the mask is drawn back to bring it into position for the mandibles to act upon it. Although there can be no question concerning the efficacy of this prehensile organ in the Odonata from an examination of its structure, the persistence of the group through long geological periods furnishes additional proof that it is well fitted to hold its own through the ability to secure abundant food.

In comparison with the structures just cited the adaptations to predatory life in some of the most highly specialized insects seem utterly inadequate. We find, for example, that the larvae of all the higher Diptera possess no mandibulate nor jaw-like mouthparts, but that some of them are nevertheless highly predaceous in spite of their evident inferiority. This deficiency is usually compensated by the selection of comparatively helpless prey, as by the aphidophagous larvae of the Syrphidae. These legless vermiform creatures (Fig. 38) occur on the leaves of plants infested by aphids, among which they circulate by a looping motion. The prey is discovered by touch, usually by more or less accidental contact with one of its appendages. The body of the prey is then located and the larva thrusts the anterior part of its body into the aphid, lifting it high

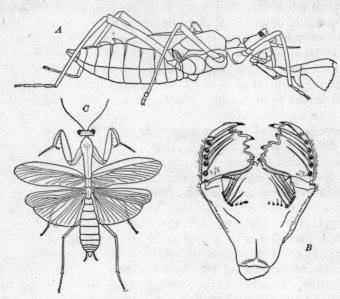

FIG. 37. Modifications for catching and holding prey. *A*, dragon-fly nymph (*Macromia*) with the labial basket extended; *B*, surface view of the opened labium of another dragon-fly (*Cordulegaster*), showing its jaw-like appendages with inner spines and bristles; *C*, a praying mantis (*Hierodula*) showing the heavily spined and raptorial front legs. *A*, redrawn from Howe, *B* from Garman, *C* from Westwood.

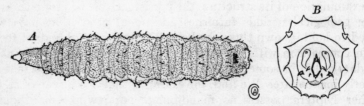

FIG. 38. The predatory larva of a syrphid fly (*Syrphus americanus*). *A*, dorsal view of larva with head at left; *B*, anterior view of head, showing the mouth hooklets. Redrawn from Metcalf.

into the air. A process of scraping and picking with the mouth hooklets now removes the tissues and the empty shell of the prey is cast aside. Thus, feeling its way about like a blind cripple, the syrphid larvae waxes fat and strangely enough fares well in the struggle for existence (*cf.* Krüger '26; Metcalf '12; '16).

Other Diptera, like the larvae of the Tabanidae or horse-flies (Pl. XV A), appear to prey generally upon the grubs of craneflies living in water or mud. The tabanid larva is even less well equipped for predaceous activity than that of the syrphid, but is more than a match for its prey once the latter is located in its bed of muck. Still more surprising is the appearance of predatism even among

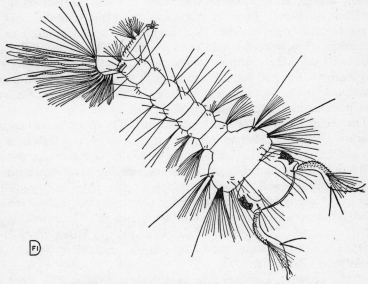

FIG. 39. Predatory larva of a mosquito, *Psorophora discolor*. Redrawn from Knab.

the "leather-jacket" larvae of certain craneflies themselves, for example, the genus *Dicranota*, discovered by Miall many years ago ('93) to be itself carnivorous. Certain mosquito larvae (Fig. 39) are also predatory and here there are clear indications of changes in the antennae and mouthparts in connection with this habit.

We have noticed that in many cases the process of feeding by predatory insects with biting mouthparts becomes modified so that the food is actually ingested in more or less liquid form. As this change has been accomplished independently by several methods and in diverse groups, it is quite evident there are advantages in previously eliminating the indigestible portions of food before it is swallowed, rather than in tearing the prey limb from limb and chewing it in the mouth. We may then as a quite natural sequence consider the modifications of the mouthparts of the haustellate or truly suctorial insects as related to predatism. Among these orders there are only two that include predaceous forms, the Hemiptera and Diptera, and of the latter group only the imagines, since as we have just seen, the larvae are not haustellate.

Occasionally leaf-hoppers will attempt to pierce the skin of the hand or arm if they happen to light thereon, and even aphids will do the same. What may be the cause of such aberrations in behavior is doubtful. Possibly it may be the same stimulus that induces certain flower-visiting flies (bombyliids and syrphids) to lap up perspiration from the skin on a hot day. This is usually attributed to its salty taste, and we might speculate that the insects long ago learned the need for extra salt on such occasions. Such a supposition appears quite fallacious, however, as there is no evidence that insects eliminate salt in any such way as we do.

The Hemiptera show only comparatively slight modifications which may be regarded as adaptations to predaceous habits, since the general type of the beak of these insects seems to be equally well suited for puncturing plant tissues, preying upon other insects or for sucking blood. Many members of the order are predatory (Fig. 34), nearly all of the suborder Cryptocerata, or waterbugs, which embraces several large families depending upon insects, or even sometimes killing small fish (Britton '11; Hungerford '17a; '17b; '19; Torre Bueno '16; Dimmock '86; Essenberg '15a; '15b; Harvey '07; Riley '18; '22).

In the Diptera, however, the mouthparts exhibit the most extreme diversity in the development of their numerous parts and as a vast series of evolutionary stages are well represented by living members of the order, it is obviously beyond the scope of the present discussion to make an even abortive attempt to correlate their structure with predatory habits. The mere fact that many predaceous types are scattered through the order, intermingled with others of entirely different nature, shows that the dipteran proboscis lends itself readily to adaptation for predatory feeding.

Associated with predatism there is an increased rigidity and sometimes, although not necessarily, an elongation of the parts, coupled commonly, as we have already indicated, with prehensile modification of the fore legs, and sometimes also the acquisition of great powers for flight.

Perhaps the most remarkable development is that of certain

dolichopodid flies culminating in the genus *Melanderia* (Fig. 40), discovered by Melander on the sea-beaches of the western United States (Aldrich '22; Snodgrass '22). In this absurd creature which has been jokingly called the mandibulate fly, the apical lobes of the proboscis, known as the labella or paraglossae, have developed a sharp chitinized lobe on the inner side. These lobes, although in no way homologous with the jaws of mandibulate insects, simulate them exactly in form and position and make equally good pincers.

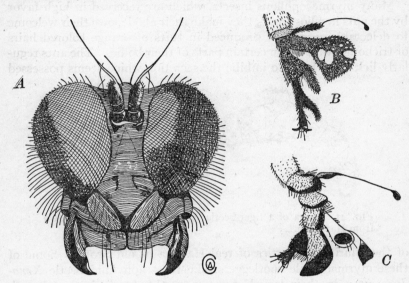

FIG. 40. Head of *Melanderia mandibulata* showing the jaw-like processes which simulate true mandibles; *B* and *C*, hind tarsi of the males of two species of *Calotarsa* (Family Platypezidae) developed as "ornamental" or "wig-wag" structures. Redrawn from Snodgrass (*A*) and Aldrich (*B* and *C*).

This case is particularly instructive as its evolution may be determined from a study of other living members of the family, including a less highly modified member of the same genus. In some other genera the epipharynx is provided with a pair of apical prongs that diverge and serve to hold the prey in a firm grip while its juices are sapped up by the hungry fly. In *Hypocorassus* these become doubled, forming two pairs, a condition retained in *Melanderia*. Finally in another species of *Melanderia* the jaw-like lobes of the labella are developed as a smooth rounded lobe, evidently the precursor of the neatly fashioned pointed "mandible."

Aside from the modification of existing organs a purely adventitious appendage has been developed in the predatory larva of the tiger-beetle, *Cicindela* (Pl. XV B and Fig. 41). This is in the form of a dorsal abdominal tubercle, tipped with a pair of curved spines, which serve to anchor the larva in a burrow which it excavates in the soil. With head protruding, the larva awaits the approach of some passing insect which, when caught, cannot dislodge its captor and is dragged ignominiously into the burrow.

Many myrmecophilous insects, which are received in high favor by the ants in whose nests they make their abode, owe their welcome to delectable secretions produced in tufts of orange colored hairs or trichomes that adorn certain parts of their bodies. The ants regularly lick these hairs to imbibe the secretion which seems possessed

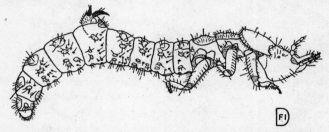

FIG. 41. Larva of a tiger-beetle (*Cicindela limbalis*). Redrawn from Hamilton.

of the alluring properties of real Havana or old Scotch. Some of these myrmecophile bootleggers, like the staphylinid beetle *Xenodusa*, prey in their larval stages upon the developing ant brood, and the golden secretory hairs of the adult are thus to be regarded as predatory adaptations, although so far as the imaginal beetles are concerned they are primarily of conciliatory function.

The climax seems to have been reached by a Javan bug, *Ptilocerus ochraceus* whose behavior as observed by Jacobson ('11) forms a sob-story that would grace the front page of a metropolitan daily or stir the narcotic squad into action, if pulled off in a highly civilized human community like that of Chicago. The *Ptilocerus* bears on the underside of its abdomen a tuft of beautiful fiery red trichomes which serve to conceal the opening of a gland and to disseminate its secretion which is fed upon by a common ant of that region, *Dolichoderus bifurcatus*. After some manoeuvers on the part of the bug the attention of an ant is attracted to this organ which is licked energetically by the unsuspecting victim. After the ant has imbibed this exudate from the surface of the trichomes for a few minutes it becomes apparent that a dose of "knock-out" drops has been included in the potion; the ant staggers and falls. The wily *Ptilocerus* now grasps her with its fore legs and thrusts its beak through a weak spot in the back of her neck. Without further ado the unconscious dupe is quickly sucked dry, its empty shell cast aside and *Ptilocerus* is ready to entice another victim.

Such a detailed account might be drawn out to great length, but there remain several other aspects of the phenomena presented by predatism that deserve brief notice. We have mentioned the fact that the prey of predatory insects consists chiefly of other insects. This is not surprising when one considers that such animals are the most abundant forms encountered in places where they customarily feed. Nevertheless we should expect to find such creatures quite generally making use of land molluscs as food. In spite of the great abundance and diversity of land snails in most regions, they are not preyed upon to any considerable extent, except by a few types of predatory insects. These are mainly certain beetles and Diptera belonging to several families. The North American Carabidae of the genus *Cychrus* (Fig. 35) and its allies devour snails, reaching into the coils of the shell with the slender head and prothorax which appears to be modified for this purpose. Similarly the larvae of several malacoderm beetles of the families Drilidae (Pl. XVI), Telephoridae and Lampyridae feed upon land snails, crawling into the shell and consuming the animal. Likewise an European species of *Silpha* belonging to another family and several dytiscid water-beetles are known to prey commonly on land and fresh-water snails. Among the Diptera, a few Phoridae, flesh flies (Sarcophagidae), and members of other muscoid families attack snails. The latter groups include also some forms which are truly parasitic.[1] Why more insects have not discovered an abundant and satisfactory source of food in the defenseless snail is, as we have said, surprising and it must depend upon some as yet undiscovered dietary or other factors.

How abundantly mites and spiders are eaten by predatory insects cannot be stated. Small or minute mites probably suffer considerable decimation, but spiders are highly immune except from some of the solitary wasps which store them as food for their young. Small crustaceans are taken abundantly by some aquatic insects and

---

[1] For detailed accounts of the predatory enemies of snails the reader is referred to the following papers: Bequaert '25; Bowell '17; Crawshay '03; Hutson and Austin '24; Keilin '19; '21; Lundbeck '23; Morris '22; Schmitz '10; '17.

even Protozoa are included in the diet of the early instar of dragon-flies.

How has the development of predatism proceeded with reference to the evolution of the several groups of insects? This will be most clearly shown by a general consideration of its present occurrence in several of those groups which exhibit it in various forms.

Odonata. Both the nymphs and imaginal Odonata are notorious gourmands partaking lavishly of a great variety of small living animals. Tillyard ('17) indicates a progressive ontogenetic change in diet; the young nymphs feed extensively on Protozoa, later water-fleas and other small crustaceans are caught, and as they become larger, nymphs of may-flies, mosquito larvae or any prey of suitable size is acceptable. The larger nymphs of the Aeschninae frequently capture the less vigorous nymphs of smaller dragon-flies and may even become cannibals, while the smaller, weaker damsel-flies (Balfour-Brown '09) consume various small organisms. The adults capture their food on the wing, taking a great variety of small insects, Diptera, Trichoptera, Hymenoptera, Coleoptera, etc., and Tillyard mentions an Australian *Telephlebia* that had crammed mosquitoes, probably more than a hundred, into its mouth till the jaws could not be closed. Adult dragonflies masticate their food so thoroughly that no very accurate determination of their food can be made from the black pasty mass that has been swallowed.

The food of the nymphs of two large Hawaiian dragon flies has been studied from dissections by Warren ('15). These contained remains of Diptera in great quantity and also of some Coleoptera, Hemiptera, Hymenoptera and of one species of dragon fly. In addition there were numerous small crustaceans, some Protozoa and a few small gastropod molluscs. The most abundant remains were those of chironomid midge larvae and the small crustacean, *Cyclops*, which together constituted fully three-fourths of the food that could be identified. An interesting case of the persistence of this larval food preference in the adult dragonfly has been observed in the case of the New Zealand *Somatochlora smithii*. This species was seen diving after midge larvae [perhaps pupae] on the surface of the water, immersing its head in order to seize them.

In the case of the members of this order it is probable that there is no particular choice in the selection of the food. It appears merely to consist of the most abundant forms of suitable size that occur within easy reach. Thus, in spite of their great age, the dragon-flies have not developed specialized dietaries.

Coleoptera. In the Coleoptera the several families forming the suborder Adephaga are generally conceded to form the most primitive group in the order. These are predominantly predatory so that it appears extremely probable that predatism prevailed generally among the earliest beetles. The same is true of the Lampyridae and related malacoderm beetles, another primitive series that are typically predators as larvae, although the adults frequently subsist to some extent upon pollen. This view of the early appearance of predatism in this order is strengthened by the fact that the saprophagous and mycetophagous families of which there are a quite considerable number are all considerably modified and could not be regarded as derived directly from the ancestral type. Furthermore, the most highly adapted phytophagous families, such as the Chrysomelidae and Rhyncophora where vegetarianism is most firmly established, exhibit more profound anatomical modifications than those found in any other Coleoptera, and are evidently of much more recent origin.

In the very extensive family Carabidae of the Adephaga we find nevertheless some phytophagous species and others that subsist upon a mixed diet. This fact was recognized many years ago by Forbes ('80; '83) and Webster ('80) from an examination of a number of common North American species. Interesting observations made by them show that the mixed diet frequently includes large quantities of fungus spores and pollen, both furnishing abundant nitrogen and probably forming a ready substitute for a meat diet rich in proteins. This combination appears to hold also in some of the highly predatory Coccinellidae, where they found fungus spores and pollen to be very generally combined with insect remains in the stomachs of several species of "lady-birds." These observations have much more recently been confirmed and extended.

The food-habits of the Coccinellidae are extremely interesting, many species are of great economic importance, and some that have been studied in considerable detail serve to illustrate the relation of predatism to phytophagy. The animal food utilized by both the larvae and imagines consists almost entirely of plant-lice and scale insects. Some are general feeders, attacking various species of either group, but a few seem regularly to combine the two. One of the coccidophagous species, the Australian lady-bird, *Rodolia cardinalis* (Howard '93; Balachowsky '32) restricts itself almost exclusively to a single genus of scales, *Icerya*, and its introduction into the United States and other countries as an enemy of the cottony cushion scale was a signal success owing to this fact. Since the family is typically carnivorous it is evident that any vegetable portion of the diet has been more recently added, and it no doubt serves to tide some species over in times of famine, for aphids are particularly erratic in their occurrence. One of our most abundant Coccinellids, *Megilla fuscilabris* (*maculata*) was found by Forbes ('80) to subsist in major part on spores and pollen, and Britton ('14) lists the eggs and larvae of beetles and a moth as represented in its dietary. Thus this mixed diet represents apparently a transition stage, no doubt quite similar to that which occurred at the time predatory insects first arose from phytophagous ones.

Still other members of the family, representing the subfamily Epilachinae have become almost purely herbivorous, and moreover the several species are each entirely dependent upon particular food plants. The association of these secondarily phytophagous beetles with their food-plants is of extreme biological interest on account of the light which it sheds on the relation of instinct to food selection, a topic which has already been dealt with in Chapter III.

Peculiar structures seen in coccinellid larvae are a series of dorsal depressions which function as wax-glands, (Böving '17), and serve to excrete the excessive amount of wax derived from their coccid food. Thus in several genera, notably *Brachyacantha* (Plate XIV) and *Hyperaspis*, the whole body is enveloped in a fluffy waxen envelope. One form, *Brachyacantha quadripunctata*, that lives upon root-coccids in ant nests of a species of *Lasius*, has been described and figured by Wheeler ('11). There seems little doubt in these cases that the wax glands of the beetle-larva have been developed as a result of its food-habits, and that they are adaptive structures which function to eliminate the wax. Observations by Simanton ('16) on *Hyperaspis binotata* serve to substantiate this view, as the white fleece is absent in the first larval instar, but forms in the second after feeding is well under way.

In another family of highly predatory beetles, the Cleridae (Beeson '26; Böving and Champlain '20) the diet is similarly restricted, here to members of the same order, as the whole family appears to prey almost exclusively upon wood-boring and bark-beetles, although the adults of a few species feed regularly upon pollen. This relationship is still further integrated by the fact that the larval clerids feed upon the eggs and brood while the imagines prey upon adults only. This choice of food with reference to the developmental stages of the prey is however not surprising when we recall that the preparatory stages of wood-borers are well protected from free-flying predators, whose larvae in turn could not cope with active prey.

The change from predatory habits to saprophagy and subsequent return to carnivorous feeding is known in the case of certain corynetid beetles, a family which includes forms closely related to and formerly included in the Cleridae. *Necrobia rufipes* is a widely distributed species living in salt hams, skins and other partly dry animal matter and appears to be a true scavenger when food is plenty. According to observations by Taschenberg ('06), however, it frequently becomes carnivorous when pressed by hunger and enters the puparia of muscoid flies, within which it completes its development, although when food is abundant, both species feed together, but independently. Similar behavior has been observed in the case of other beetles and such changes, back and forth are probably far more frequent than actual observations will warrant a statement.

That the food habits of some groups of beetles are still in a state of flux is well illustrated by the conditions that prevail in the Silphidae (Heymons and von Lengerken '26; '28–'29). Thus in the tribe Silphini, to which our common *Silpha* belongs, two genera consist of pure scavengers, while a third occurs more commonly on rotten fungi or excrement; one is phytophagous; one is predatory on caterpillars, two eat snails and one lives on a mixed diet of living plants, captured animals and carrion.

Hymenoptera. The early Hymenoptera were undoubtedly phytophagous and the more primitive living members of the orders still remain so, although the group as a whole exhibits a most varied series of food habits. This ensemble has never been satisfactorily analyzed from the standpoint of the several types of food habits which we have defined, although a number of entomologists have made progress in this direction. The greatest difficulty lies in the interpretation that we make of predatism and parasitism which appear in highly specialized guise and in such diversity that no hard and fast line can be drawn between them. The biologist who attempts to consider the predatory Hymenoptera is therefore led sooner or later to extend his definition of predatism and to regard the parasitism of many Hymenoptera as a relationship which must be distinguished from those to which the name is otherwise generally applied. Here, as is commonly the case with holometabolous insects, the widely different economy of the larva and imago renders the morphological and behavioristic modifications of the two to a large extent independent, although all form a continuous cinema in the life of every individual. If it is to bring into accord such a sequence and to have any evolutionary significance, our interpretation must seize upon the most primitive features of the individual as a starting point. All Hymenoptera are characterized by the possession of a sharp, blade-like, stiletto-like or stylet-like ovipositor, the original purpose of which is to enable the female to deposit her eggs within the tissues of plants or animals. Aside from the ants and bees, the habits, life-histories and behavior of practically the entire order have been moulded with reference to this organ.

In the more primitive Hymenoptera like the saw-flies the ovipositor is used as an instrument to permit the insertion of the eggs within the leaves of plants upon which the larvae feed. In the wood-wasps (Siricoidea) it enables the female to penetrate woody tissue and place her eggs where the larvae may bore in the wood. In the Oryssidae, Ichneumonoidea, Chalcidoidea and Serphoidea it serves to introduce the eggs into the body or upon the surface of an insect host, which serves as the larval food-supply. Finally in the bethyloids and the solitary and social wasps, it is developed, as the sting, into a paralyzing apparatus for subduing the prey which is destined to furnish the larval food. In this new rôle it has determined to a great extent the habits of the higher Hymenoptera, with the exception of the bees.

Where the egg is deposited upon or within an uninjured insect, as occurs in the vast group of Parasitic Hymenoptera, we find the larva in a sense predatory, or as Wheeler ('19) has said, showing a refinement of predatism which he calls parasitoidism. In this group the adult exhibits no predatory activity, unless we regard the feeding at oviposition punctures as an incipient, or perhaps vestigial or even incidental manifestation of predaceous tendencies. In the case of internally feeding larvae this refinement consists in at first feeding upon non-vital portions of the host until development is nearly complete, when the remainder is destroyed and the host dies. In some externally feeding forms with rapid development we find, however, that extra-intestinal digestion by the parasitoid larva occurs whereby the host succumbs quite rapidly through a liquefaction of its body quite like that which occurs in the prey of the larval *Dytiscus*. If these latter forms represent the more primitive habit, we have then a transition stage leading from gross predatism. As is well known these parasitoids are very constantly associated with

PLATE XV

A. Horse-fly and its larva (*Tabanus punctifer*). The predatory larvae live in marshy soil and the adult females are blood-sucking. Photographs by the author.

B. Larva of a tiger-beetle (*Cicindela*) with the jaws open, ready to grasp its prey. From a motion-picture film by Mrs. C. T. Brues.

PLATE XVI

Snail-eating "Trilobite" larva of a Sumatran species of Drilidae. The narrow head of these large beetle larvae enables them to pick the meat out of the snail's shell. Original photographs.

definite hosts, exhibiting again the very great fixity found so generally in insects.

If we consider the activities of the imaginal Hymenoptera in securing their own food, we find that there are practically no predatory forms. A few anthophilous saw-flies sometimes add an occasional insect to their diet, but such food is an unimportant digression. The driver ants, of the genera *Dorylus* and *Eciton*, are, however, the embodiment of insatiable predatism and some other ants depend for food upon insects which they capture, but the significance of such habits is quite obscure unless we take into consideration their relation to the rearing of the brood. This is even more clearly seen in the case of the solitary wasps, where the adults are typically anthophilous and the larvae carnivorous; but as the parent wasp provides the food supply for her offspring she develops highly predatory instincts.

The prey of the solitary wasps is killed outright or paralyzed by the sting of the female wasp, stored in a nest cell and, while in a moribund condition, gradually consumed by the wasp grub usually without further attention on the part of the mother. Here the predatory behavior of the species is separated into two acts, capture by the mother and feeding by her helpless offspring, a perfectly ordinary sequence from the standpoint of the prey, but utterly disrupted so far as the wasp is concerned, and the life of the wasp has been wonderfully modified to adapt itself to this condition. In certain wasps, notably among the social ones that feed malaxated insects to their developing larvae, predatism is restricted to the imagines, as all traces of it disappear in the larva, which is merely fed at intervals upon prepared food.

Thus necessarily omitting any discussion of the marvelous habits of the wasps and reserving those of the Parasitic Hymenoptera for later consideration we may readily realize that the distinctions between predatism and parasitism on the one hand and passive carnivorous feeding on the other are exceedingly difficult to define.

The carnivorous solitary wasps are of particular interest on account of their high specialization in the selection of prey with which they provision their nests. This provender consists mainly of insects of one kind or another, almost entirely living prey, which is stung into insensibility previous to storage. Only one native digger wasp, *Microbembex*, is known to collect dead and broken specimens for this purpose. A quite considerable number, including the members of one large family, the Psammocharidae, collect spiders and one of our common mud-dauber wasps, *Sceliphron* of the family Sphecidae fills its nests with a conglomeration of various small spiders. Other members of the superfamily Sphecoidea store insects of all sorts, large and small. Frequently they take species of sufficient size that a single individual furnishes the needed food to rear the wasp larva to maturity. Others store numerous smaller specimens which are accumulated until the brood cell is stocked with the requisite quantity to supply the complete food requirements of the developing larva. After provisioning, the wasp lays her egg and the nest is sealed. It is now left to its fate as the wasp immediately sets about preparing another cell and will not return to the scene of her previous labors. This process of mass provisioning is not varied, except in rare cases, for example, some of the sand wasps (*Bembix*) store only a part of the food when the nest is first dug, returning later on one or more occasions with another of the flies which the bembicid wasps collect for their offspring.

There is everywhere among the solitary wasps a great constancy in the kinds of prey used for provisioning. Very often only a single species is taken although frequently related insects of similar size may be collected by one species of wasp, and even a single cell may contain such a mixture. Members of each genus commonly restrict their hunting to insects of a single family, but in the aggregate most groups of insects find themselves included in the shopping list of one or several species of wasps. The wasps are such adept hunters that they ordinarily capture their preferred prey with great dispatch. When Nature is forced to ration their food supply, due to abnormal climatic conditions, they may have to spend much more time in the chase, but they are fully as persistent as the housewife seeking butter or beef, and even more fatalistic in refusing substitutes.

Some cases of monophagous wasps are known which parallel closely the vegetarian insects dealt with in Chapter III which restrict their feeding to a single species of plant. One common native wasp, *Aphilanthops frigidus*, possesses an even more old-maidish appetite, as they provision their nests only with virgin queen ants which they capture at the time these are leaving the parental nest to undertake their nuptial flight.

Larval insects, particularly caterpillars, form the prey of some wasps as do the nymphs of many Orthoptera and Hemiptera, although the great majority make use of adult insects.

The fixed habits of the solitary wasps are directly comparable to those of the vegetarian insects in that the choice of food is made by the mother preparatory to the deposition of her egg. On this account they appear possibly to represent a sort of memory and preference for the food enjoyed by the wasp during her juvenile existence, and this has been adduced as evidence in support of such an interpretation. The same difficulty appears, however, in this case, as noted with the vegetarian forms. The food preferences are clearly long-time associations, for we note whole families of wasps exhibiting the preference for closely related types of prey.

Diptera. The successful entry of the members of this order into the field of predatism has not been due to the possession of any inherently fine weapons. As we have seen, the larvae of the higher Diptera are greatly handicapped by the lack of powerful mouthparts as well as by their delicate integument which is unable to withstand the scars and scuffs which a predator may expect to receive, especially when he does not even have any legs on which to skirmish or run away. Those which have become predaceous have therefore necessarily done so by finding a type of prey still more helpless than themselves, or by the development of stealthy methods. They are consequently of erratic occurrence and scattered through the order, just as the blackmailers, porch-window sneak thieves and the like do not form a homogeneous group in human society. One Brazilian form sucks the eggs of scale insects (Borgmeier '31). Similarly the larvae of certain small acalyptrate Diptera are aphidophagous. Thus, two common European species of *Chloropisca* belonging to the family Chloropidae have recently been shown to feed upon root aphids (Mesnil '33) although they had been previously thought to be vegetarian like most other members of this group which includes several important agricultural pests.

A few of the more primitive Diptera have active aggressive larvae; thus certain mosquito larvae are predaceous (see Fig. 39) and make themselves very useful by devouring the wrigglers of other mosquitoes (Gendre '09; Paiva '10).

A number of gall midges of the family Itonididae (Cecidomyiidae) are predatory and Felt ('14) has pointed out the fact that such habits appear in widely separated groups. The prey is restricted to scale insects, aphids, small mites, other gall midges and the like. The larva of one genus, *Aphidoletes* (Fig. 42), has been observed by Davis ('16) as soon as it was hatched, to pierce the body of the aphid, usually from below between the legs. After feeding it leaves the shriveled body of the aphid to attack another and after partial growth usually inserts its mouthparts stealthily into the aphid's leg at one of the articulations. The ensuing process of blood-letting is so neatly accomplished that the aphid seldom appears to notice that it is being bled to death. Other cecidomyiids attack not only aphids, but other small inactive Homoptera and even mites (Barnes '29; '30; '33; Del Guercio '20; Voukassovitch '25; '32). Various other Diptera either as adults or larvae are similarly haematophagous but most of these must be regarded as on the way to a truly parasitic mode of life.

The scattering of predatory groups among the higher Diptera, as larvae or imagines, *e.g.*, in the Tabanidae, Mydaidae, Asilidae, Empididae, Dolichopodidae, Phoridae, Syrphidae, Anthomyiidae and

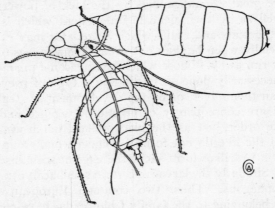

FIG. 42. The predatory larva of an itonidid midge (*Aphidoletes meridionalis*) feeding on an aphid. After Davis.

Drosophilidae, show that larger or smaller series have developed predaceous habits independently. Some larvae that prey on the eggs of spiders are only weakly carnivorous, while some adult flies like the Asilidae are accomplished predators of such piratical nature that they have become terrors in the insect world.

These robber-flies are generally large, powerful creatures and predatism is so firmly established among them that their instincts relating to the selection of prey have become quite firmly fixed (Whitfield '25; Bromley). Many observations have been made upon them, as the family is large, widely distributed, and exhibits many cases of aggressive mimicry in large, conspicuously colored species presumably associated with definite kinds of prey. These were elaborately summarized some years ago by Poulton ('06; see also Melin '23; Ruiz '25), who finds that as a whole these flies prey almost exclusively on Hymenoptera, Diptera, Coleoptera and Lepidoptera, with a frequency indicated by the order given. One surprising fact shown by Poulton's studies is the sporting spirit shown by the robber-flies in choosing worthy antagonists. Among the Hymenoptera practically all were good-sized species, over 75% were aculeates and two-thirds were wasps and bees, the hive bee leading all species by a wide margin; surely a courageous record. Honeybees appear to be especially favored as prey by the dark wasp-like European *Dasypogon diadema* which is also known to capture bumblebees. In America the large species of *Dasyllis* resemble bumblebees closely and have been seen by Plath to prey upon them. Among the dipterous prey it is noticeable that there is a predominance of Asilidae, belonging either to the same or to species other than their captors. True cannibalism thus enters into the economy of the robber-flies where it is usually, although not always, confined to pouncing upon some poor old male who has outlived his usefulness and dies by the hand, or rather the mouth, of his spouse. Observations made by others in America agree in the main with those of Poulton, for example those of Bromley ('23; '30; '42) who found the prey of the splendid large *Proctacanthus rufus* to consist almost entirely of wasps and bees, with a great preponderance of honeybee workers, while that of its smaller relative, *P. brevipennis* included beetles, ants, flies, and bugs. In feeding, the robber-flies thrust the heavily chitinized beak into their victim and suck its body dry, undoubtedly facilitating the process by the injection of a powerful digestive fluid, since Bromley found the musculature and other internal organs of the prey wholly or partly disintegrated after feeding.

Another interesting group of Diptera is the family Empididae, predatory in the imaginal stage, which feeds mainly upon various kinds of small flies. Some years ago, Poulton ('06) noticed that female empidids are much less frequently seen carrying prey than the males, while the opposite is the case among the asilids, although in both families during copulation the female usually is provided with prey. With the asilids he regards this as indicative of a timidity on the part of the male to approach the female if she is not feeding, for she frequently captures males of her own species and devours them. The female empidid does not act in this cannibalistic way, and Poulton was at a loss to account for the still more frequent possession of prey by copulating empid females. More recent observations show that the prey of the empid female has actually been presented to her by her suitor as part of an elaborate courtship (Fig. 43). Nearly a century ago Kirby and Spence suspected that such might be the case as they suggest that the prey may be "designed for the nuptial feast." In these flies we find therefore that predatism has served profoundly to influence the mating habits. Among the innumerable genera and species of muscoid Diptera, a scattering of forms have discarded their primitive vegetarian or saprophagous habits and become predatory in the larval stage. An example of this sort is the common dung-fly, *Scatophaga*. Notwith-

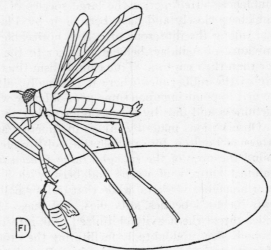

FIG. 43. Male of *Empis poplitea* carrying the "balloon" which contains the food that he is offering to his spouse. Redrawn from Aldrich and Turley.

standing the name originally applied to it, this insect develops at the expense of the larvae of other flies, including those of the housefly (Hewitt '14; Austen '21). The larvae of some still more closely related forms, like *Muscina*, belonging to the same family as the host, prey upon housefly larvae. Certain anthomyiids (*e.g. Hydrotaea*) are similarly predatory (Keilin '17).

In addition to the elaborate anatomical modifications and adaptive behavior of predatory insects, there are a few cases known where traps are actually constructed for the purpose of capturing prey. We are so accustomed to the familiar webs spun by many spiders for the capture of prey that we overlook the really marvelous nature of these structures, so admirably adapted for the purpose of entrapping insects of various kinds (*cf.* Bristowe '30). Many immature insects spin silk which is generally utilized for the construction of cocoons, or protective coverings, but only very rarely as aids in securing food.

Certain caddis-flies of the family Philopotamidae (Alm '17; '26; Wesenberg-Lund '11) utilize their power of spinning silk for the production of nets which serve to entangle food. These are commonly in the form of sacs, much larger than the larva, attached at the base or open end, but with the closed end floating freely in the swiftly moving water. In one of these, *Philopotamus nidificatus*, studied by Thienemann ('08) in Europe, the larva constructs a more unusual type of funnel-shaped net which spans the space between two adjacent stones after the fashion of the aërial webs of spiders.

Another remarkable type of trap adapted for the capture of terrestrial prey by hypogaeic insects are the sand pits excavated by the larvae of the neuropterous ant-lion, *Myrmeleon*, those of the dipterous worm-lion, *Vermileo* (Wheeler '18; '30), and the closely related *Lampromyia*. As is well-known the pit of the ant-lion is dug in dry sand in the form of an inverted cone with the sides as steep as equilibrium will allow. Ants and other insects roll down the insecure slope, and as they frantically search for a foothold, are showered with sand by a dexterous twist of the ant-lion's head, whose jaws await their arrival at the bottom. As we have already noted, these jaws are admirably suited to finish the task. *Vermileo* belongs to the dipterous family Leptidae and has developed similar habits. The pits of *Vermileo* are essentially similar to those of *Myrmeleon* and the larva, although legless and without jaws, manages to cope with its prey in a most remarkably satisfactory way (Pl. XIV).

During the course of the previous discussion we have had occasion in a number of instances to refer to the great care exercised by certain predaceous insects to select particular kinds of prey exclusively, or at least in great part. Their behavior in this respect parallels the selection of specific food-plants by phytophagous species although it is by no means identical. In the latter case the food-plant is really selected by the mother at the time of oviposition and, due to their limited activity, her offspring find it impossible, or at least very inconvenient to deviate from their mother's decision. Such conditions are reproduced exactly in the case of the larvae of solitary wasps, among aphidophagous dipterous larvae, and less rigidly among the more active predators, while others are fully able to cope with almost any type of prey that may cross their path and tempt their appetite. Moreover, the latter may easily vary their diet as fleeting fancy may dictate. We should, therefore, expect the more powerful types to tire of a continuous round of chicken and waffles, and in desperation to raid the first vendor of chile con carne that came within reach. Or, avoiding anthropomorphic bias we might foresee the interpolation of a red-peppered wasp in a protracted diet of mildly flavored butterflies. As we have seen, however, in the case of the robber-flies, a spicy diet of honey-bees, wasps and cantharid beetles has permanently perverted their tastes and ruined their appetite for simpler foods.

Some of these predatory insects have gone so completely to the dogs in this direction that they threaten to wreck some of our pet theories. They appear no longer to be able to distinguish between tasty procryptic forms and distasteful aposematic ones, and Poulton ('06) refers to observations in Africa which show "that the widely mimicked *Limnas chrysippus* has been seen to be devoured by an asilid fly, a large dragonfly and a locustid; while another species of locustid and a large wasp have been found eating the larva."

The queer appetites of certain predatory insects are also of interest in the light that they throw upon the question of aggressive mimicry. Poulton ('06) believes that the resemblance of asilids to their prey, which has often been noticed, is of value in enabling them to approach it, although he finds that only in some cases do they resemble the specific forms which they are known to capture. That is to say, the bumblebee model of *Dasyllis* or *Laphria* is not always, or perhaps only occasionally the prey chosen. Where the resemblance is so very apparent and when such flies capture other bees, wasps or beetles he finds the matter difficult of interpretation. It may, I think, be explained in a much simpler way and in no wise weaken the theory of mimicry. Resemblance to any bee or wasp should afford protection to the fly, if indeed it needs any protection, and likewise make it appear harmless to other aculeates for such do not prey upon one another. From this viewpoint there seems no reason to expect a close resemblance as the advantage of this could not be great.

Incidentally it may be mentioned in connection with the fondness of robber-flies for worker honey-bees that they frequently catch the common drone fly, *Eristalis tenax*, which as its common name denotes is apparently a mimic of the male honey-bee. This probably means, that to the insect eye *Eristalis* really looks like a bee, but on the other hand its resemblance to a bee is unfortunate when there are asilids about.

Predatory insects are almost without exception possessed of insatiable appetites and are accordingly important factors in the economy of the forms upon which they prey. In many cases accurate data are available to show just how much food is required. In the aphis-lion, *Chrysopa californica* a larva may consume thirteen aphids daily, or 141 during its development (Wildermuth '16); another, ovivorous species requires some 500 eggs of the oriental fruit moth for its larval food (Putnam '32) (see also Smith '23). A ladybird beetle, *Coccinella californica,* according to Clausen ('16) requires 475 aphids at the rate of 25 a day for development, and after transforming into an adult beetle eats 34 per day during its life, while the European *C. septempunctata* consumes on the average 267 aphids per day (Boeker '06). Nakayama ('12) found another coccinellid, *Chilocoris similis* still greedier, as its larva required 772 aphids for development and the adult destroyed 791 during its life. Further extensive observations (*e.g.,* Cressman and Dumestre '30; Miller and Thompson '26; '27; Strouhal '26) on other coccinellids show them to have equally voracious appetites. The large beetles of the genus *Calosoma,* notorious as enemies of the gipsy moth caterpillar, are less voracious, or at least their prey is much larger; the smaller native *C. frigidum* requires 14 caterpillars for its larval food and the larger imported *C. sycophanta* needs 41 to become full grown (Burgess '10; '11). The larvae of two syrphid flies reared by Curran ('20) *Allograpta obliqua* and *Syrphus americanus* disposed of 265 and 474 aphids respectively during their larval development. In California, extensive observations on the food of other syrphid larvae by Campbell and Davidson ('24), show that these have equally voracious appetites. The large toll of prey destroyed by insects is not by any means peculiar to them. Predatory mammals like the mountain lion in our western states, regularly account for a large number of prey, authentic records indicating 25–30 calves and colts destroyed by single individuals, and the same is true of wolves and coyotes, which prey extensively on domesticated animals. The importance of predatory insects in reducing the numbers of vegetarian insect pests is very great, but it has been in the past quite generally underestimated. It has been assumed that the specificity of food selection by predators is less definite than that of entomophagous parasites and that they are therefore less reliable aids in the biological control of vegetarian species. Studies of the prey of predatory species, like some of those we have noted, show that this belief is to a great extent contrary to observed facts and considering the large numbers of prey that are destroyed by a single individual during the course of its life we are forced to admit that predatory insects play a primary role in checking the increase of destructive forms. As this question may be more intelligently discussed after the behavior of parasitic insects has been considered, we must defer any more detailed statement until this has been done.

As has been indicated above, many insects which are really predatory are often referred to in a loose way as parasites. When predatory species select a particular type of prey they approach true parasites in another way, namely that a definite host relationship becomes established, something which among other animals is especially associated with parasitism. It thus becomes difficult to establish criteria to distinguish parasites from predators and there is really no sound reason to mourn the fact, for like most classes of biological phenomena there is an intergrading on all sides and there is nothing to be gained in understanding by attempts to define clear-cut distinctions which do not exist in nature.

Certain small Diptera which develop within the egg cocoons of spiders are illustrative of predators which from their relationship to definite hosts and the complete loss of an active larval life bear practically the relation of parasites. One of these, the North American *Phalacrotophora epeirae* of the family Phoridae (Brues '02) occurs commonly in the cocoons of *Aranea* (*Epeira*) and other spiders. Although the specificity of its habits has been questioned by Schwangart ('06), it is undoubtedly restricted to this habitat. Other small flies of the family Oscinidae, *Gaurax araneae* and *Siphonella oscinina*, have been reared in spider cocoons in North America by Coquillett ('98) who believed that the larvae subsist upon the egg-shells. However, observations in Europe by Schwangart ('06) upon either the same or a closely related species of *Siphonella* show that the larvae undoubtedly devour the spider eggs. Even certain caterpillars, as the pyralid, *Dicymolomia julianalis*, feed on egg masses of the bagworm moth. This species is not constant in this habit, however, and appears to be in a transition stage.

In conclusion it may be said that contrary to expectations based on the wide opportunities for varied choice we find that many of the most powerful predators belonging to highly specialized orders, like the asilids and solitary wasps, restrict themselves to a very limited diet and select the same prey throughout life and from one generation to another. As with vegetarian insects, the more primitive types, like the Mantidae and Odonata are highly polyphagous and we find in a general way that evolution has tended to narrow rather than to widen the dietary of predatory insects. In consequence it has introduced the selection of specific prey more closely to integrate the balance of nature, invoking a large number of factors to determine the abundance and fluctuations in numbers of carnivorous types. Thus again, predatory insects clearly outrank other animals of prey in the complexity of their interdependence.

REFERENCES

Aldrich, J. M.

1922 A New Genus of Two-winged Flies with Mandible-like Labella. Proc. Entom. Soc. Washington, vol. 24, pp. 145–150.

Alm, G.

1917 Till Kännedomen om de nätspinnande Trichopter-larvenas Biologi. Entom. Tidskr., vol. 38, pp. 285–297, 1 pl.

1926 Beiträge zur Kenntnis der netzspinnenden Trichopteren-Larven in Schweden. Intern. Rev. Hydrobiol., vol. 14, pp. 233–275, 2 pls., 14 figs.

Austen, E. E.

1921 The Prey of the Yellow Dung Fly, *Scatophaga stercoraria*. Ann. Mag. Nat. Hist. London (8), vol. 13, pp. 118–123.

Ayyar, T. V. R. (see Ramakrishna Ayyar)

Balachowsky, A.

1926 Note sur l'acclimatisation des prédateurs du *Parlatoria blanchardi* dans la palmérie de Colomb-Bechar. Bull. Soc. Hist. Nat. Afrique du Nord, vol. 17, pp. 93–96.

1932 Observations biologiques sur l'adaptation de *Novius cardinalis* aux dépens de *Gueriniella serrulatae*. Rev. Path. vég. Entom. agric., vol. 19, pp. 11–17.

Balduf, W. V.

1935 The Bionomics of Entomophagous Coleoptera. i + 220 pp., 108 figs. St. Louis, Mo., John S. Swift Co.

1938 The Rise of Entomophagy among Lepidoptera. American Naturalist, vol. 72, pp. 358–379.

1939 Food Habits of *Phymata pennsylvanica americana*. Canadian Entom., vol. 71, pp. 66–74.

1940a More Ambush Bug Prey Records. Bull. Brooklyn Entom. Soc., vol. 35, pp. 161–169.

1940b Ambush Bug Studies. Trans. Illinois Acad. Sci., vol. 33, pp. 206–208.

Balfour-Browne, F.

1909 The Life-history of the Agrionid Dragon Fly. Proc. Zool. Soc. London, 1909, pp. 253–285, 2 pls., 1 chart.

Banks, N.

1904　A Treatise on the Acarina or Mites.　Proc. U. S. Nat. Mus., vol. 28, 114 pp., 201 figs.

1915　The Acarina or Mites.　Rept. Off. Sec. U. S. Dept. Agric., No. 108, 153 pp., 294 figs.

Barnes, H. F.

1929　Gall Midges as Enemies of Aphids.　Bull. Entom. Res., vol. 20, pp. 433–442.

1930　Gall Midges (Cecidomyidae) as Enemies of Tingidae, Aleyrodidae and Coccidae.　*Ibid.*, vol. 21, pp. 319–329.

1933　Gall Midges as Enemies of Mites.　*Ibid.*, vol. 24, pp. 215–228.

Barth, G. P.

1907　Observations on the Nesting Habits of *Gorytes canaliculatus*.　Bull. Wisconsin Nat. Hist. Soc., vol. 5, pp. 141–149, 3 figs.

Beeson, C. F. C.

1926　Notes on the Biology of the Cleridae.　Indian Forest Records, vol. 12, pp. 217–231.

Bequaert, J.

1922　The Predaceous Enemies of Ants.　Bull. American Mus. Nat. Hist., vol. 45, pp. 271–331, 3 pls.

1925　The Arthropod Enemies of Mollusks, with Description of a New Dipterous Parasite from Brazil.　Journ. Parasitology, vol. 11, pp. 201–212, 1 fig.

Bevis, A. L.

1923　A Lepidopterous Parasite on a Coccid.　South African Journ. Nat. Hist., vol. 4, pp. 34–35.

Bhatia, M. L.

1939　Biology, Morphology and Anatomy of Aphidophagous Syrphid Larvae.　Parasitology, vol. 31, pp. 78–129, 75 figs.

Blanchard, R.

1918　Larves de Neuroptères éventuellement hématophages.　Bull. Soc. Path. exot., vol. 11, pp. 586–592, 5 figs.

Blunck, H.

1918　Die Entwicklung des *Dytiscus marginalis*.　Zeits. wiss. Zool., vol. 117, pp. 1–129.

Bordas, L.

1917　Sur le régime alimentaire de quelques Vespinae (*Vespa crabro*).　Insecta, Rennes, vol. 7, pp. 5–7.

Borgemeier, T.

1931　Eine neue Itonididengattung aus S. Paulo.　Revist. Entom., vol. 1, pp. 184–191.

Böving, A.

1917　A Generic Synopsis of the Coccinellid Larvae in the U. S. National Museum.　Proc. U. S. Nat. Mus., vol. 51, pp. 621–650, pls. 118–121.

Böving, A., and A. B. Champlain

1920　Larvae of North American Beetles of the Family Cleridae.　Proc. U. S. Nat. Mus., vol. 57, pp. 575–649, 12 pls.

Bowell, E. W.

1917   Larva of a Dipterous Fly Feeding on *Helicella itala*. Proc. Malacol. Soc. London, vol. 12, p. 308.

Bristowe, W. S.

1925   Solitary Wasps and their Prey, with Special Reference to the Mantid-hunters. Ann. Mag. Nat. Hist., (9) vol. 16, pp. 278–284.

1930   Notes on the Biology of Spiders. The Evolution of Spiders' Snares. *Ibid.* (10) vol. 6, pp. 334–342, 2 figs.

Britton, W. E.

1911   A Hemipterous Fisherman. Entom. News, vol. 22, pp. 372–373.

1914   Some Common Lady Beetles of Connecticut. Bull. Connecticut Agric. Expt. Sta., New Haven, No. 181, 24 pp., 23 figs.

Bromley, S. W.

1914   Asilids and their Prey. Psyche, vol. 21, pp. 192–198.

1923   Observations on the Feeding Habits of Robber Flies. Psyche, vol. 30, pp. 41–45.

1930   Bee-killing Robber Flies. Journ. New York Entom. Soc., vol. 38, pp. 159–176, 1 pl.

1932   Observations on the Chinese Mantid, *Paratenodera sinensis*. Bull. Brooklyn Entom. Soc., vol. 27, pp. 196–201.

1934   The Robber Flies of Texas (Diptera, Asilidae). Ann. Entom. Soc. America, vol. 27, pp. 74–110.

1942   Bee-killing Asilids in New England. Psyche, vol. 49, pp. 81–83.

Brues, C. T.

1902   Notes on the Larvae of Some Texan Diptera. Psyche, vol. 10, pp. 351–354.

1936   Aberrant Feeding Behavior among Insects and its Bearing on the Development of Specialized Food Habits. Quart. Rev. Biol., vol. 11, pp. 305–319.

Bugnion, E.

1922   Études rélatives à l'anatomie et à l'embryologie des vers luisants ou Lampyrides. Bull. Biol. France et Belgique, vol. 56, Fasc. 2, pp. 1–53.

1928   Un grand vol de Libellules. Bull. Entom. Soc. France, 1928, pp. 242–244.

Burgess, A. F.

1910   Notes on *Calosoma frigidum* Kirby, a Native Beneficial Insect. Journ. Econ. Entom., vol. 3, pp. 217–222.

1911   *Calosoma sycophanta:* Its Life-history, Behavior and Successful Colonization in New England. Bull. U. S. Dept. Agric., Bur. Entom., No. 101, 94 pp., 9 pls., 22 figs.

Burgess, A. F., and C. W. Collins

1912   The Value of Predaceous Insects in Destroying Insect Pests. Yearbook U. S. Dept. Agric. for 1911, pp. 453–466, 6 pls., 6 figs.

1915   The Calosoma Beetle (*Calosoma sycophanta*) in New England. Bull. U. S. Dept. Agric., No. 251, 40 pp.

1917   The Genus *Calosoma*. Bull. U. S. Dept. Agric., No. 417, 124 pp., 18 pls., 3 figs.

Burgess, E.

1882   The Structure of the Mouth in the Larva of *Dytiscus*. Proc. Boston Soc. Nat. Hist., vol. 21, pp. 223–228.

Camargo, F. C.

1937 Notas Taxonómicas e Biológicas sobre alguns Coccinellideos do Género Neocalvia, Predadores de Larvas do Género *Psyllobora*. Rev. Entom., vol. 7, pp. 362–377.

Campbell, R. E., and W. M. Davidson

1924 Notes on Aphidophagous Syrphidae of Southern California. Bull. S. California Acad. Sci., vol. 23, pp. 3–9; 59–71, 4 pls., 3 figs.

Campion, H.

1914– Some Dragonflies and their Prey. Ann. Mag. Nat. Hist., (8) vol. 13, pp. 496–
21   504; *ibid.*, (9) vol. 8, pp. 240–245.

Candura, G. S.

1932 Contributo alla conoscenza morfologica e biologica dello struggigrano (*Tenebrioides mauritanicus*). Boll. Lab. Zool. Portici, vol. 27, pp. 1–56, 18 figs.

Carpenter, F. M.

1936 Revision of the Nearctic Raphidiodea. Proc. American Acad. Arts and Sci., vol. 71, pp. 89–157.
1940 Revision of the Nearctic Hemerobiidae, Berothidae, Sisyridae, Polystoechotidae and Dilaridae. *Ibid.*, vol. 74, pp. 193–280.

Cherian, M. C.

1923 An Agromyzid Fly Predaceous on Aphids. Madras Agric. Journ., vol. 11, pp. 343–344.

Cherian, M. C., and V. Mahadevan

1937 A New Enemy of the Indian Honey Bee. *Ibid.*, vol. 25, pp. 65–67.

China, W. E.

1932 Reduviid Bugs (Apiomerinae) Capturing their Insect Prey by Means of Adhesive Resin-covered Fore-legs. Proc. Entom. Soc. London, vol. 7, p. 12.

Clark, A. H.

1926 Carnivorous Butterflies. Ann. Rept. Smithsonian Inst. for 1925, pp. 439–508.

Clausen, C. P.

1916 Life History and Feeding Records of a Series of California Coccinellidae. Univ. of California Publ. Tech. Bull., vol. 1, No. 6, pp. 251–299.

Cook, O. F.

1913 Web-spinning Fly Larvae in Guatemalan Caves. Journ. Washington Acad. Sci., vol. 3, pp. 190–193.

Copello, A.

1927 Biologia del moscardon cazador de abejas (*Mallophora ruficauda*). Circ. Minist. Agric. Argentina, No. 699, 19 pp., 10 figs.

Coquillett, D. W.

1891a Another Parasitic Rove Beetle. Insect Life, vol. 3, pp. 318–319.
1891b Predaceous Habit of Histeridae. *Ibid.*, vol. 4, p. 76.
1898 Habits of the Oscinidae and Agromyzidae, Reared at the United States Department of Agriculture. Bull. U. S. Dept. Agric., Bur. Entom., No. 10, pp. 70–79.

Corset, Jean

1931 Les coaptations chez les insectes. Suppl. 13, Bull. Biol. France et Belgique, 337 pp.

Cottam, R.

1922 Notes on the Bionomics of an Aphidophagous Fly of the Genus *Leucopis* in the Anglo-Egyptian Sudan. Entom. Monthly Mag., (3) vol. 8, pp. 61–64.

Couturier, A.

1938 Contribution a l'étude biologique de *Podisus maculiventris*, prédateur Américain du doryphore. Ann. Epiphyt. Phytog´nét., vol. 4, pp. 96–195.

Crawshay, L. R.

1903 On the Life-history of *Drilus flavescens*. Trans. Entom. Soc. London, 1903, pp. 39–51, 2 pls.

Cressman, A. W., and J. O. Dumestre

1930 The Feeding Rate of the Australian Lady Beetle, *Rodolia cardinalis*. Journ. Agric. Res., vol. 41, pp. 197–203.

Cuénot, L., and R. Poisson

1922 Sur le développement de quelques coaptations des insectes. C. R. Acad. Sci., Paris, vol. 175, pp. 463–466, 5 figs.

Curran, C. H.

1920 Observations on the more Common Aphidophagous Syrphid-flies. Canadian Entom., vol. 52, pp. 53–55.

Davis, J. J.

1916 *Aphidoletes meridionalis*, an Important Dipterous Enemy of Aphids. Journ. Agric. Res., vol. 6, pp. 883–888, 4 figs., 1 pl.

DeLeon, D.

1935 A Study of *Medeterus aldrichii*, a Predator of the Mountain Pine Beetle (*Dendroctonus monticolae*). Entomologia Americana, vol. 15, pp. 59–90.

Del Guercio, G.

1920 Cecidomyid Enemies of Aphids. L'Agric. Colon., Florence, vol. 13, pp. 31–62, 31 figs. (Abstract in Rev. Appl. Entom., vol. 8A, p. 159.)

Dewitz, H.

1881 Die Mundtheile der Larve von Myrmeleon. S. B. Ges. naturf. Freunde Berlin, 1881, pp. 163–166.

Didlake, Mary

1926 Observations on the Life-histories of Two Species of Praying Mantis. Entom. News, vol. 37, pp. 169–174, 2 pls.

Dimmock, G.

1886 Belostomidae and Some Other Fish-destroying Bugs. Ann. Rept. Fish and Game Comm. Massachusetts, Public Document No. 25, pp. 67–74, 1 fig.

Dodd, A. P.

1902 Contribution to the Life-history of *Liphyra brassolis*. Entomologist, vol. 35, pp. 153–156; 184–188.

Dow, Richard

1942   The Relation of the Prey of *Sphecius speciosus* to the Size and Sex of the Adult Wasp. Ann. Entom. Soc. America, vol. 35, pp. 310–317.

Edwards, F. W.

1934   The New Zealand Glowworm. Proc. Linn. Soc. London, pp. 3–10.

Edwards, W. H.

1886   On the Life-history and Preparatory Stages of *Feniseca tarquinius*. Canadian Entom., vol. 18, pp. 141–153.

Essenberg, C.

1915   The Habits and Natural History of the Back-swimmers, Notonectidae. Journ. Anim. Behavior, vol. 5, pp. 38–39.

1915   The Habits of the Water Strider *Gerris remiger*. *Ibid.*, vol. 5, pp. 397–402.

Faguiez, C.

1934   Une invasion de *Chloridea peltigera*, Lepidoptère nuisible à la sauge sclarée, arrêtée par l'intervention de *Polistes gallicus*. Rev. franç. Entom., vol. 1, pp. 27–28.

Fell, H. B.

1940   Economic Importance of the Australian Ant, *Chalcoponera metallica*. Nature, vol. 145, p. 707.

Felt, E. P.

1914   List of Zoophagous Itonididae. Journ. Econ. Entom., vol. 7, pp. 458–459.

Ferton, Ch.

1890   L'évolution de l'instinct chez les Hyménoptères. Rev. Scient., vol. 45, pp. 496–498.

Fluke, C. L.

1929   The Known Predacious and Parasitic Enemies of the Pea Aphid in North America. Res. Bull. Wisconsin Agric. Expt. Sta., No. 93, 47 pp.

Forbes, S. A.

1880   Notes on Insectivorous Coleoptera. Bull. Illinois State Lab. Nat. Hist., vol. 1, pp. 167–176.

1883   The Food Relations of the Carabidae and Coccinellidae. Bull. Illinois State Nat. Hist., vol. 1, pp. 33–64.

Fulton, B. B.

1915   The Tree Crickets of New York: Life History and Bionomics. Tech. Bull. New York Agric. Expt. Sta., No. 42, 47 pp.

1939   Lochetic Luminous Dipterous Larvae. Journ. Elisha Mitchell Sci. Soc., vol. 55, pp. 289–293.

1941   A Luminous Fly Larva with Spider Traits (Diptera; Mycetophilidae). Ann. Entom. Soc. America, vol. 34, pp. 289–302, 2 pls.

Gage, J. H.

1920   The Larvae of the Coccinellidae. Illinois Biol. Monogr., vol. 6, pp. 1–62.

Gahan, A. B.

1909   A Moth Larva Predatory on the Eggs of the Bagworm. Journ. Econ. Entom., vol. 2, pp. 236–237.

Gendre, E.

1909 Les larves carnivores de deux espèces de moustiques. Bull. Soc. Path. Exot., vol. 2, pp. 147–150.

Goodacre, W. A.

1923 A Casual Enemy of the Bee. The Dragon Fly (*Hemianax papuensis*). Agric. Gaz. New South Wales, vol. 34, pp. 373–374, 1 fig.

Grandi, G.

1937 Morfologia ed etologia comparate di insetti a regime specializzoto. Bol. Ist. Super. Agric., Lab. Entom., vol. 9, pp. 33–64.

Hadden, F. C.

1927 A List of Insects Eaten by the Mantis, *Paratenodera sinensis*. Proc. Hawaiian Entom. Soc., vol. 6, pp. 385–386.

Haddon, K.

1915 On the Methods of Feeding and the Mouthparts of the Larva of the Glow-worm. Proc. Zool. Soc. London, 1915, pp. 77–82, 1 pl.

Haddow, M. B.

1942 Note on the Predatory Larva of the Mosquito, *Culex tigripes*. Proc. Entom. Soc. London, vol. 17A, pp. 73–74.

Hamm, A. H.

1926 Biology of the British Crabronidae. Trans. Entom. Soc. London, vol. 74, pp. 297–331.

Hartman, C. G.

1905 Observations on the Habits of Some Solitary Wasps in Texas. Bull. Univ. Texas Sci. Ser., No. 6, 72 pp., 4 pls.

Hartzell, A.

1935 Histopathology of Nerve Lesions of Cicada after Paralysis by the Killer-wasp. Contrib. Boyce Thompson Inst., vol. 7, pp. 421–425.

Harvey, G. W.

1907 A Ferocious Water Bug. Proc. Entom. Soc. Washington, vol. 8, pp. 72–75.

Heiss, E. M.

1938 A Classification of the Larvae and Puparia of the Syrphidae of Illinois, Exclusive of Aquatic Forms. Illinois Biol. Monogr., vol. 16, pp. 1–142.

Hess, W. N.

1920 Notes on the Biology of some Common Lampyridae. Biol. Bull., vol. 38, pp. 39–76.

Hewitt, G. C.

1914 On the Predaceous Habits of *Scatophaga*, a New Enemy of *Musca domestica*. Canadian Entom., vol. 46, pp. 2–3.

Heymons, R., H. v. Lengerken, and M. Bayer

1926– Studien ueber die Lebenserscheinungen der Silphini. III and IV. Zeits.
32 Morph. Oekol. d. Tiere, vol. 6, p. 287; vol. 10, pp. 330–352; vol. 14, pp. 235–260; vol. 24, pp. 259–287.

Hinds, W. E.

1907   An Ant Enemy of the Cotton Boll Weevil. Bull. Bur. Entom. U. S. Dept. Agric., No. 63, pp. 45–48, 2 figs.

Hinman, E. H.

1934   Predators of the Culicidae. Journ. Trop. Med. and Hyg., vol. 10, pp. 145–150.

Hobby, B. M.

1931   The British Species of Asilidae and their Prey. Trans. Entom. Soc. So. England for 1930, pp. 1–42.

1932   The Prey of Sawflies. Proc. Entom. Soc. London, vol. 7, pp. 14–15; 35–36.

1933a   Supplementary List of the Prey of Asilidae. Journ. Entom. Soc. So. England, vol. 1, pp. 69–77.

1933b   The Prey of British Dragonflies. Proc. Entom. Soc. So. England, vol. 8, pp. 65–76.

1934   Notes on Predaceous Anthomyiidae. Entom. Monthly Mag., vol. 70, pp. 185–190.

Hollinger, A. H., and H. B. Parks

1919   *Euclemensia bassettella*, the *Kermes* Parasite. Entom. News, vol. 30, pp. 91–100.

Horsfall, W. R.

1941   Biology of the Black Blister Beetle. Ann. Entom. Soc. America, vol. 34, pp. 114–126, 1 pl.

Howard, L. O.

1893   An Important Predatory Insect. Insect Life, vol. 6, pp. 6–10, 1 fig.

Hungerford, H. B.

1917a   Notes Concerning the Food-supply of some Water Bugs. Science, vol. 45, pp. 336–337.

1917b   Food Habits of Corixids. Journ. New York Entom. Soc., vol. 25, pp. 1–5, 1 pl.

1919   The Biology and Ecology of Aquatic and Semiaquatic Hemiptera. Kansas Univ. Sci. Bull., vol. 11, pp. 3–328, 30 pls.

1936   The Mantispidae of the Douglas Lake, Michigan Region, with some Biological Observations. Entom. News, Vol. 47, pp. 69–72; 85–88, 1 pl.

Hutson, J. C., and G. D. Austin

1924   Notes on the Habits and Life-history of the Indian Glow-worm, an Enemy of the African or Kalutara Snail. Bull. Dept. Agric., Ceylon, No. 69, 16 pp., 1 pl.

Illingworth, J. F.

1929   Grasshoppers Eat Pineapple Mealy Bugs and other Pests. Proc. Hawaiian Entom. Soc., vol. 7, pp. 256–257.

Jacobson, E.

1911   Biological Notes on the Hemipteron, *Ptilocerus ochraceus*. Tijdschr. v. Entom., vol. 54, pp. 175–179.

1913   Biological Notes on the Heterocera; *Eublemma rubra, Catoblemma sumbavensis* and *Eublemma versicolora*. *Ibid.*, vol. 56, pp. 165–178.

Jordan, K.

1926   On a Pyralid Parasitic as Larva on Spiny Saturnian Caterpillars at Pará. Novitates Zool., vol. 33, pp. 367–370.

Karny, H. H.

1934   Biologie der Wasserinsekten. 311 pp. Wagner, Wien.

Kaston, B. J.

1937   Dipterous Parasites of Spider Egg Sacs. Bull. Brooklyn Entom. Soc., vol. 22, pp. 160–165, 21 figs.

Keilin, D.

1917   Recherches sur les Anthomyides à larves carnivores. Parasitology, vol. 9, 125 pp., 11 pls., 41 figs.

1919   On the Life-history and Larval Anatomy of *Melinda cognata* Parasitic in the Snail *Helicella virgata*, with an Account of Other Diptera Living upon Mollusks. *Ibid.*, vol. 11, pp. 430–455, 4 pls., 6 figs.

1921   Supplementary account of the Dipterous Larvae Feeding upon Mollusks. *Ibid.*, vol. 13, pp. 180–183.

Kershaw, J. C. W.

1905   The Life History of *Gerydus chinensis*. Trans. Entom. Soc. London, 1905, pp. 1–4, 1 pl.

Killington, F. J.

1925   Notes on the Prey of Dragonflies. Entomologist, vol. 58, pp. 181–183.

Korschelt, E.

1924   Der Gelbrand, *Dytiscus marginalis*. Bearbeitung einheimischer Tiere, Erste Monographie, vol. 2, 964 pp.

Krishnamurti, B.

1933   On the Biology and Morphology of *Epipyrops eurybrachydis*. Journ. Bombay Nat. Hist. Soc., vol. 36, pp. 944–949.

Krüger, F.

1926   Biologie und Morphologie einiger Syrphidenlarven. Zeits. Morph. Ökol. d. Tiere, vol. 6, pp. 83–149.

Lamborn, W. A.

1914   On the Relationship between Certain West African Insects, Especially Ants, Lycaenidae and Homoptera. Trans. Entom. Soc. London, 1913, pp. 436–498.

1920   The Habits of a Dipteron Predaceous on Mosquitoes in Nyasaland. Bull. Entom. Res., vol. 11, pp. 279–281.

Landis, B. J.

1937   Insect Hosts and Nymphal Development of *Podisus maculiventris* and *Perillus bioculatus*. Ohio Journ. Sci., vol. 37, pp. 252–259.

Lengerken, H. von

1921   *Carabus auratus* und seine Larve. Archiv Naturgesch., vol. 87A, No. 3, pp. 31–113.

Lopes, H. de Souza

Sobre duas especias de *Sarcophaga* cujas larvas são predadores. Revista Entom., vol. 5, pp. 470–479, 2 figs.

Lundbeck, W.

1923   Some Remarks on the Biology of the Sciomyzidae, with the Description of a New Species of *Ctenulus* from Denmark. Vidensk. Medd. Dansk Natur. Foren., vol. 76, pp. 101–109.

Lyon, M. B.

1915 The Ecology of the Dragonfly Nymphs of Cascadilla Creek. Entom. News. vol. 26, pp. 1–15.

McKenzie, H. L.

1932 The Biology and Feeding Habits of *Hyperaspis lateralis*. California Univ. Pubs., Entom., vol. 6, pp. 9–20.

Marshall, G. A. K.

1902 Experiments on Mantidae in Natal and Rhodesia. Trans. Entom. Soc. London, 1902, pp. 297–315.

Mathur, R. N., C. F. C. Beeson, and S. N. Chatterjee

1934 On the Biology of the Mantidae. Indian Forest Rec., vol. 20, No. 3, 27 pp., 1 pl.

Mattos, A. T. de

1905 The Hunting Wasps. New York, Dodd, Mead and Co.

Melander, A. L., and C. T. Brues

1906 The Chemical Nature of Some Insect Secretions. Bull. Wisconsin Nat. Hist. Soc., vol. 4, pp. 22–36.

Melin, Douglas

1923 Contributions to the Knowledge of the Biology, Metamorphosis and Distribution of the Swedish Asilids in Relation to the Whole Family of Asilids. Zool. Bidrag, vol. 8, pp. 1–317, 305 figs.

Mesnil, L.

1933 Sur deux Chloropides considérés à tort comme nuisible. Rev. Path. Vég. Entom. Agric., vol. 20, pp. 3–7, 1 pl.

Metcalf, C. L.

1912 Life Histories of Syrphidae, III, IV. Ohio Naturalist, vol. 12, pp. 477–488, 1 pl., pp. 533–541, 1 pl.

1916 Syrphidae of Maine. Bull. Maine Agric. Expt. Sta., No. 253, pp. 193–264, 9 pls., 1 fig.

Miall, L. C.

1893 *Dicranota*, a Carnivorous Tipulid Larva. Trans. Entom. Soc. London, 1893, pp. 235–253, 3 pls.

Miller, P. L., and W. L. Thompson

1926– Life Histories of Lady-beetle Predators of the Citrus Aphid. Florida Entom.,
27 vol. 10, pp. 40–46; 57–59; vol. 11, pp. 1–8.

Misra, C. S.

1924 A Preliminary Account of the Tachardiphagous Noctuid Moth, *Eublemma amabilis*. Rept. Proc. 5th. Entom. Meeting, Pusa, pp. 238–247.

Misra, M. P., P. S. Negi, and S. N. Gutpa

1930 The Noctuid Moth (*Eublemma amabilis*); a Predator of the Lac Insect and its Control. Journ. Bombay Nat. Hist. Soc., vol. 34, pp. 431–446.

Morgan, A. C.

1907 A Predatory Bug Reported as an Enemy of the Cotton Boll Weevil. Bull. Bur. Entom. U. S. Dept. Agric., No. 63, pt. 4, pp. 49–54, 9 figs.

Morris, H. M.

1922 On the Larva and Pupa of a Parasitic Phorid Fly, *Hypocera incrassata.* Parasitology, vol. 14, pp. 70–74, 1 pl., 4 figs.

Morris, K. R. S.

1938 *Eupelmella vesicularis* as a Predator of Another Chalcid. *Ibid.*, vol. 30, pp. 20–32, 5 figs.

Nechleba, A.

1929 Natürlicher Schutz von Scolytiden-Bruten gegen Raubinsekten. Anz. f. Schädlingsk., vol. 5, pp. 24–26.

Needham, J. G., and H. B. Heywood

1929 A Handbook of the Dragonflies of North America, 378 pp. C. C. Thomas, Springfield, Ill.

Noble, N. S.

1936 *Pristhesancus papuensis,* an Assassin Bug. Journ. Australian Inst. Agric. Sci., vol. 2, pp. 124–126, 1 fig.

Noyes, Alice A.

1914 Biology of the Net-spinning Trichoptera of Cascadilla Creek. Ann. Entom. Soc. America, vol. 7, pp. 251–271.

Okada, Y. K.

1928 Two Japanese Aquatic Glow-worms. Trans. Entom. Soc. London, vol. 76, pp. 101–108.

Oldenberg, L.

1924 Die bei *Eriopeltis lichtensteini* schmarotzende *Leucopis.* Deuts. Entom. Zeits., 1924, pp. 448–450.

Paiva, C. A.

1910 Notes on the Larvae of *Toxorhynchites.* Rec. Indian Mus., vol. 5, pp. 187–190.

Pariser, K.

1919 Beiträge zur Biologie und Morphologie der einheimischen Chrysopiden. Arch. Naturg., Abt. A, vol. 83, pp. 1–57, 2 pls., 26 figs.

Payne, O. G. M.

1916 On the Life History and Structure of *Telephorus lituratus.* Journ. Zool. Res., vol. 1, pp. 4–32.

Pearman, J. V.

1932 Some Coccophagous Psocids (Psocoptera) from East Africa. Stylops, vol. 1, pp. 90–96.

Peckham, G. W., and E. G. Peckham

1898 On the Instincts and Habits of the Solitary Wasps. Wisconsin Geol. Nat. Hist. Surv., Bull. No. 2, iv + 245 pp.

1899 The Instincts of Wasps as a Problem in Evolution. Nature, vol. 59, pp. 466–468, 2 figs.

1905 Wasps, Social and Solitary. Second Edition, xv + 311 pp. London.

Pemberton, C. E.

1928 Thysanuran Predatory on Eggs and Immature Forms of Termites in Borneo. Proc. Hawaiian Entom. Soc., vol. 7, p. 147.

Perkins, R. C. L.

1905 Leafhoppers and their Natural Enemies. Bull. Hawaiian Sugar Planters' Expt. Sta., No. 1, pp. 75–85.

Picard, F.

1903 Recherches sur l'éthologie du *Sphex maxillosus*. Mém. Soc. Sci. Nat., Cherbourg, vol. 33, pp. 97–130.

Plank, H. K.

1934 Some Predatory Habits of the Orange Bagworm, *Platoeketicus gloverii*. Bull. California State Dept. Agric., vol. 23, pp. 207–208.

Plummer, C. C., and B. J. Landis

1932 Records of Some Insects Predacious on *Epilachna corrupta* in Mexico. Ann. Entom. Soc. America, vol. 25, pp. 695–708, 1 fig.

Poulton, E. B.

1902 The Attacks of Predaceous Insects other than Mantidae upon Conspicuous, Specially Defended Lepidoptera. Trans. Entom. Soc. London, 1902, pp. 328–337.

1906 Predaceous Insects and Their Prey. *Ibid.*, 1906, pp. 323–409.

1924 The Relation Between the Larvae of the Asilid Genus *Hyperechia* (Laphriinae) and Those of Xylocopid Bees. *Ibid.*, pp. 121–123, 1 pl., 1 fig.

Putnam, W. L.

1932 Chrysopids as a Factor in the Natural Control of the Oriental Fruit Moth. Canadian Entom., vol. 64, pp. 121–126.

Rabaud, E.

1923 Les pattes ravisseuses et la convergence des formes. C.R. Soc. Biol., Paris, vol. 88, pp. 25–27.

1926 Sur le régime alimentaire des larves de *Chrysopa vulgaris*. Feuille Natural., vol. 47, pp. 164–167.

Ramakrishna Ayyar, T. V.

1929 Notes on Some Indian Lepidoptera with Abnormal Habits. Journ. Bombay Nat. Hist. Soc., vol. 33, pp. 668–675.

Rau, Phil

1938 Some Remarks on Prey-selection by Solitary Wasps. Ann. Entom. Soc. America, vol. 21, pp. 385–392.

Rau, P., and N. Rau

1916 The Biology of the Mud-daubing Wasps as Revealed by the Contents of Their Nests. Journ. Anim. Behavior, vol. 6, pp. 27–63, 4 pls.

1918 Wasp Studies Afield. xv + 372 pages, 68 figs. Princeton, N. J., Princeton Univ. Press.

Raymond, P. E.

1920 The Appendages, Anatomy and Relationships of Trilobites. Mem. Connecticut Acad. Arts and Sci., vol. 7, 169 pp., 11 pls., 46 figs.

Readio, P. A.

1927 Biological Notes on Phymata. Bull. Brooklyn Entom. Soc., vol. 22, pp. 256–262, 1 pl.

Reichert, A.

1905 Raubbeine bei Insekten. Entom. Jahrb., vol. 15, pp. 82–83.

Reinhard, E. G.

1924 The Life History and Habits of the Solitary Wasp, *Philanthus gibbosus*. Ann. Rept. Smithsonian Inst. for 1922, pp. 363–376.

Riley, C. F. C.

1918 Food of Aquatic Hemiptera. Science, vol. 48, pp. 545–547.

1922 Droughts and Cannibalistic Responses of the Water-strider, *Gerris marginatus*. Bull. Brooklyn Entom. Soc., vol. 17, pp. 79–87.

Riley, C. V.

1892 The Larger Digger Wasp. Insect Life, vol. 4, pp. 248–252, 7 figs.

Ripley, L. B.

1917 Notes on the Feeding Habits of Adult Chrysopidae. Entom. News, vol. 28, pp. 35–37.

Roberts, R. A.

1937 Biology of the Bordered Mantid, *Stagmomantis limbata*. Ann. Entom. Soc. America, vol. 30, pp. 96–108.

1937 Biology of the Minor Mantid, *Litaneutria minor*. *Ibid.*, pp. 111–121.

Rollinat, R.

1926 Quelques observations sur la mante réligieuse, principalement sur la nourriture pendant le premier âge. Rev. Hist. Nat. Appl. Paris, vol. 7, pp. 242–251; 270–276.

Rothschild, W.

1906 On a New Parasitic Moth from Queensland, discovered by P. F. Dood. Novitates Zool., vol. 13, pp. 162–169.

Ruiz, P. F.

1925 Voracidad de los Asílidos (Dípteros). Rev. Chil. Hist. Nat., vol. 29, pp. 220–224.

Rungius, H.

1911 Der Darmkanal (der Imago und Larve) von *Dytiscus marginalis*. Zeits. Wiss. Zool., vol. 98, pp. 179–287, 74 figs.

Schilder, F. A., and M. Schilder

1928 Die Nahrung der Coccinelliden und ihre Beziehung zur Verwandschaft der Arten. Arb. Biol. Reichanst. f. Land.- und Forstwirtsch., vol. 16, pp. 213–282.

Schmitz, H.

1910 Zur Lebensweise von *Helicobosca muscaria*. Zeits. wiss. Insektenbiol., vol. 6, pp. 107–109.

1917 Biologische Beziehungen zwischen Dipteren und Schnecken. Biol. Zentralbl., vol. 37, pp. 24–43, 7 figs.

Schwangart, F.

1906 Ueber den Parasitismus von Dipterenlarven in Spinnencocons. Zeits. wiss. Insektenbiol., vol. 2, pp. 105–107.

Sharp, D.

1882 On Aquatic Carnivorous Coleoptera. Scient. Trans. R. Dublin Soc. (II), vol. 2, p. 179.

Shelford, V. E.

1909　Life-histories and Larval Habits of the Tiger Beetles. Journ. Linn. Soc. London, vol. 30, pp. 157–184.

Simanton, F. L.

1916　*Hyperaspis binotata*, a Predatory Enemy of the Terrapin Scale. Journ. Agric. Res., vol. 6, pp. 197–205, 2 pls., 1 fig.

Smith, C. E.

1923　*Larra analis*, a Parasite of the Mole Cricket, *Gryllotalpa hexadactyla*. Proc. Entom. Soc. Washington, vol. 37, pp. 65–82.

Smith, R. C.

1920　Predacious Grasshoppers. Journ. Econ. Entom., vol. 13, p. 491.

1922　The Biology of the Chrysopidae. Mem. Cornell Agric. Expt. Sta., No. 58, pp. 1287–1372.

1923　Life Histories and Stages of Some Hemerobiids and Allied Species. Ann. Entom. Soc. America, vol. 16, pp. 128–148, 3 pls.

Snodgrass, R. E.

1922　Mandible Substitutes in the Dolichopodidae. Proc. Entom. Soc. Washington, vol. 24, pp. 148–150, 1 pl.

Speyer, W.

1924　Die Ernährungsmodificationen der Organismen unter besondere Berücksichtigung der fleishfressenden Tiere. Beitr. aus d. Tierkunde Widmungsschr. für Dr. M. Braun. Königsberg, pp. 16–20.

Strouhal, H.

1926　Die Larven der paläarktischen Coccinellini und Psylloborini. Arch. Naturg. Jahrg. 92, Abt. A, no. 3, pp. 1–63, 15 figs.

Swezey, O. H.

1905　Leaf-hoppers and Their Natural Enemies (Orthoptera, Coleoptera, Hemiptera). Bull. Hawaiian Sugar Planters' Expt. Sta., No. 1, pp. 211–238.

Taschenberg, O.

1906　Beitrag zur Lebensweise von *Necrobia ruficollis* und ihrer Larve. Zeitschr. wiss. Insektenbiol., vol. 2, pp. 13–17.

Terry, F. W.

1905　Leaf-hoppers and their Natural Enemies. Part V. Bull. Hawaiian Sugar Planters' Expt. Sta., No. 1, pp. 163–181, 3 pls.

Thienemann, A.

1908　Die Fangnetze der Larven von *Philopotamus ludificatus*. Zeits. wiss. Insektenbiol., vol. 4, pp. 378–380.

Thierwolf, W. R.

1928　The Economic Importance of *Paratenodera sinensis*. Entom. News, vol. 39, pp. 112–116; 140–145.

Thompson, R. C. M.

1937　Observations on the Biology and Larvae of the Anthomyiidae. Parasitology, vol. 29, pp. 273–358.

Thompson, W. R.

1929 On the Relative Value of Parasites and Predators in the Biological Control of Insect Pests. Bull. Entom. Res., vol. 19, pp. 343–350.

Tillyard, R. J.

1917 The Biology of Dragonflies. xii + 396 pp., 4 pls., 188 figs. Cambridge, England.

1918 Life History of *Psychopsis elegans*. Proc. Linn. Soc. New South Wales, vol. 43, pp. 787–818.

1922 The Life History of the Australian Moth-lacewing, *Ithone fusca*. Bull. Entom. Res., vol. 13, pp. 205–224.

Torre-Bueno, J. R. de la

1916 Aquatic Hemiptera, a Study in the Relation of Structure to Environment. Ann. Entom. Soc. America, vol. 9, pp. 353–365.

Trouvelot, B.

1932 Recherches sur les parasites et predateurs attaquant le doryphore en Amérique du Nord. Ann. des Epiphyt., vol. 17, pp. 408–445.

Verhoeff, C.

1892 Zur Kenntnis des biologischen Verhältnisses zwischen Wirt- und Parasiten-Bienenlarven. Zool. Anz., vol. 15, pp. 41–43.

Vogel, R.

1915 Beitrag zur Kenntnis des Baues und der Lebensweise der Larve von *Lampyris noctiluca*. Zeitschr. wiss. Zool., vol. 112, pp. 291–432.

Voukassovitch, P.

1925 Observations biologiques sur quelques insectes prédateurs des pucerons et leurs parasites et hyperparasites. Bull. Soc. Entom. France, 1925, pp. 170–172.

1932 *Isobremia kiefferi*, Diptère prédateur des pucerons. Soc. Entom. France, Livre du Centenaire, pp. 319–327, 4 figs.

Walton, W. R.

1908 Popular Fallacies Regarding Insects; and Some Insects that are Poisonous. Entom. News, vol. 18, pp. 467–473.

Warren, A.

1915 A Study of the Food Habits of the Hawaiian Dragonflies or Pinau. College of Hawaii Publ. Bull., No. 3, 45 pp., 5 pls., 1 fig.

Webster, F. M.

1880 Notes Upon the Food of Predaceous Beetles. Bull. Illinois State Lab. Nat. Hist., vol. 1, 2nd Ed., 1903, pp. 162–166.

Weiss, H. B.

1913 The Chinese Mantis, a Beneficial Insect in New Jersey. Circ. Bur. Statistics, Dept. Agric. New Jersey, No. 68, 8 pp., 3 figs.

1921 A Summary of the Food Habits of North American Hemiptera. Bull. Brooklyn Entom. Soc., vol. 16, pp. 116–118.

1922 A Summary of the Food Habits of North American Coleoptera. American Natural., vol. 56, pp. 159–165.

Wellman, F. C.

1909 Ungewöhnliche parasitäre Gewohnheiten einer afrikanischen Ephydride. Zeits. wiss. Insektenbiol., vol. 5, p. 356.

Wesenberg-Lund, C.

1911 Biologische Studien über netzspinnende Trichopteren-Larven. Internat. Rev. f. Hydrobiol. Hydrogr., vol. 3, pp. 1–64, 6 pls.

1912 Biologische Studien über Dytisciden. *Ibid.*, vol. 5, pp. 1–129, 9 pls.

Westwood, J. O.

1876 Notes on the Habits of a Lepidopterous Insect Parasitic on *Fulgora candelaria.* Trans. Entom. Soc. London, 1876, pp. 519–524.

Wheeler, W. M.

1911 An Ant-nest Coccinellid (*Brachyacantha 4-punctata*). Journ. New York Entom. Soc., vol. 19, pp. 169–174.

1913 A Solitary Wasp (*Aphilanthops frigidus*) that Provisions its Nest with Queen Ants. Journ. Anim. Behav., vol. 3, pp. 274–387.

1918 *Vermileo comstocki* sp. nov., an Interesting Leptid Fly from California. Proc. New England Zool. Club, vol. 6, pp. 83–84.

1919 The Parasitic Aculeata, A Study in Evolution. Proc. American Philos. Soc., vol. 58, pp. 1–40, 6 tables.

1923 Social Life Among the Insects. vi + 375 pp. Harcourt, Brace and Co., New York.

1930 Demons of the Dust, a Study in Insect Behavior. 378 pp., 49 figs. W. W. Norton, New York.

Whitfield, F. G. S.

1925a The Relation between the Feeding Habits and the Structure of the Mouth-parts in the Asilidae. Proc. Zool. Soc. London, 1925, pp. 299–638, 2 pls., 15 figs.

1925b The Natural Control of the Leaf-miner (*Phytomyza aconiti*) by *Tachydromia minuta*. Bull. Entom. Res., vol. 16, pp. 95–97.

Wildermuth, V. L.

1916 California Green Lacewing Fly. Journ. Agric. Res., vol. 6, pp. 515–525, 7 figs.

Williams, F. X.

1917 Notes on the Life History of Some North American Lampyridae. Journ. New York Entom. Soc., vol. 25, pp. 11–38.

1919a Philippine Wasp Studies. Bull. Hawaiian Sugar Planters' Assoc., Entom. Ser., No. 14, 186 pp., 106 figs.

1919b *Epyris extraneus* Bridwell, a Fossorial Wasp that Preys on the Larva of the Tenebrionid Beetle, *Conocephalum seriatum*. Proc. Hawaiian Entom. Soc., vol. 4, pp. 55–63, 8 figs.

1928 Studies in Tropical Wasps, Their Hosts and Associates. Bull. Hawaiian Sugar Planters' Assoc., Entom. Ser., No. 19, 186 pp., 106 figs.

1929 Notes on the Habits of Cockroach Hunting Wasps of the Genus Ampulex, with Particular Reference to *Ampulex* (*Rhinopsis*) *canaliculatus*. Proc. Hawaiian Entom. Soc., vol. 7, pp. 315–329.

Withycombe, C. L.

1925 Some Aspects of the Biology and Morphology of the Neuroptera with Special Reference to the Immature Stages and Their Possible Phylogenetic Significance. Trans. Entom. Soc. London, 1924, pp. 303–411, 6 pls.

1924 Further Notes on the Biology of some British Neuroptera. Entomologist, vol. 57, pp. 145–152, 2 figs.

Xambeau, P.

1908 Moeurs et métamorphoses des Coléoptères du groupe des Malachides. Naturaliste, Paris, vol. 30, pp. 189–192; 199–202.

Zacher, F.

1927 Eine neue Gewächsheuschrecke. Anz. Schädlingsk., vol. 3, pp. 33–34.

CHAPTER VII

PARASITISM

Parasites may properly be regarded as more advanced organisms than the predators, for they have not only had a more eventful phylogenetic career, but, during their long history, have learned to use other organisms in a very economical manner as instruments of nutrition.

W. M. WHEELER, in *Insect Parasitism and its Peculiarities*, 1911

ANY DEFINITION of parasitism sufficiently broad to encompass all the varied aspects of this common biological phenomenon is very difficult, indeed well nigh impossible to formulate. Even when the field of inquiry is simplified by restriction to the insects alone, we find a wide range of parasitic behavior, and I shall attempt to deal with its several phases and their relation to one another, rather than to squeeze them all into a single category. The prime reason for this lies in the fact that parasitic insects do not form a related natural group having a common origin in the past. On the contrary, a parasitic mode of life has made its appearance quite frequently in the history of the insects and although we may trace many similarities and parallel developments in diverse insects there are some fundamental differences which must be considered as manifestations of definitely independent phenomena.

We might regard the vegetarian insects already dealt with earlier in this book as parasites of the plants on which they feed, particularly those which weaken the plants by insidious means, such as sucking their sap or mining in their leaves. It is customary, however, among entomologists to exclude such types and apply the term parasitic only to those which affect living animals. Such a restriction is hardly logical, but it is so firmly established that only confusion could follow any deviation from this usage.

We may begin by accepting the bland statement that a parasite is an organism which lives on, or within, the body of a living host on which it depends for its sustenance (and perhaps shelter or other less vital needs as well), and which it does not immediately destroy. Any biologist may pick flaws in this definition, and we shall proceed to do so ourselves in a moment, but it is a waste of breath to safeguard its accuracy and completeness with reservations in legal form or ponderous verbiage. This is especially true of insect parasites which display varied phases of this protean type of behavior.

The classical division of parasites into external (epizoic) and internal (entozoic) ones is a highly artificial arrangement, but if we disregard numerous exceptions and intergrades it may be profitably applied to the insects since it serves at once to segregate several kinds of epizoic parasites like the fleas and lice that live on the higher vertebrates where they feed mainly by sucking the blood of their host. Such external parasites differ widely in the extent of their dependence upon the host animals. Some pass their entire life as parasites; others are free-living during the preparatory stages, like the fleas; still others, like the mosquitoes, are free-living throughout the whole life cycle and it would be stretching a point to call them parasites were it not that there is no place in an intergrading series of blood-sucking insects where it can be said that parasitism ends and some freer mode of existence supervenes. For these reasons a chapter has been reserved for haematophagous insects and other epizoic parasites, to include a more or less homogeneous series. In such an arrangement it must be freely admitted, however, that there are a number of discordant elements. One of these is, for example, a series of small flies that suck blood like mosquitoes, but affect other, and larger insects. This group includes only a very minor portion of the parasitic insects. Also, it will be recalled that some truly predatory insects are haematophagous, for example the larvae of syrphid and cecidomyiid flies that suck the juices of their prey. In a sense the same is true of predatory forms like the *Dytiscus* larva and the neuropterous larvae that inject digestive fluids into the prey to liquefy its tissues so that they can be sucked out. In these cases the prey succumbs quickly and does not survive after feeding is under way. This is all very well with reference to our definition of parasitism, but closely similar behavior is found among the larvae of certain ichneumon flies. These are regarded as true parasites because they depend upon the mother to locate their host and lay the eggs from which they will develop on its body. The host succumbs with unseemly haste, but the whole process is clearly an aberration of the quite orthodox parasitism found generally among this large family of ichneumon flies.

The second group, including internal parasites is far more extensive, as well as more varied both in morphology and behavior, and the number of unexpected modifications and adaptations disclosed by its members is so appallingly great that one is tempted to deal with it by a mere recitation of life cycles and behavior patterns.

Perhaps the most striking generalization that may be made is that the great majority of internal parasites among the insects depend upon other insects rather than upon unrelated animals and this is a relationship which does not prevail generally among other animals since hosts and their parasites are usually members of very dissimilar groups. On this account these insect-destroying forms are universally known as entomophagous parasites.

This at once introduces a peculiar set of conditions. The host and parasite are frequently of the same or quite comparable stature, although on account of the very considerable range in size among insects, there is much variation in this respect, especially as many parasitic insects are among the smaller members of their class. Nevertheless, it means quite generally that the number of individuals developing in a single host is strictly limited; frequently one host insect supports a single parasite, often several to a dozen, rarely more, and only in very exceptional cases a brood of certain very minute parasites may number into the hundreds.

Due to the similarities of host and parasite, it is reasonable to believe that insect parasites encounter less difficulty in attaining satisfactory adjustment to life in the bodies of their hosts than would be the case among radically different types of animals. The independent and sporadic appearance of parasitic forms among so many groups of insects lends support to such an idea, although direct evidence derived from the physiological reactions of parasitized insects against their invaders is as yet meagre in amount and controversial in nature.

Resistance of host animals against invasion by insect parasites is often quite ineffectual. The epizoic parasites are commonly annoying to their hosts and the latter often attempt to dislodge them by various physical means, ranging from the scratching of dogs to remove fleas, to the eating of their external parasites by monkeys and the swatting of biting green-head flies by bathers on New England beaches.

Far more important for entomophagous parasites are physiological reactions of the host which interfere with their development during the early stages. As the range of hosts by these parasites is

very rigidly maintained it appears that the host as well as the parasite is involved and that the reactions may be specific for both.
Phagocytosis is frequently observed whereby the eggs or hatching
larvae of the parasite are surrounded and encapsulated by amoeboid
cells from the blood of the host. However, immunity of some host
insects is not accompanied by phagocytosis and such resistance must
be traced to other reactions whose nature is not yet understood.

Most frequently the visible effects of insect parasites upon their
hosts are far less apparent than might be expected, especially among
the internally feeding entomophagous forms where the host is commonly consumed almost entirely by the time the parasites have
completed their growth. Indeed, the host most frequently gives
no clear indication of its impending fate till the last few hours of its
life although the growth of the contained parasite may have proceeded conjointly with that of its host for many weeks. We must
attribute this tolerance to the intense vitality of insects, to the rapid
and extensive storage of reserve foods in the body and to the con-
summate care exercised by the growing parasitic larva to confine its
feeding to non-vital structures of its host. A few of these internal
entomophagous parasites, particularly among the ichneumon flies,
have become secondarily external feeders. In such cases development
is usually greatly accelerated and the host succumbs very quickly
to powerful digestive fluids secreted by the parasites. Such a picture
recalls the extra-intestinal digestion of predatory insects and shows
the close similarity between insect parasitism and predatism.

Frequent cases are known where the presence of entomophagous
parasites induces noticeable changes in their hosts by interfering
with the development of the reproductive glands. This phenomenon
known as parasitic castration, is frequently seen in insects parasitized by nematode worms, and is perhaps best known to zoologists
in the crustacean, *Sacculina,* which affects certain decapod and
isopod crustaceans and brings about the castration of its host.
Among entomophagous parasites, the Strepsiptera produce similar
changes in the wasps and bees that they parasitize and the effects
are seen in a modification of certain secondary sexual characters,
particularly the color pattern. Among ants, worker-like females
sometimes result from parasitism by small Hymenoptera of the
family Eucharidae that prey upon the ant larvae.

Such reactions appear to have much in common with the polymorphism of ants and other social Hymenoptera where this phenomenon undoubtedly has a nutritional basis. This fact is obscured

by the use of the term nutricial castration for such cases as it implies regulatory behavior on the part of the "nurse" ants and smacks too strongly of anthropomorphism.

Another peculiarity which frequently appears in the host relations of parasitic insects is a phenomenon known as hyperparasitism, whereby certain species, mainly small Hymenoptera become secondary parasites. These secondaries attack species that are themselves primary entomophagous parasites, being thus dependent not only upon their own parasitic host but upon its host in turn. We shall return to this matter later.

One feature which is not infrequently met with among other parasitic animals is absent among the insects. This is heteroecism, a condition in which a parasite preys upon two, usually widely different, hosts which are attacked alternately either by successive generations or in sequence during the development of a single individual parasite. An analogous situation occurs among the purely vegetarian aphids which consistently migrate back and forth between two food plants during a cycle of generations. The alternation of generations in cynipid gall wasps involves a similar change in food plants, but this phenomenon does not occur among parasitic insects unless we wish to extend it to an alternation among hosts that can be traced to the exigencies of abundance, scarcity or absence and not to free choice.

We might expect that some parasitic insects would find suitable hosts among arthropods other than insects. This has happened quite frequently among the spiders as these Arachnida are affected to some extent, although far less commonly than insects. Terrestrial isopods and land crabs are rarely parasitized by insects, and except for a few snails, other invertebrates remain almost entirely unaffected.

A small, but by no means negligible number of insects develop as internal parasites of the warm-blooded vertebrates. These may be relegated to a quite distinct category, principally because they do not belong to those groups that include the entomophagous parasites, indicating that they have developed their parasitic habits independently. Also, a number of facultative parasites appear in this series, especially species which may at times invade the alimentary canal of man and other mammals. This is in clear contrast to the host relations of entomophagous parasites as these are almost invariably obligatory in their parasitic relations. Like other statements this must be qualified, although only unique or at most very

PLATE XVII

A. An example of phoresy in a parasitic hymenopteron. Females of *Lepidoscelio* attached to the body of a grasshopper. In this fashion they are transported to the place where the hopper will lay her eggs. Original photograph.

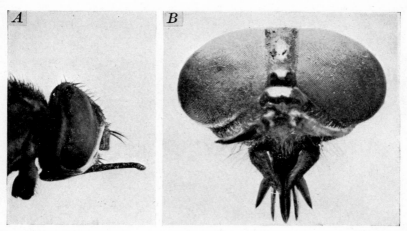

B. Head and mouthparts of two blood-sucking flies. *A*, the stable-fly (*Stomoxys calcitrans*) of the family Muscidae; *B*, a horse-fly (*Tabanus*) of the family Tabanidae. Photographs by the author.

PLATE XVIII

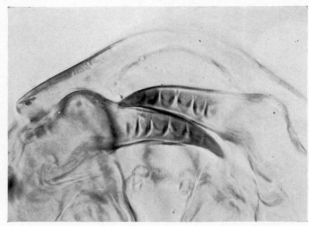

A. Mandibles of the first-stage larva or triungulin of the meloid beetle, *Horia maculata*. Original microphotograph.

B. Larvae of an ichneumon fly parasitic on a leaf-rolling caterpillar. Photograph by the author.

exceptional cases of truly facultative entomophagous parasites have been demonstrated among the thousands of species concerned.

Like their vegetarian and predatory relatives that have been treated in previous chapters, the several types of parasitic insects are very highly specialized in their host relationships and it is only very rarely that any approach to an indiscriminate selection of hosts occurs. In this respect insects are similar to all other types of parasitic animals, and plants too, for that matter. It is also fully authenticated that at least many kinds of insect parasites have retained their dependence upon certain types of hosts over long periods of time, extending at least as far back as the early Tertiary. This is true of both entomophagous and epizoic parasites. Some of the evidence relating to this remarkable persistence of host relationships will be presented in a later chapter, but we may note here that this recalls the similar retention of specific food plants by vegetarian insects over equally long periods.

Even though the several types of insect parasitism represent numerous origins, it may be asked whether all have been derived from varied ancestral stems that resembled one another in their food habits. Such is unquestionably not the case. Some of the entomophagous parasites have undoubtedly developed from insects that were saprophagous, depending on dead or moribund insects and from these a transfer to living hosts was accomplished. This transformation is still in progress among certain higher Diptera of the family Sarcophagidae, and there is good reason to believe that the very abundant and successful entomophagous parasites which constitute several related families arose in this way. The same is also quite surely true of other allied flies that are internal parasites of mammals. Some parasites have arisen from vegetarian forms. One clear-cut example of this is furnished by the parasitic gall wasps which are unquestionably derivatives of their gall-making vegetarian relatives. It may be said also that the large families of entomophagous parasites which form several major divisions of the Hymenoptera could all probably trace their ancestry back to vegetarian ancestors, although these would antedate our own Edenic forbears by uncounted centuries, or to be vaguely exact by perhaps 10^8 years. Strangely enough, a very few scattered members of this series have again more recently returned to the plants for their food supply, and we find in certain chalcid wasps and at least one ichneumon-fly, *Grotea*, that this has happened. Shifts of this sort between the several types of food habits have occurred not very infrequently among insects of various groups. We have already mentioned certain lepidopterous caterpillars that have become predatory and the reverse change from predatism to vegetarianism is illustrated by a group of coccinellid beetles. Many of these shifts appear to be in the nature of mutations in instinct since they do not necessarily involve morphological adaptations. Since parasitism appears so frequently and sporadically, its origin in this way is obvious. In many cases it may be definitely traced to a shift from either phytophagy, saprophagy or predatism, and in others its derivation must for the present remain doubtful. For a much more extended consideration of such shifts in food habits the reader is referred to an earlier account by the writer ('36).

Parasitism, of whatever type, commonly involves structural modifications. Although in insects these are not always so pronounced as might be expected, they appear quite frequently in highly exaggerated form and great variety. In the higher insects with complete metamorphosis where the parasitic period of existence is confined to the larva, this stage acquires the simplifications of structure that are usually classed as degenerative, while the imaginal stage remains almost entirely unaffected. This peculiar situation is of course due to the fact that insect larvae are able to acquire all sorts of adaptive structures which are purely transitory, as they disappear entirely on transformation, without leaving any visible impression on the imago.

A relationship known as phoresy is frequently associated with parasitism among insects. In common parlance this is really hitchhiking and embraces the behavior of certain species that use the bodies of other, usually larger insects or other animals for the purpose of transportation (Pl. XVII A). Such phoretic individuals may harbor no ulterior designs further than to bone a free ride to a desired destination but most frequently the ride is only a prelude to murderous assault, larceny or some less serious racket. Here we should pause to point out that statistically at least the human race can claim ethical superiority. Most epizoic parasites do not commonly leave the body of the host for the obvious reason that it represents their food supply. Some parasites that develop in the eggs of grasshoppers and other insects attach themselves to the body of the female and are thus carried to the place she selects to lay her eggs. The active newly hatched larvae of certain parasitic beetles belonging to the family Rhipiphoridae and those of a number of Strepsiptera that develop as parasites in the larvae of bees, social wasps and other insects attach themselves to adults of the host species and are thus carried directly to the nest where they experience little difficulty in finding the larvae within which their later development is destined to take place. Phoresy thus plays an important secondary rôle in the life of the most varied parasitic insects by affording a ready and correctly piloted transport.

It must by no means be assumed that all the structural modifications of parasitic insect larvae are simplifications. The more conspicuous peculiarities are more generally of this nature, but this has not precluded the appearance of many specialized structures. In some entomophagous parasites the minute young larva is active and seeks out its host, after which it subsides into the comfortable gluttony of its endoparasitic feeding stage. Coincident with this change in behavior, the legs and sensory structures are degraded or disappear completely, reducing the larva to an utterly helpless condition. Hypermetamorphosis of this type occurs in some parasitic beetles, in the order Strepsiptera and in several separated instances among the Hymenoptera, and numerous other forms show it in less pronounced form. Such an irregular distribution shows conclusively that we are dealing with convergence, rather than with genetic relationship, a conclusion that is further substantiated by a comparison of the structures involved, as these commonly bear the earmarks that betray their independent origin.

Among the social Hymenoptera, particularly the ants, the phenomena of parasitism may reach a higher level, transcend the parasitism of individual organisms and involve the colony. Since the ant colony is really a family in the strictly human sense, including a large group of offspring that remain with the mother, it becomes actually a sort of superorganism where the integration of individuals simulates the coördinated processes characteristic of the single organism. In the more typical cases of social parasitism an entire colony of the parasitic ant becomes dependent upon a colony of its host species, commonly through the queen insinuating herself into the host colony and destroying its queen. If she is successful in this bloodthirsty venture her brood become dependent upon the care of the workers of the host colony. This condition may be temporary and the colony may later persist as a pure group of the originally parasitic species after the workers of the host species have disappeared through natural death or accident. Among those ants that are permanent social parasites the continuance of the colony depends upon raids by the parasitic species upon other nests of the host species to replenish the supply of host workers from time to time as they begin to disappear from the population. Although this type of parasitic behavior is of deep human interest on account of the parallels that may be drawn with the social life of man, it is clearly beyond the scope of the present book to deal with it in an adequate manner. Consequently the reader must be referred to the publications of the late Professor Wheeler whose studies in this field form a large part of our present knowledge. From the humanistic standpoint, a very illuminating account is contained in a book by Haskins ('39) which deals also with the predatory and vegetarian ants.

From the foregoing brief summary it is clear that insect parasitism is extremely diverse and that it may verge on predatism, scavengerism or even on commensalism in some instances. We shall deal with its varied manifestations in the two succeeding chapters where it will be possible to present them in greater detail and more concrete form.

REFERENCES

Note: Only a few titles of general interest are included in this list. Those of more specialized content are cited in connection with the following two chapters.

Ball, G. H.
1943 Parasitism and Evolution. American Naturalist, vol. 77, pp. 345–364.

Bodenheimer, F. S.
1928 Welche Faktoren regulieren die Individuenzahl einer Insektenart in der Natur. Biol. Zentralbl., vol. 48, pp. 714–739.
1931 Zur Frühgeschichte der Erforschung des Insekten-Parasitismus. Arch. Gesch. Math. Naturwiss. Techn., vol. 13, pp. 412–416.

Brues, C. T.
1936 Aberrant Feeding Behavior among Insects and its Bearing on the Development of Specialized Food Habits. Quart. Rev. Biol., vol. 11, pp. 305–319.

Cameron, T. W. M.
1940 The Parasites of Man in Temperate Climates. xii + 182 pp. Toronto, Canada. University of Toronto Press.

Caullery, M.
1922a Le Parasitisme et la Symbiose. 400 pp., 53 figs. Paris, Gaston Doin.
1922b Parasitism and Symbiosis in their Relation to the Problem of Evolution. Trans. by G. S. Miller. Ann. Rept. Smithsonian Inst., 1922, pp. 399–409.

Chandler, A. C.
1922 Speciation and Host Relationships of Parasites. Parasitology, vol. 15, pp. 326–339.
1940 Introduction to Parasitology. xiii + 698 pp. New York, Wiley and Sons.

Clausen, C. P.

1936 Insect Parasitism and Biological Control. Ann. Entom. Soc. America, vol. 29, pp. 201–223.

Cobb, N. A.

1904 Parasites as an Aid in Determining Organic Relationship. Agric. Gaz. New South Wales, vol. 15, pp. 845–848.

Ewing, H. E.

1912 The Origin and Significance of Parasitism in the Acarina. Trans. Acad. Sci., St. Louis, vol. 21, pp. 1–70, 7 pls.

Fantham, H. B.

1936 The Evolution of Parasitism among the Protozoa. Scientia, vol. 59, pp. 316–324.

Gause, G. F.

1934 Ueber einige quantative Beziehungen in der Insekten-Epidemiologie. Zeits. angew. Entom., vol. 20, pp. 619–623.

Haskins, C. P.

1939 Of Ants and Men. 244 pp. New York, Prentice-Hall.

Hegner, R.

1926 The Biology of Host-Parasite Relationships among Protozoa Living in Man. Quart. Rev. Biol., vol. 1, pp. 393–418.

Herms, W. B.

1926 The Effects of Parasitism on the Host and on the Parasite. Journ. Econ. Entom., vol. 19, pp. 316–325.

Kaupp, B. F.

1925 Animal Parasites and Parasitic Diseases. 4th Ed. xvi + 250 pp., 81 figs. London.

Kennedy, C. H.

1933 Some Fundamental Aspects of Insect Parasitism. Trav. Congr. Intern. Entom., Paris, vol. 2, pp. 407–419.

Laloy, L.

1906 Parasitisme et mutualisme dans la nature. Paris, Félix Alcan, 284 pp., 82 figs.

Metcalf, M. M.

1929 Parasites and the Aid They Give in Problems of Taxonomy, Geographical Distribution and Palaeography. Smithsonian Misc. Coll., vol. 81, pp. 1–36.

Muir, F.

1914 [Parasites.] Presidential address. Proc. Hawaiian Entom. Soc., vol. 3, pp. 28–42.

1931 The Critical Point of Parasitism and the Law of Malthus. Bull. Entom. Res., vol. 22, pp. 249–251.

Rabaud, E.

1928 Parasitisme et Évolution. Rev. Philos., France et l'Étranger, vol. 105, pp. 48–81.

Root, F. M.

1924   Parasitism Among the Insects. Sci. Monthly, vol. 9, pp. 479–495.

Shipley, A. E.

1913   Pseudo-parasitism. Parasitology, vol. 6, pp. 351–352.
1926   Parasitism in Evolution. Sci. Prog., vol. 20, pp. 637–661.

Smith, Theobald

1934   Parasitism and Disease. xiii + 196 pp. Princeton, N. J., Princeton Univ. Press.

Stunkard, H. W.

1929   Parasitism as a Biological Phenomenon. Sci. Monthly, vol. 28, pp. 349–362.

Swellengrebel, N. H.

1940   The Efficient Parasite. Science, vol. 92, pp. 465–469.

Thompson, W. R., and H. L. Parker

1927   Études sur la biologie des insects parasites: La vie parasitaire et la notion morphologique de l'adaptation. Ann. Soc. Entom. France, vol. 96, pp. 113–146.

Wagner, F. von

1902   Schmarotzer und Schmarotzertum in der Tierwelt. Leipzig, Goeschen.

Ward, H. B.

1907   The Influence of Parasitism on the Host. Science, vol. 25, pp. 201–218.

Wheeler, W. M.

1902   A Neglected Factor in Evolution. *Ibid.*, vol. 15, pp. 766–774.
1904   A New Type of Social Parasitism in Ants. Bull. American Mus. Nat. Hist., vol. 20, pp. 347–375.
1907   The Polymorphism of Ants with an Account of Some Singular Abnormalities Due to Parasitism. *Ibid.*, vol. 23, pp. 1–93, 6 pls.
1910   The Effects of Parasitic and Other Kinds of Castration in Insects. Journ. Expt. Zool., vol. 8, pp. 377–438.
1911a   The Ant Colony as an Organism. Journ. Morph., vol. 22, pp. 307–325.
1911b   Insect Parasitism and its Peculiarities. Pop. Sci. Monthly, vol. 79, pp. 431–449.
1928   The Social Insects. xviii + 378 pp.   New York, Harcourt, Brace and Co.

CHAPTER VIII

BLOOD-SUCKING INSECTS AND OTHER EXTERNAL PARASITES

And he himself does not escape the virus secreted by these terrible stingers; mosquitoes menace him with malaria in the vicinity of marshes, tsetse flies with sleeping sickness in the damp and shady jungles of the African tropics; fleas transmit to him the germs of plague, and filthy lice the typhus fever.

L. O. HOWARD, 1922

THE HAEMATOPHAGOUS or blood-sucking insect parasites include certain members of various orders and all the representatives of two others. The latter are the fleas (Siphonaptera) and the sucking lice (Anoplura). Another extensive group of epizoic insect parasites are the biting lice or bird-lice, forming the order Mallophaga, and these are almost the only ectoparasitic insects provided with biting mouth-parts, since nearly all the others feed by sucking blood. The host animals are almost entirely terrestrial vertebrates, mainly birds and mammals, while the few invertebrate hosts are mainly other, larger insects of various kinds. A few cases of phoresy, mentioned in the previous chapter, might be interpreted as a form of transitory parasitism, for example the small Hymenoptera that attach themselves to grasshoppers while they await transport to the egg-mass to be laid by the hopper, in which their larvae will develop (Pl. XVII A). As the adult Hymenoptera probably secure food by puncturing the integument of the hopper and lunching in transit, they are thus temporarily parasitic.

Aside from these extensive groups there are to be found among the insects as a whole, scattering examples of this mode of life that appear in orders where the prevailing habits are not parasitic.

A number of generalizations may be drawn with reference to the epizoic insect parasites. There is a well marked tendency for the adults to lose the wings and became partly or usually entirely apterous in those which live on the body of the host. The most completely adapted types represented by the lice, undergo an incomplete metamorphosis and the entire life cycle takes place on the host, including the embryonic period as the eggs are deposited and undergo their development attached to its body. In the fleas the adult is wingless and highly modified to live in the fur or feathers of its host, but the eggs fall from the host and the larval stages are free-living. In several instances among the Diptera, Hemiptera and Dermaptera the insects are viviparous and larval growth is completed before birth, thus eliminating the free larva altogether. In contrast to these the free-living mosquitoes and the blood-sucking flies and bugs are essentially independent except for their adult feeding habits.

Like all insects, and like other animal parasites, the epizoic insects are closely restricted in their food relations. This holds true also quite frequently among many of the blood-sucking species that are free-living in all stages, although this fact was not fully realized until quite recently. On this account many of the insects dealt with in this chapter have assumed an important rôle as carriers of infectious diseases that affect their host animals. It is well known that many human diseases such as malarial fever, bubonic plague, yellow fever, dengue, typhus and others are insect-borne and the importance of similar insects to the health of other terrestrial vertebrates is equally great although this field has not been very completely explored even in the case of domesticated animals. Since other mammals and even birds may act as primary hosts or reservoirs for some human pathogens the range of hosts among blood-sucking insects becomes a matter of great concern to public health.

From the biological standpoint the more degenerate and highly specialized forms like the bird lice offer much interesting evidence on parallel evolution of hosts and parasites. On account of their clear-cut morphological characters these insects serve better than any other group of animal parasites to furnish reliable data concerning this phase of the evolutionary process.

In general it may be said that the development of epizoic parasitism had its origin in phlebotomy or the sucking of blood by free living insects of various types. In its most primitive form we cannot clearly distinguish between phlebotomy and the behavior of the numerous insects described in Chapter VI which suck the body juices of other insects. However, the one difference to be seized upon is that the more truly predatory type of feeding utilizes small prey, usually other insects and these are sucked dry to promptly join the spirit world. Those that we regard as blood-sucking parasites attack much larger animals, nearly always warm-blooded vertebrates which they bleed, but to a generally negligible extent. They may become an intolerable nuisance, but only under most exceptional circumstances do they bleed the patient to death. Many remain attached to their hosts during the periods between feeding, and as this attachment grows closer, epizoic life becomes obligatory, finally leading to extreme structural and developmental adaptations as well as to temperature requirements that prevent them from leaving the host for more than very brief periods.

The most familiar and most universally detested blood-sucking insects are the mosquitoes which represent a type whose developmental and adult stages show practically no modifications that may be attributed to their phlebotomic food habits. The eggs are laid either on the surface of the water, or nearby where they will be submerged, and the very active aquatic larva feeds upon plants and minute animal organisms. The pupa is likewise aquatic and able to swim about readily, a very unusual characteristic as the movement of practically all pupal insects is restricted to an occasional half-hearted wiggle. The adult males pass unnoticed as they do not suck blood while the females depend upon this medium so completely that they can rarely ripen their eggs without a meal of blood. Some mosquitoes show a decided preference for particular hosts. This relationship has been investigated most thoroughly in the anopheline mosquitoes that carry malaria in Europe, and it has been found that some prefer man while others attack domesticated animals. Until recently all were regarded as representing a single species, *Anopheles maculipennis*, but it is now established that there are several races, subspecies or probably distinct species that are practically identical structurally, except for notable differences in the egg stage. These forms show different host preferences and the occurrence of malaria in man depends upon the presence of those that seek out humans. Experiments have shown that these several forms do not interbreed freely and that the offspring of such crosses as can be made, prove partly or completely sterile. It would appear that we have to deal here with incipient species where differences in food habits, geographic range and genetic compatibility combine to influence the course of speciation. Whether we regard these populations as specifically distinct at the present time is immaterial, but they are interesting to students of evolution and are now engaging the attention of several entomologists (*cf.* Bates '39, '40). In addition to their host preferences the tolerance of their larvae to saline

water and their geographical ranges are not uniform. These peculiarities recall at once the somewhat similar cases cited in a
previous chapter of physiological races or strains among phytophagous insects. None of the anopheline mosquitoes are known to
restrict their feeding to man, but *Anopheles gambiae*, the very
dangerous malaria carrier of Africa, recently introduced into Brazil,
depends mainly on humans, while among the forms recently segre-
gated from *Anopheles maculipennis* in Europe, *A. labranchiae* attacks
humans only to an extent of 10% and *A. melanoön* only once out
of 400 feedings, as determined by precipitin tests with antisera
(Hackett '40).

It cannot be stated with certainty whether such highly specific
food preferences are characteristic of the majority of mosquitoes,
but it is probable that such a condition is much more prevalent than
was formerly believed. Without question, however, many mosquitoes show no very consistent preference for certain hosts, attacking various mammals and birds that may be available. This is
shown clearly by the fact that malaria in birds is known to be
carried by mosquitoes of several genera, including some of the very
annoying pests of man (Giovannola '39).

Numerous small midges, somewhat similar to mosquitoes, are also
vicious blood-suckers in the adult stage. Their larvae are aquatic,
some inhabiting salt marshes whence they give rise to swarms of
"sand-flies" that are particularly peppery in spite of their minute
size. They are widespread from the arctic to the equator wherever
appropriate breeding places are to be found, appearing in greatest
abundance on still nights.

There are numerous other groups of Diptera in which the adults
have similar blood-sucking habits. Indeed, the mouthparts of the
more primitive members of the order, where the mandibles and
maxillae are long and blade-shaped, are well suited for piercing the
skin of animals and consequently many forms have turned to blood
as a source of food. Among these the most conspicuous is the family
Tabanidae which includes large or medium-sized species known as
gad-flies or horse-flies. Their larvae (Pl. XVA) live in mud or damp
soil where they prey upon the larvae of crane-flies or other soft-
bodied forms. The adult flies are almost exclusively phlebotomic,
most commonly attacking the larger quadrupeds, although the
members of one genus, *Chrysops*, regularly bite human beings, particularly about the neck and ears. A few of the smaller species of
Tabanus also annoy man and among these we must mention the

"green-head," *Tabanus nigrovittatus*, a maritime species common on the Atlantic coast where its larva develops in salt marshes. The flies are a great pest along many ocean beaches where a lack of clothing renders the sun-bathers especially attractive, at least to the flies. The lancet-like mouthparts are very sharp and are dexterously thrust deep into the skin without warning when the flies alight on a vulnerable spot (Pl. XVII B). Blood quickly exudes through the gashes cut in the skin and is lapped up by the labium so that feeding requires a very brief period. Rarely the feeding is painless and unnoticed by the victim if no sensory nerve endings are directly pierced.

Another notorious family of biting flies are the black-flies or Simuliidae (Fig. 44). These are small species that pass their developmental stages in swiftly running, well aërated water. They are widely distributed, but generally most abundant in the cooler, temperate regions. In New England the black-fly season heralds the approach of the long awaited genial summer weather and it is said that their appearance in swarms was the signal to the aboriginal Indians that the time for planting corn was at hand. Further north their appearance is delayed, but wherever present in numbers they are a major pest particularly well known to those addicted to trout fishing in the northern woods. The bite of the black-fly is less painful than that of the larger horse-flies, but its effects last longer as there is local poisoning from the injection of salivary secretion. Frequently there remains a minute haemorrhagic spot due to the action of a powerful proteolytic enzyme contained in the saliva. Not all of the species bite man and some are particularly serious enemies of domesticated animals.

The other blood-sucking Diptera are mainly isolated genera scattered among several families. The females of certain American Leptidae of the genera *Symphoromyia* and *Atherix* are vicious biters and many other members of the family are predaceous on other insects. The latter habit occurs commonly among several other families like the robber flies (Asilidae), Dolichopodidae and Empididae, but none of these apparently ever suck the blood of vertebrate hosts.

Among the higher Diptera several genera of the family Muscidae are notorious biters. Practically world-wide in its distribution and most annoying to man and domesticated animals alike is the stable-fly, *Stomoxys calcitrans*. Although probably of African or Indian origin, where the other members of its genus are native, the stable fly is most abundant in temperate regions. Horses and cattle are their favorite hosts, but they bite man regularly, particularly about the shins and ankles. The piercing proboscis of the stable fly is very

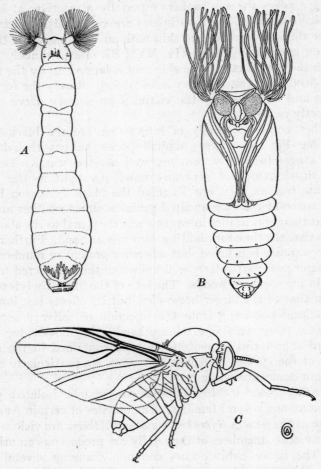

FIG. 44. Developmental stages and adult of a black-fly (Family Simuliidae). *A*, larva; *B*, pupa; *C*, adult. *A* and *B* redrawn from Cameron, *C* from Cole and Lovett.

different in structure from that of the mosquito. In the latter, as in other primitive Diptera, the mandibles, maxillae and labrum serve as piercing organs and the labium laps up the exuding blood. In the higher Diptera the mandibles and maxillae become greatly degenerated and the feeding process is restricted to the lapping of liquids.

In these biting muscids, however, the labium is secondarily modified into a thin rigid tube that is thrust directly into the skin of the host after the fashion of an hypodermic needle (Pl. XVII B). Thus the mouthparts after an almost complete transformation of parts have come to serve an identical purpose. On the other hand, there are a number of species of *Musca* whose mouthparts are not fitted for piercing the skin, yet they feed regularly on mammalian blood. This they secure in a totally unexpected manner by approaching a horse-fly or other biting species when it is in the act of feeding and lapping up the exuding blood as the real biter flies away. Patton and Cragg ('13) found that *Musca pattoni* and related species in India follow this procedure; often a number of specimens cluster about the biter and hasten its departure in order thus to "lick the plate."

Other phlebotomic Muscidae are the much dreaded tsetse flies of the Ethiopian region belonging to the genus *Glossina*. These flies serve as alternate hosts for certain flagellate protozoans known as trypanosomes that cause serious diseases in mammals, including African sleeping sickness in man. Although the adult flies are free-living like the Diptera previously mentioned, the active larval stage has been suppressed and the eggs hatch within the genital duct of the mother fly where the larva remains in a brood chamber until fully grown. During this growing period it is nourished by glandular secretions elaborated by the mother fly. The tsetse flies are thus not only viviparous, but pupiparous as the larva is grown and ready for pupation by the time it is born. After crawling into the soil, transformation is accomplished and the adult emerges. On account of its large size only a single larva may be accommodated in the mother at any one time, but "pregnancies" follow one another in continuous succession and in suitable localities the glossinas exist as large populations. Since wild animals act as reservoirs for the trypanosomes the fact that the tsetse flies may infect man after receiving the infection from such animals implies the feeding of the flies on more than one host species. Such indiscriminate feeding habits are, of course, a necessary prelude to the transfer of any animal disease to man. Conversely, where insect-borne diseases affect only one host animal as occurs with human malaria, we have already seen that the mosquito carriers which restrict their feeding most closely to man are the most efficient vectors of malaria, provided that there is no differential susceptibility among the different species of mosquitoes.

Pupiparous reproduction is characteristic of another group of Diptera in which the adults have become epizoic parasites. Some of these retain the wings and the ability to fly, but in others these organs have degenerated or disappeared and a continuous epizoic life becomes obligatory. This series, known as the Pupipara, includes three families all parasitic on birds and mammals.[1] The Hippoboscidae occur on various birds and mammals; the Streblidae nearly always on bats, and the most highly specialized, completely apterous Nycteribiidae (Fig. 45) are exclusively restricted to bats. The adap-

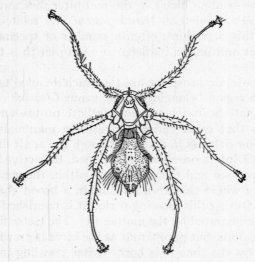

FIG. 45. A wingless parasitic fly (*Colpodia sykesi*) of the family Nycteribiidae that lives externally on bats. Redrawn from Snodgrass.

tations of these flies to parasitic life include a progressive degeneration of the wings, a hard body with leathery abdomen, culminating in the last family mentioned which has the legs enormously enlarged and the head reduced in size, folding back over the top of the thorax. In a few genera the wings develop normally, but are dehiscent, falling off at maturity. Bats are generally infested with these parasites; birds and other mammals less abundantly so. A common form is the wingless sheep-tick or sheep-ked (*Melophagus ovinus*) that infests the wool of sheep the world over.

Related to the Diptera is the order Siphonaptera, including the

---

[1] *Braula*, the bee parasite is also sometimes placed in this group.

fleas, all of which are blood-sucking epizoic parasites of birds and mammals, especially the latter. All fleas are wingless in both sexes and are otherwise beautifully adapted to their habitat in fur or feathers by the possession of a hard, polished integument, compressed body and thickly spinose legs which provide easy locomotion in such an environment. Furthermore, they are proverbially high jumpers and use their prowess in this direction to elude capture or to reach the body of their host by leaping upon it from the substratum. The sexes are similar and both feed by piercing the skin of their hosts with lancet-like mouthparts. The mandibles and also the labrum-epipharynx are needle-like and the blood is aspirated into the pharynx after being mixed with a salivary secretion that is injected at the point of puncture. In the case of human beings, some are greatly plagued by fleas and show a violent reaction by the development of local oedematous spots or welts, while others are practically immune to their bites. This difference is not related to sex, but has never been further elucidated. The reaction is undoubtedly due to the injected salivary liquid. Although the adult flea is tremendously modified, parasitic life has made no impress on the developmental stages. The eggs are laid in the hair or feathers of the host whence they fall to the ground, usually into the nest. The larvae are vermiform and feed on waste organic matter. As such material is most readily available in the case of mammals like rodents which commonly occupy permanent nesting sites, the flea fauna of these animals is unusually extensive. After metamorphosis the adult flea then seeks its warm-blooded host where it remains during adult life. Fleas are commonly restricted to specific hosts and the correlation of host and parasite within the order is quite close although not generally so specific as in the case of entomophagous parasites and many vegetarian insects. Closer attention has been given to the host preferences of the species that infest rats than to those of any other animals since fleas serve as the vectors of bubonic plague which they transfer from rats and some other rodents to man, thus giving rise to this disease in the human species often in the form of widespread and disastrous epidemics. Both the tropical rat flea, *Xenopsylla cheopis*, and the common rat flea of temperate regions, *Nosopsyllus fasciatus*, readily bite man. The same is true of the cat and dog flea, both of which are related members of another genus, *Ctenocephalides*. In our region these two are the most annoying species, whose larvae occasionally develop abundantly in houses. Another species, *Pulex irritans*, known as

the human flea is a pest of man on our own Pacific coast and in
Europe.

The Mallophaga (biting or bird lice) and Anoplura (sucking lice)
are the most highly specialized of all the ectoparasitic insects both
structurally and in habits. In general appearance the members of
these two orders are similar in many respects, except that the former
possess biting mouthparts and the latter have true suctorial ones.
It is generally believed that the Mallophaga are related to the
Corrodentia and that the Anoplura are derived from forms related
to the large order Hemiptera. Nevertheless it is thought by some
that the biting and sucking lice are more closely related than this
in spite of the fundamental differences in the structure of their
mouthparts. Both undergo an incomplete metamorphosis, the
entire life cycle taking place on the body of the host. They are
flattened, entirely without wings, blind or with greatly reduced
eyes, with the legs modified for clinging to the fur or feathers of
their hosts. Moreover, they are unable to survive for more than a
very short period when deprived of the warmth that they enjoy in
their natural environment and have few opportunities to migrate
from one host animal to another, more particularly from one species
to another.

On account of these peculiar conditions of life they are very closely
associated with specific hosts and it was soon evident to students
who gave close attention to these degraded and loathsome parasites
that they were of great interest from an evolutionary standpoint.
They illustrate very clearly the phenomenon of parallel, or more
properly, concurrent evolution of host and parasite to which nearly
all post-Darwinian parasitologists have at least given passing
notice.

The Mallophaga (Fig. 46) feed to a large extent upon the barbules
of the feathers of birds and the small number that infest mammals
depend mainly upon the fatty secretions of the skin and the outer,
scaly layer of the epidermis. On the other hand, the Anoplura are
entirely dependent upon the blood of their hosts.

The American entomologist, Vernon Kellogg, was the first to
make any detailed study of the significance of bird lice in tracing
the phylogeny and relationships of their hosts ('96; '13; '14). Several other students, notably Harrison and Ewing, have greatly extended our knowledge, dealing also with the sucking lice. It is clear
from the work of these observers that there are a great many instances of close relationship between species of parasites infesting

obviously related birds. This is partly understandable on the basis
that the Mallophaga of non-gregarious birds have little opportunity to pass from one species of bird to another, although, on the
other hand, transfer between birds of the same species may easily
be effected at the time of mating or still more readily while the parent
birds are in close contact with their nestlings, thus passing from one
generation to another in perpetuity. Another wholly unexpected

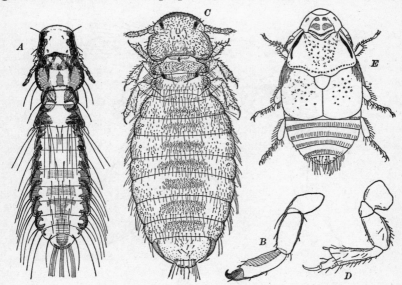

FIG. 46. Epizoic parasites. *A, Lipeurus mississippiensis, B,* hind
leg of same; *C, Trichodectes hermsi; D,* hind leg of same (Mallophaga). *E, Platypsyllus castoris* (Coleoptera). *A* and *B* after
MacGregor, *C* and *D* after Kellogg and Nakayama, *E* after
Westwood.

The *Lipeurus* is a parasite of the flicker, the *Trichodectes* of the
goat and the *Platypsyllus* lives on the beaver.

method of migration is indicated by MacAttee ('22) who found a
bird louse attached to a hippoboscid fly which served as transportation to a new host sought by the likewise parasitic fly. Similarly
the large oestrid fly (*Dermatobia*) that lives in the subcutaneous
tissues of man regularly lays its eggs on mosquitoes and other blood-
sucking Diptera (Fig. 47) whence the larva is able to reach the
human host when its carrier is feeding. Kellogg noted in the case
of the birds of the Galapagos islands that there is an unusual amount
of "straggling," or the occurrence of certain species of Mallophaga

on more than one and often on unrelated kinds of birds. He attributed this condition to the close association of various birds in very limited areas on these islands. While there is evidently good reason why straggling should ordinarily be greatly limited it is clear that this alone cannot account for the persistence of these Mallophaga in restricting themselves to specific hosts. This persistence is shown by a great many species and is in general illustrated by the occur-

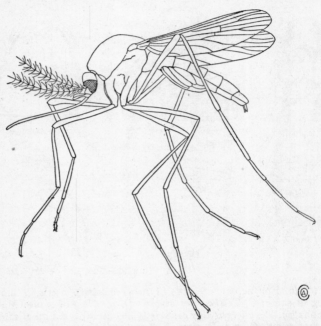

FIG. 47. A mosquito, *Janthinosoma lutzi*, bearing attached to the underside of the abdomen, a cluster of eggs of the oestrid fly, *Dermatobia hominis*, which is parasitic on mammals, including the human species. Redrawn from Sambon.

rence of related species on the several members of certain families of birds that enjoy wide geographical ranges. In a number of cases a single species of bird louse occurs on two or more related birds whose habitats are so widely separated that contact between these birds is now utterly out of the question. We must conclude then that the attachment of these Mallophaga to such related species dates at least from the time when the birds had not become specifically distinct or possibly in some cases where their ranges may have overlapped. Still more striking cases are seen in identical mallophagan parasites affecting much more distantly related birds. Thus, among the Australian cassowary, the African ostrich and the South American rheas there are species of bird lice on the rheas, one of which occurs on each of the other two giant old-world birds. This condition bespeaks a most persistent attachment to hosts over a long period of evolution as the ancient separation of these continents is incontrovertible. This condition is not alone true of their insect parasites, for Harrison ('28) has pointed out that the same relationship is shown by their parasitic helminths.

Some observations made by Ewing ('33) show that the host relationships of the bird lice are not matters of chance migration, but that they have a physiological basis, at least in some cases. It will be remembered that the cowbirds comprise a series of parasitic birds that lay their eggs in the nests of other species. Among these, the common *Molothrus ater* is known to lay its eggs in the nests of "no less than 158 species of birds. These belong to eight orders, 25 families and 103 genera." Here then is a condition where there is every opportunity for the transfer of foreign species to the nestling cowbirds and practically no chance for them to acquire parasites from their own parents. Ewing finds this cowbird comparatively free of Mallophaga, yet those that do infest it are forms belonging to genera found on blackbirds belonging to the same family (Icteridae) as the cowbird. Here we have definite proof that there is a specificity of host relationship prevalent among these birds and their parasites, for we do not find them infested with lice derived from their foster parents in whose nests they have been reared. Bagnall has come to the same conclusion from a study of the European cuckoo. Later, Geist in America noted several species of lice on cowbirds. He found these to fall into two groups; one includes species that occur only on icterid hosts, a second those that lack host specificity, and a third made up of those that do not regularly parasitize cowbirds. The third group is much the smallest.

Other evidence on host specificity among the sucking lice (Anoplura) is contained in the paper by Ewing cited above. He undertook the rather heroic experiment of feeding lice from a spider monkey, another baboon-like monkey and also the common dog-louse on his own arm. Although they accepted the strange human host more or less readily, all those that fed died promptly of acute indigestion. Aside from showing that entomologists are not so innocuous as some people take them to be, these experiments demonstrate conclusively that the lice are unable to utilize the blood of an abnormal host animal, even though their instincts do not deter them from making the attempt. This does not imply, however, that changes to a new host never occur and become permanent. Such an accidental shift and establishment on a new, permanent host has probably occurred a few times in the course of every million years. From what has been said in previous chapters concerning the food preferences of other kinds of insects, particularly vegetarian species, it is at once evident that we are here dealing with different conditions. In these lice there can be a rigid elimination of individuals that partake of food which their instincts do not forbid them to try, whereas most vegetarian forms absolutely refuse to taste strange plants which we have every reason to believe would prove fully acceptable to their digestive apparatus.

Other blood-sucking insects are included among the true bugs of the large order Hemiptera, most of whose members depend upon the sap of plants for food. Some predatory members of this group which suck the juices of other insects have already been mentioned in a previous chapter. The acquisition of blood sucking habits by such insects may be easily accomplished by transferring their attention from insect to vertebrate hosts and examples of such a transition in process may be cited. Certain water-bugs of several families, *e.g.*, *Pelocoris*, *Lethocerus* (Fig. 48), *Belostoma*, feed normally on other insects, but all of these will bite man if the opportunity to do so presents itself. They inflict an extremely painful wound, injecting into the skin a digestive enzyme so powerful that it causes a localized destruction of the tissue at the point where the beak is inserted. Members of another family of terrestrial Hemiptera, the Reduviidae, known as assassin-bugs will behave in the same way and the notorious "kissing-bugs" are familiar examples. One of these, *Reduvius personatus*, often called also the "big bedbug," is a case of "the biter bitten," as one of its principal sources of food is the bedbug from which it acquires human blood second-hand by stabbing and draining the body of the primary thief. Other reduviids, notably species of *Conorhinus* or *Triatoma* (Fig. 48), are obligate parasites of mammals, including man, obtaining their blood meals directly from the vertebrate hosts. They are of major importance in the tropics where they act as carriers of a disease due to trypanosomes related to those causing African sleeping sickness.

The bedbug depends similarly on blood for food and will feed upon a great variety of animals, often becoming a pest on animals raised in laboratories for experimental purposes. It is nevertheless primarily dependent on the human species and practically cosmopolitan although represented by two common species, *Cimex lectularius* and *C. rotundatus*, the latter distributed generally throughout tropical countries. A small number of related forms affect birds and bats. The effects of parasitism are evident in the loss of wings, but the bugs are free-living, visiting the host only for feeding, so that in this

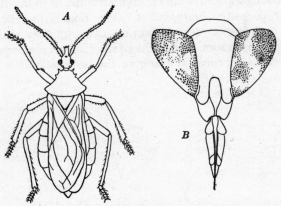

FIG. 48. *A*, a blood-sucking bug, *Triatoma; B*, mouthparts of the predatory bug, *Lethocerus.* Original.

respect they are intermediate between the mosquitoes and fleas, although on account of their incomplete metamorphosis the younger stages suck blood like the adults.

The few remaining epizoic parasites among the insects are scattered through various groups, but no account would be complete without reference to some of these as they illustrate very beautifully the great power of adaptation exhibited by insects in general.

One of the most remarkable of these is a minute beetle (*Platypsyllus*) (Fig. 46) parasitic in the fur of the beaver in Europe and North America. Only a single species is known, resembling a cockroach, entirely blind, with flattened body, no wings and greatly modified antennae. The exact nature of its food is not known. A few other beetles belonging to the enormous family Staphylinidae are true ectoparasites of various marsupials and rodents (Seevers '44).

Bats are for some reason particularly susceptible to insect ectoparasites. In addition to the pupiparous flies mentioned on a previous page, there are members of two other orders that infest these animals. Some of the most remarkable Hemiptera, forming the family Polyctenidae, are widespread, occurring on various tropical bats (Jordan '12). They resemble the Nycteribiid parasites of bats and like them are viviparous, producing young differing only slightly from the adults as they undergo an incomplete metamorphosis. They are blind, with flattened body, flightless, and have a series of combs, or ctenidia on the underside of the head and thorax (Fig. 49). These ctenidia represent a most peculiar modification which, curiously enough appears also in nycteribiids, in many fleas and in the beaver parasite, *Platypsyllus*. The reason for such striking structural convergence seems obscure, although they must serve to anchor their possessors, and incidentally, in the mammal-infesting species, perhaps to comb the hair of the host as a little gesture of

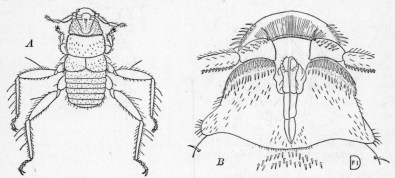

Fig. 49. *Polyctenes fumarius*, a West Indian hemipterous insect that lives as an external parasite of bats. *A*, dorsal view; *B*, ventral view of head and thorax, showing the rows of comb-like bristles. Redrawn from Westwood.

appreciation for their bed and board. Other bat parasites are species of *Arixenia* (Jordan '09), a Malayan genus of earwigs, but parasitism is here apparently not obligatory as they have been found also frequenting the guano beds in bat-caves. Related to these are species of *Hemimerus* that live as ectoparasites in the fur of giant African rats of the genus *Cricetomys*. These are also greatly degenerate earwigs, entirely wingless, totally blind and viviparous. They feed on the epidermal scales of the host and also on a fungus that is likewise parasitic on the rats.

There are a few species of Diptera whose larvae feed by sucking the blood of birds and mammals, in spite of the handicap imposed upon them by the entire absence of legs and elaborate sense organs. In North America several species of *Protocalliphora* develop in the nests of a great variety of birds where they attack the nestlings, and in Europe, *P. azurea* has similar habits. The flies are often numerous and exact a large toll from the bird population. The larvae puncture the skin by rasping movements of the sharp mouth hooklets and the margin of the first thoracic segment forms a sucker that enables the larva to adhere to the skin of its victim while feeding.

The African *Passeromyia heterochaeta* attacks birds in the same way (Rodhain and Bequaert '16). Larvae of this sort must of course depend upon sedentary hosts, a condition admirably fulfilled by nestling birds. Others, however, live on the blood of mammals like the African genus *Choeromyia*, attacking species with naked skin such as the aard-vark (*Orycteropus*) and the wart-hog. One species even attacks man. This is *Aucheromyia luteola*, known as the Congo floor maggot, whose larvae occur in the soil of huts where they are able to reach the bodies of sleeping natives and feed stealthily at night when their victims lie upon the ground.

During the past several decades numerous cases of small midges belonging to *Ceratopogon* and other genera of the family Ceratopogonidae have come to light in which the adults have been found sucking the blood of other insects. Several of these relate to attacks upon caterpillars; one of the earliest (Baker '07) was a species of *Forcipomyia* in Cuba and other instances have been observed in Florida (Knab '14), Sumatra (Jacobson '21), Java (Roepke '21) and in South America, involving related forms. In India they have been found sucking the blood of mosquitoes (Gravely '11). One species of *Forcipomyia* attaches itself to the wing veins of lacewing flies, *Chrysopa* (Tjeder '36) and a Danish *Ceratopogon* occurs on the wings of moths with the proboscis inserted into the wing veins (Kryger '14). A very interesting account of a number of species belonging to several genera has been given by MacFie ('32), especially of some that occur on the wings of dragonflies where they feed by inserting the proboscis into the veins (Fig. 50). Some of these lack the tarsal claws, clinging to the wings of the host by means of the remarkably modified empodium at the tip of the last tarsal joint. Likewise the aquatic larva of a remarkable midge of the family Chironomidae develops as an ectoparasite of mayfly larvae (Dorier '38) attached to the body beneath the wing cases.

These cases of insect hosts for adult midges are particularly interesting as practically all the blood sucking Ceratopogonidae attack warm-blooded vertebrates. An engorged mosquito is of course really a reservoir of vertebrate blood, and furthermore it appears that at least one caterpillar-infesting *Forcipomyia* may also bite human beings, so it appears probable that the habit of attacking other insects is a secondary one of erratic although not rare occurrence. With some of these midges, at least, there can be no specific association with hosts.

Another minute fly, *Braula coeca*, known as the "bee louse" lives in the adult stage as an ectoparasite on the body of the honeybee. Until recently it was thought to suck the blood of the bees, but is now known to beg food by tickling the mouthparts of its host and inducing the regurgitation of drops of liquid which it drinks up. The eggs of the fly are laid on the comb and the larvae secure food

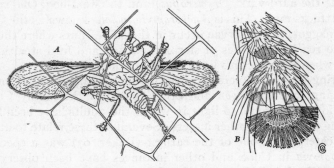

Fig. 50. A blood-sucking midge, *Pterobosca adhesipes*, which attaches itself to the wings of dragonflies. *A*, adult in position on the wing of *Agrionoptera*; *B*, tip of tarsus, showing the sucker-like process.

from the store provided by the bees for their own brood. *Braula* is therefore really a commensal and neither a true parasite nor predator at any stage of its life-history. It lives only with the honeybee.

Still other fly larvae have been found in South America mining beneath the scaly covering on the wings of butterflies. These have never been bred so far as I know, although first noted many years ago by Schulz ('04), who believed them to be Phoridae.

The epizoic parasites are certainly a lousy lot, and we have allotted them more space than will be agreeable to most of our readers. Nevertheless they are really of deep interest to anyone concerned with the problems of general biology and evolution.

REFERENCES

Aitken, T. H. G.

1939 The *Anopheles maculipennis* Complex of Western North America. Pan-Pacific Entom., vol. 15, pp. 191–192.

Barber, M. A., and T. B. Hayne

1924 Some Notes on the Relation of Domestic Animals to *Anopheles*. United States Public Health Reports, vol. 39, p. 141.

Bates, M.

1939a The Use of Salt Solutions for the Demonstration of Physiological Differences Between the Larvae of Certain European Anopheline Mosquitoes. American Jour. Trop. Med., vol. 19, pp. 357–384, 4 figs.

1939b Hybridization Experiments with *Anopheles maculipennis*. American Journ. Hygiene, vol. 29, Sec. C, pp. 1–6.

1940 The Nomenclature and Taxonomic Status of the Mosquitoes of the *Anopheles maculipennis* Complex. Ann. Entom. Soc. America, vol. 33, pp. 343–356.

Bates, M., and L. W. Hackett

1939 The Distinguishing Characteristics of the Populations of *Anopheles maculipennis* Found in Southern Europe. Verh. VIIten Int. Kongr. Entom., Berlin, 1938, vol. 3, pp. 1555–1569, 3 figs.

Bishopp, F. C.

1933 Mosquitoes Kill Live Stock. Science, vol. 77, pp. 115–116.

Bruck, C.

1911 Ueber das Gift der Stechmücke. Deutsche Mediz. Wochens., vol. 37, pp. 1787–1790.

Bull, C. G., and W. V. King

1923 The Identification of the Blood Meal of Mosquitoes by Means of the Precipitin Test. American Journ. Hyg., vol. 3, pp. 491–496.

Bull, C. G., and B. D. Reynolds

1924 Preferential Feeding Experiments with Anopheline Mosquitoes. II. American Journ. Hyg., vol. 4, pp. 109–118.

Bull, C. G., and F. M. Root

1923 Preferential Feeding Experiments with Anopheline Mosquitoes. I. *Ibid.*, vol. 3, pp. 514–520.

Genna, M.

1923   Richerche sulla nutrizione dell'*Anopheles claviger*.   Arch. Zool. Ital. Napoli, vol. 10, pp. 15–34.

Giovannola, Arnaldo

1939   I plasmodi aviari. Riv. Parasitol., vol. 3, pp. 222–276.

Hackett, L. W.

1937   Malaria in Europe — an Ecological Study. xvi + 366 pp. London, Oxford Univ. Press.

1940   Some Obscure Facts in the Epidemiology of Malaria. American Journ. Pub. Health, vol. 30, pp. 589–594.

Hackett, L. W., and A. Missiroli

1935   The Varieties of *Anopheles maculipennis* and Their Relation to the Distribution of Malaria in Europe. Riv. Malariol., vol. 14, pp. 45–109, 4 pls.

Hinman, E. H.

1930   A Study of the Food of Mosquito Larvae. American Journ. Hyg., vol. 12, pp. 238–270.

1941   The Problem of Races of *Anopheles quadrimaculatus* in the United States. Proc. 6th Intern. Congr. Entom., Madrid, vol. 2, pp. 937–942.

Howard, L. O.

1924   On Zoophilism with *Anopheles*. A Review. Journ. Parasitol., vol. 10, pp. 191–198.

Howard, L. O., H. G. Dyar, and F. Knab

1912   The Mosquitoes of North and Central America and the West Indies. I. A General Consideration of Mosquitoes, Their Habits and Their Relations to the Human Species. Carnegie Inst. Washington, Pub., No. 159, 520 pp.

Huff, C. G.

1929   Ovulation Requirements of *Culex pipiens*. Biol. Bull., vol. 56, pp. 347–350.

Knab, F.

1907   Mosquitoes as Flower Visitors. Journ. New York Entom. Soc., vol. 15, pp. 215–219.

McKinley, C. B.

1929   The Salivary Gland Poison of *Aëdes aegypti*. Proc. Soc. Expt. Biol. Med., vol. 26, pp. 806–809.

Roubaud, E.

1933   Sur la vie du moustique commun (*Culex pipiens*). Ann. Sci. Nat. Zool., (10) vol. 16, No. 1, 108 pp., 8 pls., 32 figs.

Roy, D. N.

1927   The Physiology and Function of the Oesophageal Diverticula and of the Salivary Glands in Mosquitoes. Indian Journ. Med. Res., vol. 14, pp. 995–1004.

Rudolfs, W.

1925   The Food of Mosquito Larvae. Proc. 12th Ann. Meeting New Jersey Mosq. Extermination Assoc., pp. 25–33.

Tate, P., and M. Vincent

1932 Influence of Light on the Gorging of *Culex pipiens*. Nature, vol. 130, pp. 336–337.

Thiel, P. van, and L. Bévère

1939 Preuve expérimentale de l'anthrophilie d'*Anopheles maculipennis labranchiae* et *elutus*. Bull. Soc. Path. exot., 32, pp. 103–109.

Vargas, A. R.

1939 The Occurrence of Anophelines in the Absence of Malaria in Mexico. Medicina, vol. 19, pp. 333–339.

Wesenberg-Lund, C.

1921 Sur les causes du changement intervenu dans le mode de nourriture de *l'Anopheles maculipennis*. C. R. Soc. Biol., France, vol. 85, pp. 383–386.

Yorke, W., and J. W. S. Macfie

1924 The Action of the Salivary Secretion of Mosquitoes and of *Glossina tachinoides* on Human Blood. Ann. Trop. Med. Parasitol., vol. 18, pp. 103–108.

OTHER BLOOD–SUCKING FLIES

Austen, E. E.

1906 British Blood-sucking Flies. London, British Museum, pp. 1–74, 34 pls.

1909 Illustrations of African Blood-sucking Flies other than Mosquitoes and Tsetse-flies. London, British Museum, xv + 221 pp. 13 pls.

1911 A Handbook of the Tsetse Flies (Genus *Glossina*). London, British Museum, x + 110 pp., 10 pls.

Bequaert, J. C.

1934 Notes on the Black Flies or Simuliidae. Contrib. Dept. Trop. Med., Harvard Univ., No. 6, pp. 175–224. Cambridge, Mass., Harvard Univ. Press.

Bezzi, M.

1911 Études systématiques sur les muscides hématophages du genre *Lyperosia*. Arch. Parasitol., vol. 15, pp. 110–143, 1 pl., 15 figs.

Bishopp, F. C.

1913 The Stable Fly (*Stomoxys calcitrans*) an Important Live Stock Pest. Journ. Econ. Entom., vol. 6, pp. 112–126.

Bromley, S. W.

1926 The External Anatomy of the Black Horse-fly, *Tabanus atratus*. Ann. Entom. Soc. America, vol. 19, pp. 440–460, 12 figs.

Brues, C. T.

1913 The Geographical Distribution of the Stable-fly *Stomoxys calcitrans*. Journ. Econ. Entom., vol. 6, pp. 459–477.

Cameron, A. E.

1922 The Morphology and Biology of *Simulium simile*. Bull. Canada Dept. Agric., n.s. No. 5, 26 pp., 9 figs.

Cornwall, J. W., and W. S. Patton

1914 Some Observations on the Salivary Secretion of the Commoner Blood-sucking Insects and Ticks. Indian Journ. Med. Res., vol. 2, pp. 569–593.

Dyar, H. G.

1927   The North American Simuliidae. Proc. United States Nat. Mus., vol. 69, art. 10, 54 pp., 7 pls.

Friedrichs, K.

1919–   Untersuchungen über Simuliiden. Zeits. angew. Entom., vol. 6, pp. 61–83, 21   15 figs.; vol. 8, pp. 31–92, 11 figs.

Grünberg, K.

1907   Die blutsaugenden Dipteren. vi + 188 pp., 127 figs. Jena, Gustav Fischer.

Jobling, B.

1928   The Structure of the Head and Mouth Parts of *Culicoides pulicaria.* Bull. Entom. Res., vol. 18, pp. 211–236.

Knab, F.

1912   Blood-sucking and Supposedly Blood-sucking Leptidae. Proc. Entom. Soc. Washington, vol. 14, pp. 108–110.

1915   Dipterological Miscellany: Evolution of the Blood-sucking Habit in *Symphoromyia. Ibid.,* vol. 17, pp. 38–40.

Knab, F., and R. A. Cooley

1912   Symphoromyia as a Blood-sucker. *Ibid.,* vol. 14, pp. 161–162.

Krijgsman, B. J.

1933   The Host Preference of *Lyperosia exigua.* Australian Council Sci. and Industr. Res., Pamphlet 43, pp. 7–8.

1933   The Relation Between the Adult *Lyperosia exigua* and Mammalian Faeces. *Ibid.,* Pamphlet 43, pp. 9–19.

Lester, H. M. O., and L. Lloyd

1928   Notes on the Process of Digestion in Tsetse Flies. Bull. Entom. Res., vol. 19, pp. 39–60.

Mitzmain, M. B.

1913   The Bionomics of *Stomoxys calcitrans,* a Preliminary Account. Philippine Journ. Sci., vol. 8B, pp. 29–48.

Newstead, R.

1911   The Papatici Flies (*Phlebotomus*) of the Maltese Islands. Bull. Entom. Res., vol. 2, pp. 47–78.

Newstead, R., A. M. Evans, and W. H. Potts

1924   Guide to the Study of Tsetse-flies. Med. Mem. Liverpool School Trop. Med., n.s. 1, xi + 268 pp., 4 maps, 28 pls., 59 figs.

Painter, R. H.

1927   The Biology, Immature Stages and Control of the Sand-flies (Ceratopogoninae) in Honduras. Ann. Rept. Med. Dept. United Fruit Co. for 1926, pp. 245–262.

Patton, W. S., and F. W. Cragg

1913   On Certain Haematophagous Species of the Genus *Musca.* Indian Journ. Med. Res., vol. 1, pp. 11–23.

Petersen, A.

1924 Bidrag til de Danske Simuliers Naturhistorie. D. kgl. Danske Vidensk. Skrifter Naturv. Math. (8), vol. 5, pp. 237–339, 2 pls., 54 figs.

Roubaud, E.

1909 La *Glossina palpalis*. Sa biologie, son rôle dans l'étiologie des trypanosomiases. Rapport de la Mission d'études de la maladie du sommeil au Congo Français (1906–1908). Paris, pp. 383–632.

Snyder, T. E.

1917 Notes on Horse-flies as a pest in Southern Florida. Proc. Entom. Soc. Washington, vol. 18, pp. 208–211.

Stokes, J. H.

1914 A Clinical, Pathological and Experimental Study of the Lesions Produced by the Bite of the Black Fly (*Simulium venustum*). Journ. Cutan. Diseases, Nov.–Dec. 1914.

Stuhlmann, F.

1907 Beiträge zur Kenntnis der Tsetse-fliege (*Glossina fusca* und *G. tachinoides*). Arbeit. K. Gesundheitsamte, Berlin, vol. 26, pp. 301–383.

Swynnerton, C. F. M.

1936 The Tsetse Flies of East Africa. Trans. Roy. Entom. Soc. London, vol. 84, iv + 597 pp., 22 pls., 33 figs., 8 maps.

Wilhelmi, J.

1917 Die gemeine Stechfliege; Untersuchungen über die Biologie der *Stomoxys calcitrans*. Zeits. angew. Entom., No. 2, 110 pp.

Windred, G. L.

1933 Some Food Reactions of the Larvae of *Lyperosia exigua*. Australian Council Sci. and Industr. Res., Pamphlet 43, pp. 20–22.

PUPIPAROUS DIPTERA

Bezzi, M.

1916 Riduzione e scomparsa delle ali negli insetti Ditteri. Riv. Sci. Nat. "Natura," vol. 7, pp. 85–182, 11 figs.

Falcoz, L.

1923 Biospeologica, XLIX. Pupipara (Diptères). Première Série. Arch. zool. Paris, vol. 61, pp. 521–552.

Jobling, B.

1926 A Comparative Study of the Head and Mouth-parts of the Hippoboscidae. Parasitology, vol. 18, pp. 319–449, 4 pls., 4 figs.

1928 The Structure of the Head and Mouthparts of the Nycteribiidae. *Ibid.*, vol. 20, pp. 234–272, 3 pls., 4 figs.

Keilin, D.

1916 Sur la viviparité chez les Diptères et sur les larves des Diptères vivipares. Arch. Zool. Expér., vol. 55, pp. 393–415.

Massonat, E.

1909 Contribution a l'étude des Pupipares. Ann. Univ. Lyon, n.s., vol. 128, 356 pp.

Phillips, W. W. A.

1924   Ectoparasites of Ceylon Bats. Spolia Zeylandica, vol. 13, pp. 65–70.

Rodhain, J., and J. Bequaert

1915   Observations sur la biologie de *Cyclopodia greffi*, Nycteribiid parasite d'un chauve-souris congolaise. Bull. Soc. Zool. Paris, vol. 40, pp. 248–262.

Ryberg, O.

1939   Fortpflanzungsbiologie und Metamorphose der Fledermausfliegen, Nycteribiidae. Verh. 7ten internat. Kong. Entom., Berlin, vol. 2, pp. 1285–1299.

Speiser, P.

1899   Ueber die Nycteribiiden. Arch. f. Naturg., Jahrg. 67, pp. 11–78, 1 pl., 4 figs.

1900   Ueber die Strebliden, Fledermansparasiten. Arch. f. Naturg., Jahrg. 66, pp. 31–70, 2 pls., 2 figs.

1908   Die geographische Verbreitung der Diptera Pupipara und ihre Phylogenie. Zeits. wiss. Insektenbiol., vol. 4, pp. 241, 301, 420, 437.

FLEAS

Bacot, A.

1914   A Study of the Bionomics of the Rat Fleas and Other Species. Journ. Hyg., Plague Supplement, vol. 3, pp. 447–655.

Baker, C. F.

1904   A Revision of the American Siphonaptera or Fleas. Proc. United States Nat. Mus., vol. 27, pp. 365–470.

Doane, R. W.

1908   Notes on Fleas Collected on Rat and Human Hosts in San Francisco and Elsewhere. Canadian Entom., vol. 40, pp. 303–304.

Dunn, L. H.

1923   Fleas of Panama, their Hosts and their Importance. Journ. Trop. Med., vol. 3, pp. 335–344.

Fox, I.

1940   Fleas of Eastern United States. vii + 191 pp., 168 figs. Ames, Iowa, Iowa State College Press.

Hicks, E. P.

1930   The Early Stages of the Jigger, *Tunga penetrans*. Ann. Trop. Med. Parasitol., vol. 24, pp. 575–586.

Mitzmain, M. B.

1910   General Observations on the Bionomics of Rodent and Human Fleas. Bull. United States Public Health Serv., vol. 38, pp. 1–34.

Sharif, M.

1937   On the Life History and the Biology of the Rat-flea (*Nosopsyllus fasciatus*). Parasitology, vol. 29, pp. 225–238.

Wagner, F. von

1926   Hosts and Phylogeny of Fleas. Acta Soc. Entom. Serb. 1926, pp. 29–37.

MALLOPHAGA AND ANOPLURA

Bagnall, R. S.

1931 Some Problems connected with the Cuckoo and its Ectoparasites. Scottish Naturalist, No. 191, pp. 145–147.

Buxton, P. A.

1939 The Louse. ix + 115 pp. London, Edward Arnold and Co.

Ewing, H. E.

1924a On the Taxonomy, Biology and Distribution of the Biting Lice of the Family Gyropidae. Proc. United States Nat. Mus., vol. 63, art. 20, 42 pp., 1 pl., 18 figs.

1924b Lice from Human Mummies. Science, n.s., vol. 60, pp. 389–390.

1926 A Revision of the American Lice of the Genus *Pediculus*, Together with a Consideration of the Significance of the Geographical and Host Distribution. Proc. United States Nat. Mus., vol. 68, art. 19, 30 pp., 3 pls., 8 figs.

1933 Some Peculiar Relationships Between Ectoparasites and their Hosts. American Naturalist, vol. 67, pp. 365–372.

Fahrenholz, H. von

1920 Bibliographie der Läuse (Anopluren). Zeits. angew. Entom., vol. 6, pp. 106–160.

Ferris, G. F.

1916a Mallophaga and Anoplura from South Africa, with a List of Mammalian Hosts of African Species. Ann. Durban Mus., vol. 1, pp. 245–252.

1916b A Catalogue and Host List of the Anoplura. Proc. California Acad. Sci., vol. 6, pp. 129–213.

Geist, R. M.

1935 Notes on the Infestation of Field Birds by Mallophaga. Ohio Journ. Sci., vol. 35, pp. 93–100.

Grimshaw, P. H.

1917 The British Lice (Anoplura) and their Hosts. Scottish Naturalist, 1917, pp. 13–17; 65–68.

Harrison, L.

1916a Bird-Parasites and Bird-Phylogeny. Ibis (10), vol. 4, pp. 254–263.

1916b A Preliminary Account of the Structure of the Mallophaga. Parasitology, vol. 9, pp. 1–156.

1916c A Preliminary Account of the Structure of the Mouthparts in the Body-louse. Proc. Cambridge Philos. Soc., vol. 18, pp. 207–226, 1 pl., 7 figs.

1928 Host and Parasite. Proc. Linn. Soc., New South Wales, vol. 35, pp. ix–xxxi.

Jellison, W. L.

1942 Host Distribution of Lice on Native American Rodents. Journ. Mammal., vol. 23, pp. 245–250.

Kellogg, V. L.

1896 Mallophaga of North American Birds. Zool. Anz., vol. 19, pp. 121–123.

1902 Mallophaga from Birds [of Galapagos]. Proc. Washington Acad. Sci., vol. 4, pp. 457–499, 4 pls.

1910 Mallophagan Parasites of the California Condor. Science, vol. 31, pp. 33–34.

1913a Distribution and Species Forming of Ectoparasites. American Naturalist, vol. 47, pp. 129–158.
1913b Ectoparasites of Monkeys, Apes and Man. Science, vol. 38, pp. 601–602.
1914 Ectoparasites of Mammals. American Naturalist, vol. 48, pp. 257–279.

Kellogg, V. L., and G. F. Ferris

1915 The Anoplura and Mallophaga of North American Mammals. Leland Stanford Univ. Publ., Univ. Ser., 74 pp., 8 pls.

Kotlan, A.

1923 Ueber die Blutaufnahme als Nahrung bei den Mallophagen. Zool. Anz., vol. 56, pp. 231–233.

McAtee, W. L.

1922 Bird Lice (Mallophaga) attaching Themselves to Bird Flies (Hippoboscidae). Entom. News, vol. 33, p. 90.

Martini, E.

1918 Zur Kenntnis des Verhaltens der Läuse gegenüber Wärme. Zeitschr. f. Angew. Entom., vol. 4, pp. 34–70.

Neiva, A., and J. F. Gomes

1917 Biologia da Mosca do Berne (*Dermatobia hominis*). Ann. Paulist. Med. Cirug., vol. 8, pp. 197–209. [Phoresy.]

Newstead, R., and W. H. Potts

1925 Some Characteristics of the First Stage Larva of *Dermatobia hominis*. Ann. Trop. Med. Parasitol., vol. 19, pp. 247–260. [Phoresy.]

Nuttall, G. H. F.

1917a The Biology of *Pediculus humanus*. Parasitology, vol. 10, pp. 80–185, 2 pls.
1917b Bibliography of *Pediculus* and *Phthirius*. Ibid., vol. 10, pp. 1–42.
1918 The Biology of *Phthirius pubis*. Ibid., vol. 10, pp. 383–405.

Oppenheim, M.

1923 Zur Entstehung der maculae coeruleae. Dermat. Wochenschr., vol. 76, pp. 170–173.

Osborn, H.

1892 Origin and Development of the Parasitic Habit in Mallophaga and Pediculidae. Insect Life, vol. 5, pp. 187–191.

Pavlovsky, E. N., and A. K. Stein

1924a Ueber die Wirkung des Speichels des *Pediculus* auf die Integumenta des Menschen. Zeits. ges. exper. Med., vol. 42, pp. 15–24, 3 figs.
1924b *Maculae coeruleae* and *Phthirius pubis*. Parasitology, vol. 16, pp. 145–149.

Peacock, A. D.

1918 The Structure of the Mouthparts and Mechanism of Feeding in *Pediculus humanus*. Ibid., vol. 11, pp. 98–117.

Snodgrass, R. E.

1899 The Anatomy of the Mallophaga. Occas. Pap. California Acad. Sci., vol. 6, pp. 145–224.

Waterston, J.

1926a On the Crop Contents of Certain Mallophaga. Proc. Zool. Soc. London, 1926, pp. 1017–1020, 1 fig.

1926b The Mallophaga and Anoplura and their Host-relations. Verh. 3ten Intern. Entom. Kongr. Zurich, vol. 2, p. 576.

Werneck, F. L.

1936 A Contribution to the Knowledge of the Mallophaga of the Mammals of South America. Mem. Inst. Oswaldo Cruz, vol. 31, pp. 391–590, 1 pl., 227 figs.

Zinsser, Hans

1935 Rats, Lice and History. xii + 301 pp. Boston, Mass., Little Brown & Co.

HAEMATOPHAGOUS HEMIPTERA

Bergevin, E. de

1924 Nouvelles observations sur les Hémiptères suceurs de sang humain. Bull. Soc. Hist. Nat. Afrique du Nord, vol. 15, pp. 259–262, 1 fig.

Bruner, S. C.

1914 Importance du Cannibalisme et de la Coprophagie chez lez Réduviides hématophages. Bull. Soc. Path. exot., vol. 7, p. 702.

Buxton, P. A.

1930 The Biology of a Blood-sucking Bug, *Rhodnius prolixus*. Trans. Entom. Soc. London, vol. 78, pp. 227–236.

Galliard, H.

1935 Recherches sur les Réduvides hématophages, *Rhodnius* et *Triatoma*. Ann. Parasitol., vol. 13, pp. 289–306; 401–423; 497–527.

Girault, A. A.

1905 The Bedbug, *Clinocoris lectularia*, Part I. Life History at Paris, Texas, with Biological Notes, and some Considerations on the Present State of our Knowledge Concerning it. Psyche, vol. 12, pp. 61–74.

Hase, A.

1917 Die Bettwanze (*Cimex lectularius*) ihr Leben und ihre Bekämpfung. Zeits. angew. Entom., vol. 4, vi + 144 pp.

Horvath, G. von

1911 Les Polycténides et leur adaptation à la vie parasitaire. Mém. 1er Congr. Int. d'Entom., Bruxelles, vol. 1, pp. 249–256, 1 pl.

Jordan, K.

1912 Contribution to Our Knowledge of the Polyctenidae. Novit. Zool., vol. 18, pp. 555–579.

Myers, J. G.

1929 Facultative Blood-sucking in Phytophagous Hemiptera. Parasitology, vol. 21, pp. 472–480.

Readio, P. A.

1927 Studies on the Biology of the Reduviidae of America. Univ. Kansas Sci. Bull., vol. 17, pp. 5–291.

Reuter, O. M.

1913 Die Familie der Bett- oder Hauswanzen (Cimicidae), ihre Phylogenie, Systematik, Oekologie und Verbreitung. Zeits. wiss. Insektenbiol., vol. 9, pp. 251–255; 303–306.

Usinger, R. L.

1934 Blood sucking among Phytophagous Hemiptera. Canadian Entom., vol. 66, pp. 97–100.

INSECTS ECTOPARASITIC ON OTHER INSECTS

Arnhart, Ludwig

1923 Die Larve der Bienenlaus in den Wachsdeckeln der Honigzellen. Bienen-Vater, vol. 55, pp. 136–137.

Baker, C. F.

1907 Remarkable Habits of an Important Predaceous Fly. Bull. United States Dept. Agric., Bur. Entom., no. 67, pp. 117–118.

Claassen, P. W.

1922 The Larva of a Chironomid which is Parasitic on a Mayfly Nymph. Univ. Kansas Sci. Bull., vol. 14, pp. 395–405.

Codreanu, R.

1925 Sur la larve d'un Chironomide ectoparasite des nymphes d'une Éphémérine. C. R. Soc. Biol., Paris, vol. 93, pp. 731–732.

Dorier, A.

1938 Une nouvelle station de *Dactylocladus brevipalpis*, Chironomide parasite de nymphes d'Éphémères. Bull. Soc. Entom. France, pp. 45–46.

Edwards, F. W.

1920 Some Records of Predaceous Ceratopogoninae. Entom. Monthly Mag., vol. 56, pp. 203–205.

1923 New and Old Observations on Ceratopogonine Midges Attacking Other Insects. Ann. Trop. Med. Parasitol., vol. 17, pp. 19–29.

Forsius, R.

1924 On Ceratopogoninae as Ectoparasites of Neuroptera. Notul. Entom., vol. 4, pp. 98–99.

Gravely, F. H.

1911 Mosquito Sucked by a Midge. Rec. Indian Mus., vol. 6, p. 45.

Herrod-Hempsall, W.

1931 The Blind Louse (*Braula*) of the Honey-bee. Journ. Ministry Agric. London, vol. 37, pp. 1176–1184, 4 pls.

Jacobson, E.

1921 Bloedzuigende Diptera op eene Rups. Tijdschr. v. Entom., vol. 64, no. 121, p. 5.

1923 Micro-Dipteren als Ectoparasiten anderer Insekten. *Ibid.*, vol. 66, pp. 135–136.

Knab, F.

1914 Ceratopogoninae Sucking the Blood of Caterpillars. Proc. Ent. Soc. Washington, vol. 16, pp. 63–66.

Kryger, I. P.

1914   En Myg, der Angriber en Sommerfugl. Entom. Meddel, 1914, pp. 83–88.

Macfie, J. W. S.

1932   Ceratopogonidae from the Wings of Dragonflies. Tijdschr. v. Entom., vol. 75, pp. 265–283.

Meijere, J. C. H. de

1923   Ceratopogon-Arten als Ectoparasiten anderer Insekten. *Ibid.*, vol. 66, pp. 137–142, 2 figs.

Patel, P. G.

1921   Note on the Life History of *Culicoides oxystoma*. Rept. Proc. 4th Entom. Meeting, Pusa, pp. 272–278, 1 pl.

Peyerimhoff, P. de

1917   *Ceratopogon* et *Meloë*. Bull. Soc. Entom. France, 1917, pp. 250–253.

Phillips, E. F.

1925   The Bee-louse, *Braula coeca*, in the United States. Circ. United States Dept. Agric., no. 334, 11 pp.

Roepke, W.

1921   Rupsenbloed-zuigende Dipteren op Java. Tijdschr. Entom., vol. 6, No. 123, pp. 39–40.

Schulz, W. A.

1904   Dipteren als Ektoparasiten an südamerikanischen Tagfaltern. Zool. Anz., vol. 28, pp. 42–43.

Skaife, S. H.

1921   On *Braula coeca*, a Dipterous Parasite of the Honeybee. Trans. Roy. Soc. South Africa, vol. 10, pp. 41–48, 11 figs.

Suire, J.

1931   Contribution a l'étude de *Braula coeca*. Rev. Zool. Agric., vol. 30, pp. 85–89; 101–104, 4 figs.

Tjeder, B.

1936   Contributions to the Knowledge of *Forcipomyia eques*. Notul. Entom., vol. 16, pp. 85–88.

HAEMATOPHAGOUS LARVAE

Bezzi, M.

1922   On the Dipterous Genera *Passeromyia* and *Ornithomusca*, with Notes and Bibliography on the Non-Pupiparous Myiodaria Parasitic on Birds. Parasitology, vol. 14, pp. 29–46.

Coutant, A. F.

1915   Habits, Life History and Structure of a Blood-sucking Muscid Larva. Journ. Parasitol., vol. 1, pp. 135–150.

Dobroscky, I. D.

1925   External Parasites of Birds and the Fauna of Birds' Nests. Biol. Bull., vol. 48, pp. 274–281.

Jellison, W. L., and C. B. Philip

1933 Faunae of Nests of the Magpie and Crow in Western Montana. Canadian
       Entom., vol. 65, pp. 26–31.

Johnson, C. W.

1929 The Injury to Nestling Birds by the Larvae of *Protocalliphora*. Ann. Entom.
       Soc. America, vol. 22, pp. 131–135.

Miller, C. W.

1909 The Occurrence of the Larvae of a Parasitic Fly (*Protocalliphora chrysorrhoea*
       in Bird Nests. Washington Soc. Study Bird Life, vol. 2, 8 pp., 4 pls.

Plath, O. E.

1919a Parasitism of Nestling Birds by Fly Larvae. Condor, vol. 21, pp. 30–38.
1919b The Prevalence of *Phormia azurea* in the Puget Sound Region. Ann. Entom.
       Soc. America, vol. 12, pp. 373–381.

Rodhain, J., and J. Bequaert

1916 Histoire de *Passeromyia heterochaeta* et de *Stasisia* (*Cordylobia*) *rodhaini*. Bull.
       Sci. France et Belgique, vol. 49, pp. 236–289, 1 pl., 13 figs.

Roubaud, E.

1913a Études biologiques sur les Aucheromyiés. Bull. Soc. Path. exot., vol. 6, pp.
       128–130.
1913b Recherches sur les Aucheromyiés, Calliphorines à larves suceuses de sang de
       l'Afrique tropicale. Bull. Sci. France et Belgique, vol. 47, pp. 105–202, 2 pls.,
       32 figs.
1915a Les Muscides à larves piqueuses et suceuses de sang. C. R. Soc. Biol., Paris,
       vol. 78, pp. 92–97, 2 figs.
1915b Hématophagie larvaire et affinités parasitaire d'une mouche Calliphorine,
       Phormia sordida, parasite des jeunes oiseaux. Bull. Soc. Path. exot., vol. 8,
       pp. 77–79.
1917 Précisions sur *Phormia azurea*, Muscide à larves hémophages parasites des
       oiseaux d'Europe. Bull. Biol. France et Belgique, vol. 51, pp. 420–430, 1 pl.

Shannon, R. C., and J. D. Dobroscky

1924 North American Bird Parasites of the Genus *Protocalliphora*. Journ. Wash-
       ington Acad. Sci., vol. 14, pp. 247–253.

Wellman, F. C.

1906 Observations on the Bionomics of *Aucheromyia luteola*. Entom. News, vol.
       17, pp. 64–67, 3 figs.

MISCELLANEOUS EPIZOIC PARASITES AND HAEMATOPHAGS

Bugnion, E., and H. Du Buysson

1924 Le *Platypsyllus castoris*. Ann. Sci. Nat. Zool., vol. 7, pp. 83–130, 21 figs.

Burr, M., and K. Jordan

1913 On Arixenia, a Suborder of Dermaptera. Trans. 2nd Intern. Entom. Congr.,
       vol. 2, pp. 398–421.

Champion, G. C.

1926 Coleoptera in Nests of the Marmot. Entom. Month. Mag., (3) vol. 12, p. 140.

Ewing, H. E.

1929   A Manual of External Parasites. xiv + 225 pp. Springfield, Illinois, C. C. Thomas.

Hagan, H. R.

1935   Vivparity in Insects. Journ. New York Entom. Soc., vol. 43, pp. 251.

Jordan, K.

1909   Description of a New Kind of Apterous Earwig Apparently Parasitic on a Bat. Novitates Zool., vol. 16, pp. 313–326, 3 pls.

Kolbe, H. J.

1911   Uber ekto- und entoparasitiche Coleopteren. Entom. Nation. Bibliot., vol. 2, p. 116.

Lawson, P. B.

1926   Some "Biting" Leafhoppers. Ann. Entom. Soc. America, vol. 19, pp. 73–74.

Matheson, Robert

1932   Medical Entomology. xii + 489 pp. C. C. Thomas, Springfield, Illinois.

Mönnig, H. O.

1934   Veterinary Helminthology and Entomology. xvi + 402 pp. 264 figs. London, Bailière, Tindall and Cox.

Muir, F.

1912   Two New Species of *Ascodipteron*. Bull. Mus. Comp. Zool., Harvard, vol. 53, pp. 349–366, 3 pls.

Notman, Howard

1923   A New Genus and Species of Staphylinidae Parasitic on a South American Opossum. American Mus. Novitates, No. 68, 3 pp.

Patton, W. S., and F. W. Cragg

1913   A Textbook of Medical Entomology. xxiii + 768 pp. London, Madras & Calcutta, Christian Lit. Soc.

Patton, W. S., and A. M. Evans

1929   Insects, Ticks, Mites and Venomous Animals. x + 786 pp. Croydon, H. R. Grubb.

Rehn, J. A. G., and J. W. H. Rehn

1936   A Study of the Genus *Hemimerus*. Proc. Acad. Nat. Sci. Philadelphia, vol. 87, pp. 457–508.

Riley, W. A., and O. A. Johannsen

1938   Medical Entomology. xiii + 483 pp. New York, McGraw-Hill.

Seevers, C. H.

1944   A New Subfamily of Beetles Parasitic on Mammals. Zool. Ser., Field Mus. Nat. Hist., vol. 28, pp. 155–172.

Williams, C. B.

1921   A Blood-sucking Thrips. Entomologist, vol. 54, p. 163.

SPECIATION, ETC. IN EPIZOIC PARASITES

Chandler, A. C.

1923   Speciation and Host Relationships of Parasites. Parasitology, vol. 15, pp. 326–329.

Rothschild, N. C.

1917   Convergent Development Among Certain Ectoparasites. Proc. Ent. Soc. London, 1916, pp. 141–156.

Schwalbe, G.

1915   Ueber die Bedeutung der äusseren Parasiten für die Phylogenie der Säugetiere und des Menschen. Zeits. Morph. Anthropol., vol. 27, pp. 585–590.

Note: Other References are to be found under "Mallophaga and Anoplura."

CHAPTER IX

ENTOMOPHAGOUS PARASITES AND OTHER INTERNAL PARASITES

L'insecte n'appartient pas à notre monde. Les autres animaux, les plantes même, en dépit de leur vie muette et des grands secrets qu'ils nourissent, ne nous semblent pas totalement étrangers. Malgré tout, nous sentons en eux une certaine fraternité terrestre. Ils surprennent, émerveillent souvent, mais ne bouleversent point de fond en comble notre pensée. L'insecte lui, apporte quelque chose qui n'a pas l'air d'appartenir aux habitudes, à la morale, à la psychologie de notre globe. On dirait qu'il vient d'une autre planète, plus monstrueuse, plus énergique, plus insensée, plus atroce, plus infernale que la nôtre.

GEORGES MAETERLINCK, 1911

IT HAS BEEN indicated earlier that the term Parasitism is applied to a number of similar, but clearly distinguishable modes of life among insects. Moreover, any attempt to define a number of categories soon leads us into difficulties, even in dealing with the entomophagous parasites alone since they represent a polyphyletic series. Some are clearly derived from predators, others from scavengers and still others from phytophagous ancestors, as nearly as we are able to judge on the basis of available criteria. Extreme specialization and convergence in both structure and development are commonly encountered throughout the entire series and this serves further to obscure the actual relationships between the numerous groups of insects concerned.

The entomophagous parasites include members of a number of the natural orders of insects, but only a single, highly specialized one, the Strepsiptera, is composed entirely of such forms. By far the greatest number belong to the large order Hymenoptera, of which certainly well over half of the existing species are parasites of other insects. These include three entire superfamilies, Ichneumonoidea, Chalcidoidea, and Serphoidea, with the exception of a few genera of chalcidoids which are secondarily phytophagous, feeding in seeds and other vegetable tissues, and a very few most exceptional ichneumonids that are known to subsist mainly on food stored in the nests of solitary bees, although these destroy the bee-egg or young larva also. The superfamily Cynipoidea, to which the gall-making wasps belong, includes one large family of parasites. Also, among the wasps and bees there are a small number of forms parasitic on species closely related to themselves. These wasps and bees exhibit a different form of parasitism, however, as they usually kill the young host larva and complete their development by consuming the store of food provided by the mother bee or wasp and intended for her own offspring. Owing to the very close similarity between host and parasite in a number of isolated instances it appears probable that this parasitic relationship originally arose through the appearance of parasitic groups of individuals comparable to the gunmen, racketeers, brain trusters, political grafters and pirates that human society has produced either as new developments or atavisms reminiscent of its early beginnings. In the case of our own species only vague and illusory morphological differences are as yet associated with such behavior, but another million years should permit of definite speciation if the host organisms do not become extinct in the meantime. In the bees the parasitic species are distinguished mainly by a degeneration of the pollen-collecting apparatus, and in the digger wasps by a loss of the bristling on the legs that is used in excavating their nests.

In some parasites of social wasps differentiation has hardly begun and in each case the parasite is structurally very closely similar to its host, so extremely like it, in fact, that we can hardly believe the parasitic form to be other than a very recent and direct derivative of the host species.

Several species of hornets of the genus *Vespula* are known to be thus parasitic. This condition is true of the European *Vespula rufa* and its parasite, *V. austriaca*. In North America *V. arctica* lives as a parasite of *V. arenaria*.

The behavior of these wasps is of such deep interest that I have been tempted to treat it at greater length than many of my readers may deem proper. All the parasitic species have lost the worker, or infertile female caste and consist only of males and fertile females, the latter corresponding to the "queens" in the host species. The parasitic females enter the nests built and maintained by their hosts where they lay their own eggs in the paper cells already provided. Their young are then fed and reared in cells, intermingled with those containing the brood of the host, all the feeding being done by workers of the host species.

So far the relation of host and parasite seems perfectly natural. It is simply the coincident development of several Parasitic Wasp Associations or "P.W.A.'s." Each is dependent upon the parent species from which it is not as yet morphologically distinct, except for the disappearance of the worker caste and the enjoyment of greater opportunities for reproduction. Both modifications are always popping up among social parasites, and are to be expected by anyone other than an economist or American politician. This disappearance of the worker caste is purely a question of food as the brood of the parasites do not mature until late in the summer at which time only males and true females are produced by the host species also.

However, if we assume that each parasitic species has arisen independently from its host species, we encounter one difficulty. In all of these several parasitic vespulas, the sting is of a different form from that of the non-parasitic ones. It is larger and shows a striking curvature, so that we may distinguish any of the parasitic forms by this structural peculiarity which is common to all of them, but does not occur in other members of the genus. If we assume that this modified sting arose only once in the evolution of the genus, then all the parasitic forms must have had a common ancestry. We can hardly assume that it has just happened to appear each time a parasitic strain has originated and this is strong evidence for the common origin and later differentiation of the several parasitic vespulas. On the other hand, each parasite is so obviously closest to its own host in structure that we cannot believe the similarities to have arisen in each case through modifications in a common stock coming to resemble in each case its own particular host.

Is there any possible explanation for such apparently contradictory taxonomic relationships among these wasps? We have seen that habits, at least with direct reference to choice of food, appear to have undergone sudden changes or mutations which have become permanent and persisted for vast periods of time. The appearance of parasitic habits in these vespulas must have originated in a similar way, although undoubtedly at some rather recent time as they are scarcely distinguishable from their parent species. It appears here that the change in habits is linked with the differently formed sting, perhaps in two genes that always remain together. The objection may be made that the longer and more powerful sting may be adaptive, or necessary for the parasite and that there is a reason for its development. However, we find similar parasitic species in the related paper wasps, *Polistes*, and here the only morphological character that separates them is a peculiar groove on the outer side of the mandible, obviously a structure that is not open to interpretation as particularly related to parasitic habits. Otherwise the taxonomic relations between hosts and parasites appear to be the same in *Polistes* as in *Vespula*.

I believe, therefore, that we are dealing in *Vespula* and *Polistes* with mutations involving both a behavioristic and a structural character which are invariably associated through some genetic mechanism.

We may then construct thus the history of the parasitic vespulas: Once upon a time in the late Pleistocene one summer after another had been long and warm, and nature had produced an abundant supply of flies and caterpillars. Many populous colonies of hornets were busily engaged in bringing these home and feeding them to their hungry brood, but food was so abundant and life was so easy that there was not enough work for all of the wasps and there were no labor leaders exhorting them to limit their hours of work. Many of them found their services unnecessary and allowed themselves to sponge on the industry of their fellows. Thus "on relief" the unemployed wasps had to find something to do so they took to laying a lot of extra eggs of their own in the nest for the already overburdened workers to rear and care for. This worked beautifully at first and it looked as though real work was out of date and that Recent times would dawn on a population of wasps with no responsibilities beyond the production of fertile eggs. However, such a New Deal couldn't last, for it is a biological axiom that parasites are strictly limited by the numbers of their hosts and what we now find is a reduced population of parasitic vespulas, scarcely distinguishable from their hosts but entirely dependent upon them and multiplying as rapidly as the responsible members of the colony can feed and care for them.

Although either the parasitic bees or the parasitic hornets just mentioned are truly parasitic their behavior is not that of the true entomophagous parasite. However, it is referred to here because we shall wish to return to it later in connection with hyperparasitism for the two phenomena are in some respects similar.

All of the primitive families of Hymenoptera are phytophagous, feeding either in plant tissues or consuming the leaves. We can be sure, therefore, that the parasitic Hymenoptera are derived directly from vegetarian ancestors, excluding the parasitic bees and some of the parasitic wasps which may have passed through a predatory stage in their evolution. These primitive Hymenoptera form a suborder, Chalastogastra, characterized by their broadly sessile abdomen and caterpillar-like larvae. However, one family of this series, the Oryssidae, has already acquired parasitic habits and it is evidently near to the stem from which the enormous superfamily Ichneumonoidea has been derived. The most primitive Ichneumonoids, forming the family Stephanidae are clearly related to the Oryssidae. The origin of the Chalcidoidea, Serphoidea and Cyni-

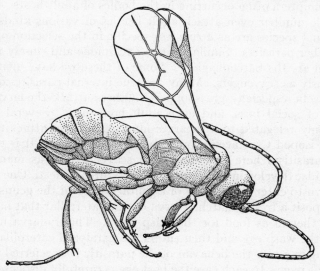

FIG. 51. An undescribed species of ichneumon-fly, preserved as a fossil in Baltic amber of Oligocene age. Original.

poidea is not so clear, although the Serphoidea may also be derivatives of stephanid-like insects. Both ichneumonoids and serphoids are known to be very ancient as they were represented in the Jurassic by forms close to living genera (Fig. 51). Many hyperparasites belong to this order.

Entomophagous parasitism is much more sporadic in the Diptera, as it is encountered in numerous families scattered throughout the order. It is not so firmly established since there are a number of families where this habit appears only in some genera. Also, most of the insect parasites of other invertebrates are Diptera, although as has already been indicated, this is unusual among parasitic insects in general. The largest series of Diptera are the tachina-flies,

forming the old family Tachinidae, now variously divided into
several or even thirty or more closely related families by some of
the rabid "splitters" among taxonomists. These flies are one of
the most recently developed types of insects and are evidently undergoing rapid evolution at the present time. Their maggot-like larvae
are internal parasites, occurring frequently in lepidopterous caterpillars, although many attack the grubs of beetles, larvae of Hymenoptera or rarely other Diptera. Many attack insects with
incomplete metamorphosis such as grasshoppers, walking sticks
and Hemiptera often occurring in the bodies of adult hosts. A considerable number even attack adult beetles of various kinds. The
individual species are as a rule less specific in the selection of hosts
than other parasites. Similar to the Tachinidae and closely related
to them are the Sarcophagidae. Some of these we have mentioned
previously as scavengers. Many are true internal parasites of various insects, especially grasshoppers. Others attack the larvae and
pupae of social bees, and even adult honeybees. Several genera
commonly relegated to a separate family, Dexiidae, attack the terrestrial isopod Crustacea commonly called sow-bugs. It is evident
that parasitism here is arising from scavengerism, as many Sarcophagidae develop in the bodies of insects already dead. One species
is known to enter the burrow of a solitary wasp of the genus *Sphex*
and deposit a freshly hatched larva on the caterpillar that has been
placed therein as food for the wasp larva. The fly-larva first destroys the wasp egg and then enters the paralyzed caterpillar, thus
recalling in a way the behavior of the parasitic bees referred to on a
previous page. In each case the host egg is carefully dispensed with
to avoid any controversy later as to ownership of the food-basket.
This is a really fine racket and the wasps sense that something is
wrong as they obviously make efforts to keep the alert and pestiferous flies from approaching their burrows.

Other dipterous parasites belong to the Bombyliidae, attacking
insect larvae and frequently the egg-masses of grasshoppers; to the
Cyrtidae, internal parasites of spiders, and to several other families
with habits too varied to deal with briefly.

The Coleoptera, or beetles, form the largest order of insects, and
although a great many are predators, almost none are parasites.
The parasitic forms include the members of two families, the
Meloidae and Rhipiphoridae and a few scattered genera in several
other families. Many Meloidae develop in the cells of solitary bees,
their larvae first destroying the host egg and later consuming the

stored food supply, like the parasitic bees. Other meloids feed within the egg-masses of grasshoppers and are thus probably to be regarded as purely predatory. The Rhipiphoridae are entirely parasitic, affecting larval wasps and bees and in the case of one highly degenerate genus, *Rhipidius*, living in cockroaches.

The order Strepsiptera includes a number of highly aberrant insects that develop internally as parasites of various wasps, bees and Hemiptera. The males are active, winged, very short-lived insects, but the larviform females of practically all species never leave the body of the host (Fig. 52). They simply protrude the tip of the body to permit mating with the male and to allow the later escape of the

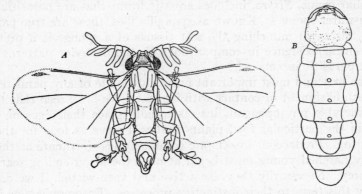

FIG. 52. Sexual dimorphism in Strepsiptera. *A*, male of *Stenocranoplus quadratus; B*, female of *Xenos peckii* (diagrammatic). *A* redrawn from Pierce, *B* original.

young which are born alive. These minute larvae are active and seek the host, commonly resorting to some sort of hitch-hiking to accomplish their purpose. The relationships of the Strepsiptera have long been a bone of contention among entomologists, but it is now generally conceded that they are closely allied to the Coleoptera. Their active first-stage larvae are very similar to those of meloid and rhipiphorid beetles, and some highly degenerate females of the latter family are strikingly like some recently discovered Strepsiptera that become free-living at maturity.

Other scattering cases of parasitism are found among certain moths, although most of these verge closely on predatism, for example, caterpillars that feed on scale insects or within the egg-masses of spiders. Apparently, the larvae of some moths of the family Epipyropidae are true external parasites of various Homoptera, especially lantern-flies (Fulgoroidea). The same is true of a remarkable Australian moth, *Cyclotorna*, whose larvae attack leaf-hoppers, also as external feeders, during the first larval stage. Later, the caterpillars become myrmecophilous, ingratiating themselves with foraging ants of the genus *Iridomyrmex* which carry them into their nests where they suck the juices of the ants' larvae. This change of food during development is most remarkable and recalls the behavior of the larvae of the butterfly, *Lycaena arion*, which are at first vegetarian, but later complete their development as predators in ant-nests where they devour the larvae and pupae of the ants.

The larvae of most Neuroptera are highly predaceous, but one peculiar genus, *Sisyra*, includes aquatic forms that are parasitic in fresh-water sponges. Known as spongilla flies, these are true parasites, although munching the soft tissues of a sponge is a pretty hum-drum existence in comparison with the behavior patterns of most parasitic insects.

The first and most important event in the life of any parasite is the establishment of contact with its host. We have seen that the innumerable phytophagous insects usually place their eggs on or within the particular food plants that will serve as food for their offspring. Predaceous insects are commonly less fortunate as their newly hatched young must begin at once the never-ending search for prey. Necessarily they are active and keen-witted, if we dare apply this term to their instinctive processes. Entomophagous insect parasites avail themselves of both methods with many variations and modifications that often involve remarkable and unexpected behavior on the part of both the mother and new-born larva.

The most positive method of approach is the insertion of the egg directly into the host. This occurs generally among those parasitic Hymenoptera whose larvae feed internally. The sharp stiletto-like ovipositor of the female is beautifully adapted to this purpose, acting like an hypodermic needle (Fig. 53). Through this, the very small, flexible egg may be squeezed into the haemocele of the host, or sometimes into some particular organ such as a nerve ganglion or even attached to the body wall by a stalk. The passage of the egg through the very small lumen of the ovipositor has been carefully observed in a chalcid wasp, *Habrocytus*, by Fulton ('33). The tip of the egg is worked through the ovipositor by means of minute stylets, then the contents of the egg flow through the greatly constricted portion and the protoplasm accumulates within the host to be followed finally by the remainder of the rubber-like chorion that envelops the egg. Among the minute Hymenoptera which mature on the contents of a single egg of some much larger insect, the parasite's egg is introduced directly into that of the host (see Fig. 60). The same procedure is followed in the case of some larger parasites where the larva does not eat the contents of the egg or the embryo, directly, but awaits the hatching of the host larva before proceeding to feed. The development of the parasite is then continued within the body of the host which does not succumb until much later, sometimes not until the pupal stage has been attained.

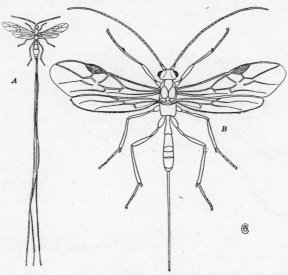

FIG. 53. Two ichneumon flies in which the ovipositor of the female is enormously lengthened. *A, Euurobracon penetrator; B, Macrocentrus gifuensis. A* after Ashmead; *B* after Parker.

The injection of the parasite's egg is frequently accompanied by a peculiar process of feeding by the female at the puncture made by the ovipositor (Fig. 54). This little diversion furnishes droplets of food at intervals during the period of egg laying and no doubt serves also as a stimulus that may quite likely have a psychological as well as a nutritional basis. That food is thus secured is demonstrated by the fact that the life of females deprived of the opportunity to feed in this way is greatly shortened. This phenomenon has been observed among a number of different parasitic Hymenoptera. It is commonly practiced quite independently of oviposition and the host animal may in the case of aphids be killed by the fierce stabbing and blood-letting. In some cases where direct contact between the mouth of the parasite and the body of the host cannot be made, a feeding tube is built up by a secretion from the ovipositor. This hardens and serves like a lemonade straw as a conduit through which the liquid food may be drawn into the mouth. Fulton ('33) has described and figured this process in the chalcidid wasp, *Habrocytus*, where the wasp has to puncture a grain of corn in order to reach the cavity within which the host caterpillar is feeding.

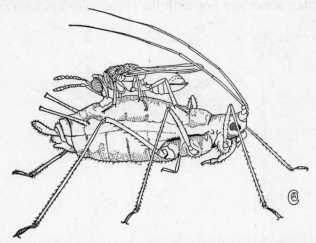

FIG. 54. A chalcid-wasp (*Aphidencyrtus inquisitor*) feeding at an oviposition puncture in the abdomen of its host aphid, *Macrosiphum cornelli*. After Griswold.

The opportunity to feed on the juices of the host is, of course, withheld if hosts are absent or scarce, and this happens frequently in the natural course of events. Under such circumstances many parasitic Hymenoptera undergo a period of phasic castration, due to the disintegration and absorbtion of eggs before they become mature. This process of oösorbtion is obviously an adaptation for producing ripe eggs only at times that hosts are available, and to avoid the necessity for oviposition when hosts cannot be found. It appears frequently to be conditioned by failure to feed on the juices of the host (Flanders, '42).

Many Diptera are also furnished with a sharp and pointed or blade-like ovipositor that may be thrust directly into the body of the host. Some Tachinidae have such instruments and a considerable variety in shape is seen in certain minute flies of the family Phoridae some of which actually insert their eggs into the bodies of adult ants. In one remarkable species of *Apocephalus*, the fly larva feeds within the head of the ant, finally decapitating the hapless victim on completing its growth.

A much simpler method practiced by a great variety of parasites is the attachment of the eggs to the body of the host by a glutinous secretion. Many such forms feed externally, but in tachina flies with this habit the larva bores its way through the integument as soon as it hatches and immediately finds itself within the body of the host.

There are numerous instances known among both Hymenoptera and Diptera where the eggs are produced in excessively large numbers and laid in pretty much hit or miss fashion, but nevertheless contact with the insect host is made. One of these involves the actual seeking of the host by the newly hatched parasitic larva and we shall return to this in a moment. Frequently, however, the host unwittingly swallows the eggs when these have been placed on its food plant. Once they are thus taken into the alimentary canal prompt hatching is induced by the digestive fluids and the young larva has only to burrow through the epithelium of the gut to find itself within the body cavity where its future growth is destined to take place. This method is of frequent occurrence among the Tachinidae. Many other flies of this family place their eggs directly on foliage where the host caterpillars are feeding, laying minute or microtype eggs which are produced in excessively large quantity. As many as 13,000 eggs have been reported in one species, although each female ordinarily produces only several thousand. This in itself is a surprisingly large number for an insect and is obviously a response to the slight expectancy of life for eggs scattered in this manner. Strangely enough, a closely similar method of placing the eggs on foliage occurs in the hymenopterous family Trigonalidae. Of the thousands laid by a single female, some are eaten by the caterpillars or sawfly host larvae. These hatch after ingestion and the parasitic larvae enter the body cavity of the host.

The development of a structurally complex, highly active first stage larva is an innovation which appears in parasitic insects of many groups and naturally plays an important part in enabling them to contact their hosts (Fig. 55). This involves the phenomenon of hypermetamorphosis to a high degree, as the later larval stages promptly discard such embellishments which are no longer necessary once they are snugly established in the comfortable interior of a nice fat grub or caterpillar. There is a great similarity in these first-stage larvae in the three great orders of insects where they occur (Coleoptera, Hymenoptera and Diptera), and those of the Strepsiptera are also essentially of the pattern found in the Coleop-

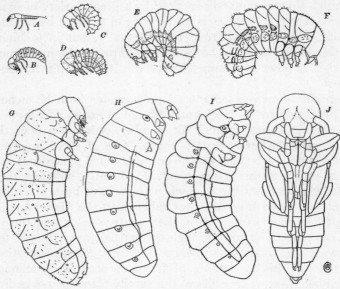

FIG. 55. Developmental stages of a hypermetamorphic blister-beetle, *Epicauta pennsylvanica* (Family Meloidae). *A*, unfed first stage larva; *B*, engorged first stage larva; *C*, newly molted second stage larva; *D*, newly molted third stage larva; *E*, gorged fourth stage larva; *F*, newly molted fifth stage larva; *G*, gorged fifth stage larva; *H*, sixth stage larva; *I*, active seventh stage larva; *J*, pupa. Redrawn to the same scale after Horsfall.

tera (Fig. 56). The two latter are provided with well developed legs and resemble the larvae of certain predatory beetles. In the Hymenoptera and Diptera they are derived from forms in which the legs have been lost, but they have made up for this pretty hopeless deficiency by developing spines, bristles and projecting corners from the body segments that enable them to crawl about with quite respectable proficiency. The larvae of this type are very minute and are produced in great numbers. Those with legs are known as triungulins and those with makeshift appendages as planidia. Both are generally encased in a tough, heavy integument and, in spite of their minute size, will live for considerable periods without food. Those of many meloid (Pl. XVIII A) and rhipiphorid beetles are

commonly found on flowers where they may lie in wait for a chance
to hop a ride on the back of a bee or other insect that will later visit
its nest where the larva will undergo its later development. In the

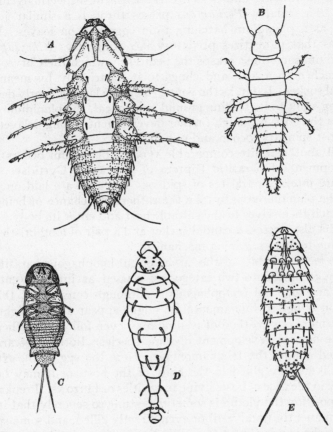

FIG. 56. Active, first-stage larvae of several diverse parasitic
insects that undergo hypermetamorphic development. *A*, *Rhipiphorus* (Coleoptera); *B*, *Horia* (Coleoptera); *C*, *Xenos* (Strepsiptera); *D*, *Perilampus* (Hymenoptera); *E*, *Pterodontia* (Diptera).
Redrawn; *A* after Pyle; *B* and *C* after Brues; *D* after King; *E*
after Smith.

Strepsiptera the triungulins issue alive from the body of the grub-
like female and enter the body of a new host by some devious path
which finally leads a select few to the proper place.
    Still more remarkable are the Hymenoptera of the family Eucha-

ridae all of which are parasites of the larvae and pupae of ants. Here the eggs are laid in masses in plant tissue or scattered over foliage and the planidia on hatching manage to attach themselves to the bodies of foraging ants which carry them unsuspectingly into the nest. The related *Perilampus* passes through a similar juvenile stage as a planidium hatching from eggs laid on leaves. In one species that parasitizes predatory lace-wing flies (*Chrysopa*) the planidium sometimes grasps the pedicel by which the Chrysopa egg is attached to a leaf, and clings to it tenaciously by means of a caudal sucker. Later, as the young *Chrysopa* larva gingerly descends the egg stalk to its feeding ground on the leaf, the planidium transfers to the host, although further development is long delayed until the host spins its cocoon and pupates.

Still another quite comparable type of planidium occurs in the development of parasitic Diptera of the family Cyrtidae. These flies are internal parasites of spiders. The eggs are laid on plants and the planidia have to take the rather slim chance of being able to attach themselves to the suitable host and enter its body. These planidia also possess a caudal sucker and a pair of long bristles that combine to form a leaping mechanism.

The feeding habits of the larvae of entomophagous parasites are readily grouped into two categories, classed as internal (entophagous) and external (ectophagous). Although convenient, this distinction is not so fundamental as might appear at first sight, and furthermore one method of feeding may even follow the other during the normal development of some species. Both types are represented among the Ichneumonidae. Here the egg of the external feeder is usually glued to the body of the host, or it may be laid nearby in the case of hosts living in cavities or burrows. Preparatory to oviposition the victim is sometimes stung so severely that it may be permanently paralyzed, or even directly killed, and some species when picked up in the fingers can inflict a quite painful sting. In one genus of spider parasites, *Polysphincta*, the host is paralyzed, but soon recovers and leads its normal life, and in many ichneumonids only very temporary effects of the sting are noted, Among the internally parasitic species the reactions of the host to oviposition and subsequent feeding are less noticeable or at least, their clinical expression is greatly reduced. This condition prevails generally among the several types of entomophagous parasites although by no means invariably. It is evident, however, that the internal parasites are on the whole more highly adapted with reference to their hosts. This will be particularly evident in a moment when we consider the several subterfuges to which they have resorted to fit their respiratory and other functions to the unusual environment encountered in the body of their hosts.

In general the endoparasitic larvae restrict their early feeding to the less important and non-vital tissues of the host, particularly the blood, blood cells and fat-body, the latter representing an accumulation of energy-supplying material. This is destined normally to provide the requirements for final metamorphosis and frequently also a major part or even the whole supply of food that is required

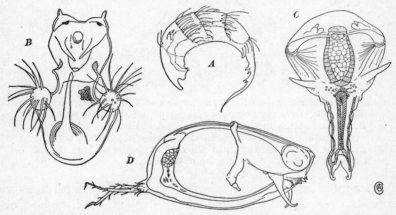

FIG. 57. Several strangely modified first stage larvae of parasitic Hymenoptera. *A, Polynema* (Family Mymaridae); *B, Teleas* (Scelionidae); *C, Trichasis* (Platygastridae); *D, Synopeas* (Platygastridae). The last three are known as cyclopoid larvae. *A* and *B* after Ayers; *C* and *D* after Marchal.

to form the eggs. The parasitic larva is at first diminutive (Fig. 57) and necessarily fastidious in consuming minute and delicate morsels, but as it grows it still feeds with great caution and does not blunder by nipping off or perforating any essential structures that might spell sudden disaster to its host and promptly seal its own fate. Such self-imposed continence is cast aside only as the last stages of development are reached. In a final orgy of feeding nothing is spared and the host may be quite thoroughly consumed by the time the parasite is ready for pupation. The parasitic grub then usually makes its way out although it may sometimes transform to the adult before emergence (Fig. 58). The behavior of some parasites is far less seemly and circumspect, and the host may succumb more quickly, particularly when attacked by species that undergo a rapid development.

Among the external or ectophagous feeders, the larva punctures the integument of its host and inserts its mouthparts into the body cavity (Pl. XVIII B). Feeding and growth in such forms is commonly more rapid than among the internal parasites and a liquefaction of the body contents about the mouthparts indicates a process

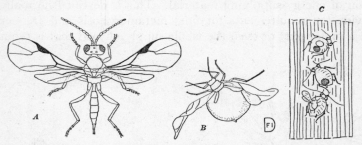

FIG. 58. Small parasitic hymenopteron, *Lysiphlebus*, whose larvae develop in the bodies of aphids. *A*, adult female parasite; *B*, aphid, swollen and deformed by the contained parasitic larva; *C*, three aphids on a leaf with emergence holes of parasites and one *Lysiphlebus* just crawling out through the circular hole it had cut with its jaws. Redrawn from Webster.

of extra-intestinal digestion effected by the injection of digestive ferments. When such rapid and violent dissolution of the host occurs, its death promptly supervenes.

There is thus a great range in the visible effects of the parasite on the host, but certain more subtle changes are often induced by the presence of internal parasites. These may involve its behavior or appetite, although such disturbances are far less than might be expected and serve to emphasize the tremendous vitality of insects which give scarcely any indication that the greater part of their bulk consists of a living, gnawing grub which will soon consume them completely. The presence of internal parasites does, however, frequently produce striking changes, giving rise to a condition known as parasitic castration where certain secondary sexual characters of the host fail to develop. In ants such abnormal individuals may be intermediate between the normal female and worker caste, and bees parasitized by Strepsiptera commonly develop a color pattern approaching that of the opposite sex. Such phenomena are, of course, not confined to insects as similar occurrences are well known in other animals. It has been generally assumed that such changes are due directly to injuries suffered by the sexual glands, but the discovery of hormonal substances in insects not formed by the gonads will make necessary a re-investigation of the whole problem. Recent studies by Beard ('40) on parasitic castration of the common squash bug by the tachinid parasite, *Trichopoda*, show a progressive atrophy of the ovaries.

Although the majority of entomophagous parasites develop at the expense of the eggs, larvae or nymphs of their hosts, there are many that attack them after they have reached the pupal or adult stage. These may be either internal or external parasites, in the latter case often within the cocoon spun by the host or within the shell-like puparium which invests the pupae of the higher Diptera. Certain chalcid wasps have such habits and may be commonly bred from the pupae of the housefly and related Diptera.

Various insects, notably those undergoing an incomplete metamorphosis, and some beetles, are attacked by entomophagous parasites that affect them after they have become adult. Such parasites of beetles are comparatively rare, including mainly flies of the family Tachinidae and Hymenoptera belonging to several highly specialized genera of Braconidae that affect Coccinellidae, Chrysomelidae and weevils. Strangely enough the extremely hard shells of the hosts fail to serve as a deterrent to these parasites. One interesting fact concerning the braconid parasites of ladybird beetles is the frequent recovery of the beetle after the emergence of the parasite. Timberlake ('16) was able actually to rear two generations of a species of *Dinocampus* from the same individual host beetle. However, the poor ladybird did not survive her second experience.

As we have already noted, the selection of particular hosts is rigidly adhered to by nearly all entomophagous parasites just as we have seen to be the case with vegetarian insects and many predatory species also. There are, however, other factors that enter into the association of parasite and host which are not encountered among the non-parasitic forms. This is particularly true of internal parasites, because of the resistance of the host as manifested by the activities of phagocytic cells in its blood. These appear to be frequently highly detrimental, especially to the early stages of the parasite. Workers are not entirely in agreement concerning the general occurrence and importance of phagocytosis. The significance of the reactions of phagocytes was early referred to by Timberlake ('12) who noted in the case of an ichneumon-fly, *Eulimneria valida*, that this species will oviposit in various caterpillars, but that in some of these, to which it seems not to be adapted, the parasite is unable to develop. This is associated with an accumulation of phagocytes about the eggs or newly hatched larvae that results in the death of the young parasites. In species which are natural hosts, like the fall webworm (*Hyphantria cunea*) the phagocytes do not accumulate about the parasites and there appears to be no defensive reaction of this sort. Whether the parasites are directly killed by phagocytic action, or whether they are enclosed in a phagocytic cyst and later destroyed after death cannot be stated. This idea has been combated by Thompson in several papers and supported by Strickland ('30) and others who have apparently brought forward much in support of the supposition that phagocytes play an important rôle in protecting hosts to which the parasites are not well adapted. This adaptation may depend upon their failure to irritate the tissues of the natural host and attract the phagocytes, or it may be due to the production of defensive substances by the parasites. At any rate it appears that all host-parasite relations are not uniform and many more studies will be required to clarify the matter completely. That the host may exercise a profound influence on the development of a parasite is shown by the observation of Salt ('38) that apterous males of a species of the minute chalcid, *Trichogramma*, are produced when they develop in the eggs of one host, but not in those of another. It seems, however, that there is no clear-cut distinction between immunity and susceptibility of hosts to specific parasites, but that the differences are quantitative rather than qualitative (*cf.* Bess '39).

However, differences in the food plants of the same species of host insect may influence susceptibility to certain parasites by inhibiting normal development of the parasitic larvae. Such differences would appear possibly to depend upon specific proteins derived from plants utilized by the hosts (*cf.* Flanders '42).

As insect larvae are primarily air-breathing, life in a liquid medium such as confronts the internal parasite involves an adjustment of respiratory function. This is commonly accomplished by a decrease in the thickness of the integument to permit a more rapid exchange of gases through the body wall. This suffices in many cases, but a remarkable method of tapping the tracheal trunks of the host is practiced by certain tachina flies. In these larvae the spiracles at the posterior end of the body open at the tip of a projection which the larva thrusts directly into a large trachea (Fig. 59). Thus a connection between its own tracheal system and that of its host is established, enabling it to breathe almost as though its spiracles were protruded directly into the air as happens in the aquatic larvae of many Diptera. In other tachina flies, and even in some of the parasitic Hymenoptera, contact is established with the outside air through perforations in the integument of the host.

Another interesting peculiarity of insect parasitism is the frequent occurrence of secondary parasitism or hyperparasitism where some entomophagous parasites attack only hosts that are in turn parasitic on a primary host. Although these primaries and secondaries ordinarily form two distinct categories there are a few species which may develop either as one or the other so that we cannot be too dogmatic

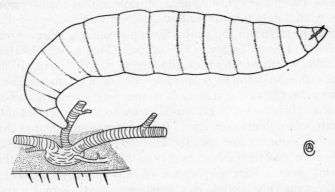

FIG. 59. Larva of a parasitic tachina fly (*Prosena sibirita*) with its respiratory funnel inserted into a trachea of its host. Redrawn from Clausen, King and Teranishi.

in our classification. Frequently, although by no means necessarily, the secondary is a close relative of the primary parasite.

Most secondary parasites begin their life as primary ones, inasmuch as they develop from eggs laid in the primary host and later enter the body of the secondary one within which their development will proceed. In one remarkable chalcid-fly, *Perilampus hyalinus*, the first-stage planidium larva penetrates the larva of a tachinid fly in the body of its host caterpillar or locust and remains there without development till the parasitic tachinid larva pupates. It then makes its way to the surface of the pupa and begins to feed and grow as an external parasite of this pupa. It is evident in such cases that the secondary parasite lives a pretty precarious existence in that it has gone up a blind alley if it chooses a host free from the primary parasite. Forms that are able to develop either as primary or secondary parasites are, of course, less liable to get into such a hopeless situation. One of this type is *Monodontomerus aëreus*, a common chalcid parasite of the brown-tail moth. This may attack certain tachinid flies and parasitic Hymenoptera as well as the caterpillars themselves.

It sometimes happens that two species of primary parasites may attack the same individual host. In such cases, known as multiple parasitism, it is a fight to the death to select the winner, much like the condition described in the case of the parasitic bees that destroy the rightful owner before they settle down to gorge on the spoils.

It may happen also that two parasitic larvae of the same species may find themselves in a similar situation, like the shipwrecked pirates whose supply of water and food will not suffice for division among all members of the gang. Such an excess is known as superparasitism. This difficulty is partially avoided by the instincts of the mother parasite which enable her to restrict oviposition and forestall wholesale fratricide on the part of her prospective little offspring. So far as the species is concerned, the case may be even worse than this as all the individuals may be so weakened that they succumb before final emergence. It has been demonstrated that the ovipositing females are usually able to tell whether eggs have already been laid in the host and that they avoid placing others in the same host. The phenomena of superparasitism have been studied extensively by Salt in a series of papers. Females avoid to some extent the deposition of eggs in hosts already attacked, but not entirely so as oviposition cannot be too long delayed. Under such stress of urgency additional eggs are added even though this endangers the progeny.

The matter has gone farther, with an added complication in some chalcid flies where the larvae destined to become males are actually parasitic on female larvae of the same species. This has been described by Flanders ('36; '39) for certain chalcid-wasps of the family Aphelinidae. The males are thus reduced to the condition of abject hyperparasites on their girl-friends or perhaps even sisters although they are still indispensable for the production of female progeny by the next generation.

In most of the parasitic Hymenoptera, and also among the aculeate members of this order, females result from the development of fertilized eggs and males develop from those that have not been fertilized. This condition does not prevail universally, however, for it has been shown at least in one genus, *Habrobracon* by the Speichers ('38) that some ova are tetraploid and give rise to normal diploid females without fertilization. Such cases are presumably quite unusual, although a few species of parasitic Hymenoptera regularly reproduce parthenogenetically for many generations without the appearance of any males whatsoever. Generally then, it is evident that when eggs are laid by a female supplied with sperm, these will be female if they receive sperm, and male if they do not. As fertilized females are capable of producing offspring of either sex, the emission or withholding of sperm in the spermatheca of the mother is the determining factor, unless it happens that her supply of sperm has been exhausted in which case only males can result. During the normal course of oviposition mixed broods occur and it is believed that possibly the rate of oviposition or the size of the host may determine whether individual eggs will be fertilized or not. Flanders ('39) believes rapid oviposition produces males by inhibiting the emission of sperms, and that among parasites which oviposit singly that a small host has the same effect since male egg are normally laid when in a small host. It is evident in the case of social insects like the honeybee that the fertilization of the egg is conditioned by the size of the cell in which it is laid since the drone ($\female$) and the worker ($\female$) cells of the honeycomb differ in the diameter of the orifice. Attempts have been made to explain the emission or retention of sperms on the basis of a carbon dioxide gradient in the sperm duct, but it is difficult to believe that a high correlation could result in this way. In the past it has generally been thought that muscular control is exercised by the female but this is not wholly in accord with the anatomical structure of the organs involved. In the case of the honeybee, it is likewise clear that the emission of sperm is regulated with great accuracy as the queen bee so carefully doles out her precious supply of sperm that she can fertilize hundreds of thousands of eggs with the product of a single mating. If such wonderful economy could be applied to the human species as seems now within the bounds of possibility, the indispensability of males in our communities may be due for serious reconsideration.

Superparasitism commonly becomes a very complex association when numerous parasites and hyperparasites are combined into an interdependent association or biocoenosis. Thus many of our common insects, especially those that are so numerous as to become severe agricultural or forest pests, are trailed by diverse parasites that account for a consistently heavy mortality among their hosts. Theoretically the abundance of parasites must remain below the level of that of their hosts, but practically there is great fluctuation from one year to another and a flat hundred percent mortality is sometimes achieved. This often involves a long series of parasites, really a fauna of such species, if we may legitimately apply the term thus. From the practical standpoint this is a very important phase of insect parasitism, and numerous papers dealing with it will be found in the references appended to this chapter.

As we have already noted briefly, the specificity of entomophagous parasites in selecting particular hosts is generally very great. There

FIG. 60. A minute parasitic chalcid-fly, *Trichogramma*, in the act of laying her egg in the comparatively large egg of the brown-tail moth within which the larva of the *Trichogramma* will undergo its development. Redrawn from Howard and Fiske.

are notable exceptions, however, For example, some of the most minute chalcid-wasps belonging to the genus *Trichogramma* (Fig. 60) which develop within the eggs of larger insects exhibit a most extraordinary range. A single species has been bred from dozens of host species, and when the full story is told this will undoubtedly increase to many hundreds. Ordinarily all the parasitic forms have a select clientèle and do not form new associations readily. This again is highly significant from a practical standpoint, as it means that noxious pests brought from foreign lands find themselves suddenly freed from the restraint exercised by such enemies and multiply far beyond the fondest hopes of the most optimistic census-taker. Any consideration of this important matter is utterly beyond the scope of the present discussion and must be regretfully dismissed without further comment.

There are, however, several matters relating to host selection by entomophagous parasites that should receive at least scant attention. The host appears to be generally selected at least to some extent by odor although its surroundings may have a profound influence upon the ovipositing female. Thus, hosts normally enclosed within a cocoon, or concealed in plant tissues, will sometimes be free from attack when such accessories are removed. Again, species in low herbage may be attacked while others on trees may escape. These appear to be quite similar to environmental influences such as we have already noted among vegetarian insects. There is also necessarily a correlation between the life cycles of host and parasite, and these are in some cases obviously prime factors in affecting host selection. A common chalcid parasite of the gipsy moth, *Anastatus bifasciatus*, develops in the eggs of this moth. The gipsy moth hibernates in the egg stage and the caterpillars hatch in the early spring. The *Anastatus* delay their emergence, however, until the eggs of the moths of the next generation have been laid in the late summer when they find fresh eggs in which to deposit their own. Upon the proper timing of its developmental cycle the future of this and similar parasites depends absolutely, unless as often happens another acceptable host is available to fill in the gap.

Many such factors relating to competition in one way or another have undoubtedly guided the evolution of parasitic habits with reference to superparasitism, multiple parasitism and hyperparasitism. Some of the preliminary stages in the development of hyperparasitism appear to be illustrated by a few species that are facultative primary or secondary parasites. Such forms have a better chance for survival, other things being equal, than the more specialized types, but the advantage is lessened, of course, by the highly purposeful instincts that go with more restricted habits. Nevertheless, it would seem in this case as among many other types of parasites that host restriction has gone farther than is advantageous to the species, and that survival has been possible only when the behavior pattern could be sufficiently elaborated. The remarkable fact is that this has been accomplished in so many surprising ways.

Attention has been called to the scarcity of insect parasites that attack invertebrates other than insects. This is, of course, due in part to the fact that most other groups of invertebrates are marine, wholly or in great part, and we have seen that this habitat is taboo for practically all insects. Nevertheless, insect parasites of terrestrial annelids as represented by the ever-present earthworms are almost unknown. Molluscs are also neglected, except for a very few sporadic cases. Among other groups of Arthropoda, the Crustacea harbor a few parasites restricted to terrestrial isopods or sowbugs, and the same is true of the ticks among the Arachnida. Spiders are somewhat more favored as hosts, although their parasites include only a scattering of insects.

A very common large muscid fly, *Pollenia rudis*, often seen in dwellings which it enters for hibernation, develops as an internal parasite of earthworms, a fact which strangely enough escaped notice until recently. As this family of flies is among the most recently evolved groups of insects it is obvious that such a habit is, geologically speaking, a very modern development and that it does not represent an association of long standing which we might expect if earthworms had ever been attractive to insect parasites in general. The known parasites of sowbugs are likewise somewhat similar flies quite closely related to the common tachina-fly parasites of insects already mentioned. Spider parasites include several groups of Diptera, one of which (Cyrtidae) has already been mentioned as having a remarkable first-stage planidium larva. They are attacked also by a few related genera of ichneumon flies, such as *Polysphincta*. Spiders' eggs are however commonly infested with true internal parasites in the form of minute serphoid Hymenoptera of the genus *Baeus* and its allies. This association must be of long standing as this group of parasites was abundantly represented in the lower Oligocene (Baltic amber) and is known even from the Cretaceous (Canadian amber) at which time the fossil species evidently had similar habits. The pimpline Ichneumonidae to which *Polysphincta* belongs are likewise an ancient group associated in great part with wood-boring beetles which are in turn among the earlier holometabolous insects. One surprising spider parasite is *Mantispa*, a member of the order Neuroptera, most of which are very active predators. The *Mantispa* larva feeds within the egg-mass of a spider and may thus be regarded as a predator. Its active first stage seeks out the egg-cocoon, but the later, helpless, legless larva is typically a parasitic type.

Parasites of ticks are extremely rare, although there are a couple of genera of minute chalcid-wasps of the family Encyrtidae whose larvae develop within the bodies of various partly grown ticks.

Molluscs, as represented by land snails, are frequently subject to attack by a number of insects in addition to the predatory forms mentioned earlier. It is difficult to group these as scavengers or parasites because a number of them are of more or less uncertain status. Some, however, are clearly parasitic on living snails, although it appears probable that most if not all have been derived quite directly from true scavengers that fed on non-living hosts. Certain minute, wingless flies of the family Phoridae (*Wandolleckia*) frequent the bodies of large African *Achatina* snails feeding apparently on the slime that covers the surface of their host. Other flies of the same family have been bred from snails in which the larvae develop. In many cases, probably all, these flies do not attack living snails and this may be true also of other small flies of the family Sciomyzidae. Two large muscoid flies are known, however, one in Europe (*Melinda*) and another in South America (*Malacophagula*) which are true internal parasites of land snails. The larvae of some chironomid midges are thought also, but with some doubt, to be parasitic in the mantle cavity of aquatic snails. These are more probably commensals.

It will be remembered that we have considered in the previous chapter, devoted to predatory insects, the solitary wasps which provision their nests with insects or spiders that form the food for their larvae. Various other wasps sting their prey into insensibility, lay an egg on its body and leave it without further molestation. The wasp larva on hatching pierces the body of the paralyzed prey and feeds to maturity quite after the fashion of the externally feeding parasites just described. Here the behavior of the mother and the growth of the larva is essentially similar in the two groups, for we find that the differences relate only to the severity of the stinging with the ovipositor by the wasp and by the parasitic Hymenoptera. Between these wasps and the predatory ones the difference lies only in the transport of the prey after stinging and the preparation of a nest, so that we cannot draw very clear distinctions between the members of the three series. Nevertheless, the highly developed instincts and behavior of the digger wasp as manifested by the mother are something entirely lacking in the other two groups. On morphological grounds there is reason to believe that the predatory wasps have probably arisen from wasps that simply stung their prey, but that the parasitic Hymenoptera, including the families mentioned earlier in this chapter have had an independent origin.

Insects which develop as internal parasites of vertebrates are not numerous, and as might be expected, are practically restricted to purely terrestrial hosts. Moreover, they are mainly a few highly specialized Diptera whose larvae develop in the alimentary canal or subcutaneous tissues of their hosts, or more rarely in open wounds, or in natural spaces that open externally, such as the nasal cavities. Most remarkable among these are the bot-flies, representatives of two families, the Oestridae and Gastrophilidae. These resemble one another in having the mouthparts so completely vestigial that the adults cannot feed after completing their metamorphosis. However they differ greatly in other characters and are now regarded as having had an independent origin in spite of the similarly atrophied mouthparts.

The Oestridae affect a considerable variety of mammals. *Oestrus ovis* is a widespread parasite of sheep which deposits minute larvae about the nostrils of its host. The tiny grubs enter the nasal cavities, often penetrating the frontal sinuses to complete their growth. Two species of *Hypoderma* affect cattle, laying their eggs on the hairs of the legs below the knee. The newly hatched larvae first penetrate the skin, then migrate through the tissues into the body cavity and accumulate mainly in the neighborhood of the gullet. After moulting they again migrate, going this time to the connective tissues underlying the skin on the back of the cattle. Growth is completed here after perforating the hide in order to extrude the tip of the abdomen that bears the spiracles, thus securing air for respiration. The larva is covered with minute chitinous prickles which serve apparently to stimulate the flow of serum and rasp the tissues to augment the food supply. On attaining full growth, the larva works its way out through the skin, drops to the ground and digs in to pupate.

The species of *Cuterebra* affect various rodents such as squirrels and chipmunks in a similar way, some of them occurring only in male hosts where they develop within the scrotum. As a rule the oestrid parasites are highly specific in relation to their hosts, restricting themselves closely, for example, to reindeer, elephants, hippopotami, elk or in a more general way to ruminants, rodents, etc. Nevertheless there are occasional references to man serving as an accidental host, at least for the earlier larval stages of several common forms.

One species, *Dermatobia hominis*, commonly referred to as the "ver macaque," is known to attack man quite consistently in the American tropics although it is more commonly a parasite of cattle.

This species frequently attaches its eggs to the bodies of mosquitoes and other blood-sucking insects. When these feed, the eggs hatch immediately, due it is believed, to the warmth from the mammalian body. These larvae (Pl. XIX A) then penetrate the skin directly where they produce a boil-like swelling within which they feed to maturity. Finally they work their way out and pupate in the soil. That the squirming of these spiny larvae beneath the skin is very painful is attested by Dunn ('30) who chose to offer himself as an experimental animal and has written a most interesting account of his success in breeding the species through its life-cycle.

The bot-flies of the genus *Gastrophilus* lay their eggs on the hairs of the fore legs or about the lips of horses, and their larvae develop in the alimentary canal of the host, attached by means of their mouth hooks. When full-grown they allow themselves to be voided with the dung of the animal whence they enter the soil for pupation.

Other parasitic fly larvae with habits more or less similar to those of the true bot-flies occur in several genera of Anthomyiidae, Calliphoridae and Sarcophagidae, affecting mammals, including man, and birds. Even batrachians and reptiles very rarely fall prey to certain sarcophagids, and a species of green-bottle fly, *Lucilia bufonivora* is a true parasite of toads.

Best known of these is the African tumbu fly, *Cordylobia anthropophaga*, sometimes parasitic on man and *Staisia rodhaini* which affects antelopes and rodents, but may on occasion affect man. More recently a Malayan genus, *Booponus*, has been found to infest the water-buffalo, its larva developing in the hoofs. The best known bird parasites are species of *Philornis* which undergo their larval development in superficial or subcutaneous tumors. They are frequently encountered in the warmer parts of the Americas. As with the larger muscoid flies parasitic on insects, there is a wide range in host specificity among those that attack vertebrates, and even the distinction between parasitism and necrophagy cannot be clearly drawn in the case of many forms.

The occurrence of various larvae of this kind in decaying animal material has already been mentioned (p. 199). Among the flies whose maggots develop in open wounds, numerous species belonging to several genera may be mentioned, but most of these are attracted by conditions similar to those present in the carcasses of dead animals. Some of these are, however, primary invaders which do not wait for decay to set in before they deposit their eggs. Notable among these is the common American screwworm (*Cochliomyia americana*). Following infestation by this species, various carrion-eating members of related genera appear, but the occurrence of these others is secondary as they appear only in the presence of the true screwworm. Others, like the larvae of *Wohlfartia*, produce lesions under the skin of animals, penetrating through the uninjured skin.

Among species with such habits we must include also the blowfly larvae that have been turned to a beneficial purpose by using them surgically to clean out and hasten the healing of wounds. Used in this way on human patients the larvae are reared under aseptic conditions so that they remain sterile. They may then be introduced without danger of contaminating the tissues with additional pathogenic organisms. Interesting details relating to such treatments are contained in several papers by Robinson. Due to the recent discovery of more ordinary therapeutic agents serving the same purpose, the use of living maggots has been discontinued. Nevertheless they served a necessary purpose, as it was through studies of their physiological action that the non-living remedies were discovered. The species mainly concerned were two blowflies, *Lucilia sericata* and *Phormia regina*.

The larvae of many Diptera are able to withstand the disintegrating action of the digestive juices of mammals and when they are taken into the alimentary canal with food or water they may survive for periods of varying length. Such accidental occurrences are comparatively infrequent but they may cause considerable discomfort. The species most commonly concerned are those that may be present on salads or other foods that are eaten raw, or without heating just previous to ingestion. Ordinarily such larvae cannot complete their transformation to the adult stage, but there is one small phorid fly, *Megaselia xanthina* which has been known to develop into winged flies in the alimentary tract of man. Such individuals of course never survive as adults in this environment.

A few most extremely aberrant insects of several groups fall also into the category of internal parasites of mammals. Thus *Ascodipteron*, a streblid fly, which never develops the wings or other adult structures in the female, lives as a parasite in the skin of bats. In a highly modified flea, the chigoe (*Tunga penetrans*) the adult female bores her way beneath the claws of dogs or under the toenails of human subjects to complete her final engorgement and mature her complement of eggs.

Although not numerous in comparison the entomophagous parasites, those that actually penetrate the tissues of the higher vertebrates illustrate very beautifully the versatility of insects in adapting themselves to most unusual and seemingly precarious situations in their search for food.

REFERENCES

NOTE: In the following list the references have been separated arbitrarily into a number of groups. The first several headings refer to specific orders of insects and include titles difficult to place easily in one of the other groups. The later ones are arranged so that they refer in a general way to the matters specifically discussed in the text. This has resulted in a considerable amount of overlapping and in the placing of some papers with reference to only a part of their contents. It seems advisable, however, to adopt such an arrangement as this, for the list is quite extensive, even though it represents only a small fraction of the really enormous literature on entomophagous insect parasites. On account of the similarity between parasites and certain predators, some references of interest will be found in the list appended to Chapter VI.

HYMENOPTERA

Adler, H. F.
1918 Zur Biologie von *Apanteles glomeratus*. Zeitschr. Insektenbiol., vol. 14, pp. 182–186, 2 figs.

Ainslie, C. N.
1920 Notes on *Gonatopus ombrodes*, a Parasite of Jassids. Entom. News, vol. 31, pp. 169–173; 187–190.

Armstrong, T.
1936 Two Parasites of the White Apple Leafhopper (*Typhlocyba pomaria*). Ann. Rept. Ontario Entom. Soc., vol. 66, pp. 16–31.

Baker, W. A., and L. G. Jones
1934 Studies of *Exeristes roborator*, a Parasite of the European Corn Borer, in the Lake Erie Area. Tech. Bull. United States Dept. Agric., No. 460, 27 pp., 2 pls., 7 figs.

Bischoff, H.
1923 Biologie der Hymenopteren. In Biol. Tiere Deutschlands, Lief. 7–8, Teil 42, 156 pp.
1927 Biologie der Hymenopteren. 571 pp., 224 figs., Berlin, J. Springer.

Bradley, W. G., and E. D. Burgess

1934 The Biology of *Cremastus flavo-orbitalis*, an Ichneumonid Parasite of the European Corn Borer. Tech. Bull. United States Dept. Agric., No. 441, 15 pp.

Brues, C. T.

1919 A New Chalcid-fly Parasitic on the Australian Bull-dog Ant. Ann. Entom. Soc. America, vol. 12, pp. 13–21, 2 pls.

Buckell, E. R.

1928 Notes on the Life History and Habits of *Melittobia chalybii* Ashmead. Pan-Pacific Entom., vol. 5, pp. 14–22.

Cendana, S. M.

1937 Studies on the Biology of *Coccophagus*, a Genus Parasitic on Nondiaspine Coccidae. California Univ. Pubs. Entom., vol. 6, pp. 337–400.

Cheeseman, L. E.

1922 *Rhyssa persuasoria;* its Oviposition and Larval Habits. Proc. South London Entom. and Nat. Hist. Soc., 1921–22, pp. 1–2.

Clancy, D. W.

1934 The Biology of *Tetracnemus pretiosus*. California Univ. Pubs. Entom., vol. 6, pp. 231–248.

Clausen, C. P.

1924 The Parasites of *Pseudococcus maritimus* in California. Calif. Univ. Pubs. Entom., vol. 3, pp. 253–288.

Cox, J. A.

1932 *Ascogaster carpocapsae*, an Important Larval Parasite of the Codling Moth and Oriental Fruit Moth. Tech. Bull. New York State Agr. Expt. Sta., No. 188, 26 pp.

Crawford, A. W.

1933 *Glypta rufiscutellaris*, an Ichneumonid Larval Parasite of the Oriental Fruit Moth. Tech. Bull. New York State Agr. Expt. Sta., No. 217, 29 pp.

Cros, A.

1935 Biologie du *Trichopria stratiomyiae* Kieffer. Bull. Soc. Hist. Nat. Afrique du Nord, vol. 21, pp. 133–160.

Crossman, S. S.

1922 *Apanteles melanoscelus*, an Imported Parasite of the Gypsy Moth. Bull. United States Dept. Agr., no. 1028, 25 pp.

DeLeon, D.

1935 The Biology of *Coeloides dendroctoni*, an Important Parasite of the Mountain Pine Beetle (*Dendroctonus monticolae*). Ann. Entom. Soc. America, vol. 28, pp. 411–424.

Duncan, C. D.

1940 The Biology of North American Vespine Wasps. 272 pp., 54 pls. Stanford Univ. Press.

Evans, A. C.

1933 Comparative Observations on the Morphology and Biology of Some Hymenopterous Parasites of Carrion-infesting Diptera. Bull. Entom. Res., vol. 24, pp. 285–405, 12 figs.

Fink, D. E.

1926 The Biology of *Macrocentrus ancylivorus*, an Important Parasite of the Strawberry Leaf Roller (*Ancylis comptana*). Jour. Agr. Res., vol. 32, pp. 1121–1134.

Fisher, K.

1932 *Agriotypus armatus* and its Relation with its Hosts. Proc. Zool. Soc. London, 1932, pp. 451–461.

Flanders, S. E.

1938 Cocoon Formation in Endoparasitic Chalcidoids. Ann. Entom. Soc. America, vol. 31, pp. 167–180, 6 figs.

Fox, J. H.

1927 The Life-history of *Exeristes roborator*, a Parasite of the European Corn Borer. Rept. Nat. Res. Council, No. 21, 58 pp., 14 pls.

Genieys, P.

1924 Contributions à l'étude des Évaniidae: *Zeuxevania splendidula* Costa. Bull. Biol. France et Belgique, vol. 58, pp. 482–494.

1925 *Habrobracon brevicornis* Wesm. (Translation from French by L. O. Howard.) Ann. Entom. Soc. America, vol. 18, pp. 143–202, 35 figs.

Graenicher, S.

1905a Some Observations on the Life-History and Habits of Parasitic Bees. Bull. Wisconsin Nat. Hist. Soc., vol. 3, pp. 153–167, 1 pl.

1905b On the Habits of Two Ichneumonid Parasites of the Bee, *Ceratina dupla*. Entom. News, vol. 16, pp. 43–49.

1906 The Habits and Life-history of *Leucospis affinis*, a Parasite of Bees. Bull. Wisconsin Nat. Hist. Soc., vol. 4, pp. 153–159.

Griswold, G. H.

1925 A Study of the Oyster-shell Scale, *Lepidosaphes ulmi* and one of its Parasites, *Aphelinus mytilaspidis*. Mem. Cornell Agr. Expt. Sta., no. 93, 67 pp.

Haeussler, G. J.

1932 *Macrocentrus ancylivorus*, an Important Parasite of the Oriental Fruit Moth. Jour. Agric. Res., vol. 45, pp. 79–100.

Hase, A.

1922 Biologie der Schlupfwespe *Habrobracon brevicornis*. Arb. Biol. Reichsanst. f. Land u. Forstw., vol. 11, pp. 95–168.

Haviland, M. D.

1920 On the Bionomics and Development of *Lygocerus testaceimanus* and *Lygocerus cameroni*, Parasites of *Aphidius*. Quart. Jour. Micr. Sci., vol. 65, pp. 101–127.

Henriksen, K. L.

1918 The Aquatic Hymenoptera of Europe and their Biology. Entom. Meddel., vol. 7, pp. 137–251.

1922 Notes upon some aquatic Hymemoptera. Ann. biol. lacustre, vol. 11, pp. 19–37.

Hill, C. C., and H. D. Smith

1931 *Heterospilus cephi*, a Parasite of the European Wheat Sawfly, *Cephus pygmaeus*. Jour. Agr. Res., vol. 43, pp. 597–609.

Hoffer, E.

1886 Zur Biologie der *Mutilla europaea* L. Zool. Jahrb., Abth f. Syst., vol. 1, pp. 679–686.

Howard, L. O.

1891 The Biology of the Hymenopterous Insects of the Family Chalcididae. Proc. United States Nat. Mus., vol. 14, pp. 567–588.

Imms, A. D.

1918 Observations on *Pimpla pomorum*, a Parasite of the Apple Blossom Weevil. Ann. Appl. Biol., vol. 4, pp. 211–227.

Ishii, T.

1932 The Encyrtidae of Japan. II. Studies on Morphology and Biology. Bull. Imp. Agric. Expt. Sta., Japan, vol. 3, pp. 161–202, 8 pls.

King, J. L., and J. K. Holloway

1930 *Tiphia popilliavora*, a Parasite of the Japanese Beetle. Circ. United States Dept. Agric., no. 145, 11 pp., 2 figs., 4 pls.

Lathrop, F. H., and R. C. Newton

1933 The Biology of *Opius melleus*, a Parasite of the Blueberry Maggot. Jour. Agr. Res., vol. 46, pp. 143–160.

Lundie, A. E.

1924 A Biological Study of *Aphelinus mali*, a Parasite of the Woolly Apple Aphid, *Eriosoma lanigerum*. Mem. Cornell Agric. Expt. Sta., No. 79, 27 pp.

Maple, J. D.

1937 The Biology of *Ooencyrtus johnsoni* and the Role of the Egg Shell in the Respiration of Certain Encyrtid Larvae. Ann. Entom. Soc. America, vol. 30, pp. 123–145.

Marsh, F. L.

1937 Biology of the Ichneumonid *Spilocryptus extrematis*. Ann. Entom. Soc. America, vol. 30, pp. 40–42.

Martin, C. H.

1928 Biological Studies of two Hymenopterous Parasites of Aquatic Insect Eggs. Entom. Americana, vol. 8, pp. 105–151, 5 pls.

Morris, K. R. S.

1938 *Eupelmella vesicularis* as a Predator of Another Chalcid, *Microplectron fuscipennis*. Parasitology, vol. 30, pp. 20–32.

Muesebeck, C. F. W., and D. L. Parker

1933 *Hyposoter disparis*, an Introduced Ichneumonid Parasite of the Gipsy Moth. Jour. Agric. Res., vol. 46, pp. 335–347.

Myers, J. G.

1929 Further Notes on *Alysia manducator* and Other Parasites of Muscoid Flies. Bull. Entom. Res., vol. 19, pp. 357–360.

1930 *Carabunis myersi* (Hym., Encyrtidae), a Parasite of Nymphal Froghoppers. *Ibid.*, vol. 21, pp. 341–351.

1932 Biological Observations on Some Neotropical Parasitic Hymenoptera. Trans. Entom. Soc. London, vol. 80, pp. 121–136.

Nielsen, E.

1923 Contributions to the Life-history of the Pimpline Spider Parasites (*Polysphincta, Zaglyptus, Tromatobia*). Entom. Meddel., vol. 14, pp. 137–205.

Parker, D. E.

1936 *Chrysis shanghaiensis*, a Parasite of the Oriental Moth. Jour. Agric. Res., vol. 52, pp. 449–458.

Parker, H. L., and H. D. Smith

1933 *Eulophus viridulus*, a Parasite of *Pyrausta nubilalis*. Ann. Entom. Soc. America, vol. 26, pp. 21–36.

Perkins, R. C. L.

1905 Leaf-hoppers and their Natural Enemies (Mymaridae, Platygasteridae). Bull. Hawaiian Sugar Planters' Expt. Sta., No. 1, pp. 187–203.

Phillips, W. J., and F. W. Poos

1927 Two Hymenopterous Parasites of American Joint-worms. Jour. Agr. Res., vol. 34, pp. 473–488.

Picard, F.

1913 Sur le genre *Zeuxevania* Kieffer et sur les moeurs du *Zeuxevania splendidula*. Bull. Soc. Entom. France, 1913, pp. 301–304, 1 fig.

1921 Sur la biologie du *Tetrastichus rapo*. *Ibid.*, 1921, pp. 206–208.

1922 Note sur la biologie de *Melittobia acasta*. *Ibid.*, 1922, pp. 301–304.

Proper, A. B.

1931 *Eupteromalus nidulans*, a Parasite of the Brown-tail and Satin Moths. Jour. Agr. Res., vol. 43, pp. 37–56.

Ramme, W.

1920 Zur Lebensweise von *Pseudagenia*. SB. Ges. Naturforsch. Freunde, Berlin, 1920, pp. 130–132.

Raynaud, P.

1935 *Phaenoserphus viator*, parasite de larves de Carabidae. Misc. Entom., vol. 36, pp. 97–100.

Riley, C. V.

1888 The Habits of *Thalessa* and *Tremex*. Insect Life, vol. 1, pp. 168–179.

Roberts, R. A.

1933 Biology of *Brachymeria fonscolombei*, a Hymenopterous Parasite of Blowfly Larvae. United States Dept. Agric. Tech. Bull., No. 365, pp. 1–21.

Sanders, G. E.

1911 Notes on the Breeding of *Tropidopria conica*. Canadian Entom., vol. 43, pp. 48–50.

Seurat, L. G.

1899 Contributions à l'étude des Hymenoptères entomophages. Ann. Sci. Nat. Zool., vol. 10, pp. 1–159, 5 pls.

Silvestri, F.

1908 Contribuzioni alla conoscenza biologica degli Imenotteri parassiti. II. Sviluppo dell' *Ageniaspis fuscicollis*. III. Sviluppo dell' *Encyrtus aphidivorus*. IV. Sviluppo dell' *Oöphthora semblidis*. Bol. R. Scuola Super. Agr., vol. 3, pp. 29–84.

Smith, H. D.

1930 The Bionomics of *Dibrachoides dynastes*, a Parasite of the Alfalfa Weevil. Ann. Entom. Soc. America, vol. 23, pp. 577–593.

Stellwag, F.

1921 Die Schmarotzerwespen (Schlupswespen) als Parasiten. Monog. angew. Entom., No. 6 (Beiheft to vol. 2 of Zeits. angew. Entom.), 100 pp., 37 figs.

Swezey, O. H.

1903 Observations of Hymenopterous Parasites of Certain Fulgoridae. Ohio Naturalist, vol. 3, pp. 444–451.

Thienemann, A.

1916 Über Wasserhymenopteren. Zeits. f. wiss. Insektenbiol., vol. 12, pp. 49–54.

Thomsen, M.

1928 Some Observations on the Biology and Anatomy of a Cocoon-making Chalcid Larva, *Euplectrus bicolor*. Vidensk. Meddel., Dansk. Naturhist. För. Kjøbenhavn, vol. 84, pp. 73–89.

Tower, D. G.

1915 Biology of *Apanteles militaris*. Jour. Agr. Res., vol. 5, pp. 495–506.

Vance, A. M.

1927 On the Biology of some Ichneumonids of the Genus *Paniscus*. Ann. Entom. Soc. America, vol. 20, pp. 405–416.

1932 The Biology and Morphology of the Braconid, *Chelonus annulipes*, a Parasite of the European Corn Borer. Tech. Bull. United States Dept. Agr., No. 294, 48 pp.

Vance, A. M., and H. L. Parker

1932 *Laelius anthrenivorus*, an Interesting Bethylid Parasite of *Anthrenus verbasci* in France. Proc. Ent. Soc. Washington, vol. 34, pp. 1–7.

Voukassovitch, P.

1925 Sur la biologie de *Goniozus claripennis*, parasite d'*Oenophthira pilleriana*. Bull. Soc. Hist. Nat. Toulouse, vol. 53, pp. 225–246, 10 figs.

Wheeler, W. M., and L. H. Taylor

1921 *Vespa arctica*, a Parasite of *Vespa diabolica*. Psyche, vol. 28, pp. 135–144.

Willard, H. F.

1920 *Opius fletcheri* as a Parasite of the Melon Fly in Hawaii. Jour. Agr. Res., vol. 20, pp. 423–438.

DIPTERA

Aldrich, J. M.

1915 The Economic Relations of the Sarcophagidae. Jour. Econ. Entom., vol. 7, pp. 242–247.

Autuori, M.

1928 *Syneura infraposita*, un Novo Parasita da *Icerya purchasi*. Arch. Inst. Biol., São Paulo, vol. 1, pp. 193–200.

Baer, W.

1920 Die Tachinen als Schwarotzer der schädlichen Insekten, ihre Lebensweise, wirtschaftliche Bedeutung und systematische Kennzeichnung. Zeits. f. angew. Entom., vol. 6, pp. 185–246, 63 figs. 2nd. part, *ibid.*, vol. 7, pp. 97–163 and 349–423, 30 figs.

Baerg, W. J.

1920 An unusual Case of Parasitism on *Clastoptera obtusa*. Entom. News, vol. 31, pp. 20–21, 1 fig.

Beard, R. L.

1940 The Biology of *Anasa tristis* with Particular Reference to the Tachinid Parasite *Trichopoda pennipes*. Bull. Connecticut Agr. Expt. Sta., No. 440, pp. 593–679, 3 pls., 18 figs.

1942 On the Formation of the Tracheal Funnel in *Anasa tristis* by the Parasite, *Trichopoda pennipes*. Ann. Entom. Soc. America, vol. 25, pp. 68–72, 2 figs.

Berg, V. L.

1940 The External morphology of the Immature Stages of the Bee Fly, *Systoechus vulgaris*, a Predator of Grasshopper Egg Pods. Canadian Entom., vol. 72, pp. 169–178, 6 figs.

Berland, L.

1933 Sur le Parasitisme des Phorides. Bull. Soc. Zool. France, vol. 57, pp. 529–530.

Borgmeier, T.

1931 Sobre alguns phorideos que parasitam a suava e outras formigas contadeiras Arch. Inst. Biol., São Paulo, vol. 4, pp. 209–228.

Bougey, E.

1935 Observations sur l'*Ammophila hirsuta* et sur *Hilarella stictica*, son parasite. Rev. Franç. d'Entom., vol. 2, pp. 19–27.

Branch, H. E.

1920 A Webspinning Sarcophagid, Parasitic upon a Mantis. Entom. News, vol. 31, p. 276.

Bussart, J. E.

1937 The Bionomics of *Chaetophleps setosa*. Ann. Entom. Soc. America, vol. 30, pp. 285–295.

Buysson, H. du

1917 Observations sur des nymphes de *Coccinella septempunctata*, parasitées par le *Phora fasciata*. Bull. Soc. Entom. France, 1917, pp. 249–250.

Cleare, L. D.

1939 The Amazon Fly (*Metagonistylum minense*) in British Guiana. Bull. Entom. Res., vol. 30, pp. 85–102.

Dowden, P. B.

1933   *Lydella nigripes* and *L. piniariae*, Fly Parasites of Certain Tree-defoliating Caterpillars. Jour. Agr. Res., vol. 48, pp. 97–114.

1934   *Zenillia libatrix*, a Tachinid Parasite of the Gipsy Moth and the Brown-tail Moth. *Ibid.*, vol. 48, pp. 97–114.

Fage, L.

1933   A Propos du Parasitisme des Phorides. Bull. Soc. Zool. de France, vol. 58, pp. 90–92.

Felt, E. P.

1911   Summary of Food Habits of American Gall Midges. Ann. Entom. Soc. America, vol. 4, pp. 55–62.

1914   Adaptation in the Gall Midges. 44th Rept. Entom. Soc. Ontario, pp. 76–82, 2 figs.

Ferrière, Ch.

1922   Le parasitisme externe des Oncophanes. Schweiz. entom. Anz., vol. 11, pp. 3–4.

Fuller, M. E.

1938   Notes on *Trichopsidea oestracea* (Nemestrinidae) and *Cyrtomorpha flaviscutellaris* (Bombyliidae)—Two Dipterous Enemies of Grasshoppers. Proc. Linn. Soc. New South Wales, vol. 63, pp. 95–104.

Greene, C. T.

1921   Dipterous Parasites of Saw-flies. Proc. Entom. Soc. Washington, vol. 23, pp. 41–43.

Hendel, F.

1933   Ueber das Auftreten der in Schildläusen parasitisch lebend Diptera-Gattung *Cryptochaetum* in Deutschland. Zeits. f. Pflanzenkr., vol. 43, pp. 97–103.

Keilin, D., and W. R. Thompson

1915   Sur le cycle évolutif des Pipunculides (Diptères), parasites intracoelomiques des Typhlocybes (Homoptères). C. R. Soc. Biol., Paris, vol. 78, pp. 9–12.

Kelly, E. O. G.

1914   A New Sarcophagid Parasite of Grasshoppers. Jour. Agr. Res., vol. 2, pp. 435–446.

Knab, F.

1914   Drosophilidae with Parasitic Larvae. Insecutor Ins. Mens., vol. 2, pp. 165–169.

Künckel d'Herculais, J.

1905   Les Lépidoptères Limacodides et leurs Diptères parasites, Bombylides du genre *Systropus*. Bull. Sci. France et Belgique, vol. 39, pp. 141–151.

Lamb, C. G.

1918   On a Parasitic Drosophila from Trinidad. Bull. Entom. Res., vol. 9, pp. 157–162, 4 figs.

Landis, B. J.

1940   *Paradexodes epilachnae*, a Tachinid Parasite of the Mexican Bean Beetle. Tech. Bull. United States Dept. Agr., No. 721, 31 pp.

Lepiney, J. de, and J. M. Mimeur

1930   Sur *Glossista infuscata* et *Anastoechus nitidulus*, parasites Maroccains de *Dociostaurus maroccanus*. Rev. Path. Vég. Entom. Agric. France, vol. 17, pp. 419–430.

Lloyd, J. T.

1919   An Aquatic Dipterous Parasite, *Ginglymyia acrirostris* and Additional Notes on its Lepidopterous Host, *Elophila fulicalis*. Jour. New York Entom. Soc., vol. 27, pp. 263–265, 1 pl.

Marriner, T. F.

1932   A Coccinellidae Parasite. Naturalist, London, No. 906, pp. 221–222.

Meijere, J. C. H. de

1904   Beiträge zur Kenntnis der Biologie und der Systematischen Verwandschaft der Conopiden. Tijdschr. v. Entom., vol. 46, pp. 144–224, 4 pls.

Menozzi, C.

1927   Contributo alla biologia della *Phalacrotophora fasciata*, parassita di coccinellidi. Bol. Soc. Entom. Italiana, vol. 59, pp. 72–78, 1 fig.

Morris, H. M.

1922   On the Larva and Pupa of a Parasitic Phorid Fly—*Hypocera incrassata*. Parasitology, vol. 14, pp. 70–74, 1 pl., 4 figs.

Nielsen, J. C.

1903   Ueber die Entwicklung von *Bombylius pumilus*, eine Fliege, welche bei *Colletes daviesana* schmarotzt. Zool. Jahrb., Abt. f. System., vol. 18, pp. 647–658.

Pergande, T.

1901   The Ant Decapitating Fly. Proc. Entom. Soc. Washington, vol. 4, pp. 497–502.

Perkins, R. C. L.

1905   Leaf-hoppers and their Natural Enemies (Pipunculidae). Bull. Hawaiian Sugar Planters' Expt. Sta., No. 1, pp. 123–157.

Rennie, J., and C. H. Sutherland

1920   On the Life History of *Bucentes* (*Siphona*) *geniculata*, a Parasite of *Tipula paludosa* and Other Species. Parasitology, vol. 12, pp. 199–211.

Ronna, A.

1937   *Melaloncha ronnai* (Phoridae) endoparasita de *Apis mellifica*. Rev. Entom., vol. 6, pp. 1–9.

Roubaud, E.

1906   Biologie larvaire et métamorphoses de *Siphona cristata*; adaptation d'une Tachinide à un hôte aquatique diptère; un nouveau cas d'ectoparasitisme interne. C. R. Acad. Sci., Paris, vol. 142, pp. 1438–1439.

Scaramuzza, L. C.

1930   Preliminary Report on the Biology of *Lixophaga diatraeae*. Jour. Econ. Entom., vol. 23, pp. 999–1004.

Schmitz, H.

1930   Phoriden aus Eipaketen von *Locusta migratoria* in Daghestan.

Severin, H. C.

1937   *Zodion fulvifrons* (Conopidae), a Parasite of the Honey Bee. Ent. News, vol. 48, pp. 243–244.

Shelford, V. E.

1913   The Life History of a Bee-fly (*Spogostylum anale*). Ann. Entom. Soc. America, vol. 6, pp. 213–225.

Smith, H. S., and H. Compère

1916   Observations on the *Lestophonus*, a Dipterous Parasite of the Cottony Cushion Scale. Month. Bull. California Comm. Hortic. Sacramento, vol. 5, pp. 384–390.

Smith, M. R.

1928   *Plastophora crawfordi* and *Plastophora spatulata*, Parasitic on *Solenopsis geminata*. Proc. Entom. Soc. Washington, vol. 30, pp. 105–108.

Strickland, E. H.

1923   Biological Notes on Parasites of Prairie Cutworms. Bull. Canada Dept. Agric., n.s., No. 26, 40 pp., 20 figs.

Thompson, W. R.

1915   Sur un Diptère parasite du larve d'un Mycetophilide. C. R. Soc. Biol. Paris, vol. 78, pp. 87–89.

1921   Recherches sur les Diptères parasites. I. Les larves des Sarcophagidae. Bull. Biol. France et Belgique, vol. 54, pp. 313–463.

1923   Observations sur quelques espèces biologiques dans le groupe des tachinaires. Bull. Soc. Zool. France, vol. 48, pp. 165–170.

Thorpe, W. H.

1930   The Biology, Post-embryonic Development, and Economic Importance of *Cryptochaetum iceryae* Parasitic on *Icerya purchasi*. Proc. Zool. Soc. London, 1930, pp. 929–971, 5 pls., 23 figs.

Timon-David, J.

1938   Sur un Phoride parasite (*Megaselia giraudii* Egger) de la sauterelle verte (*Phasgoneura viridissima*). Ann. Parasitol. Humaine et Compar., vol. 16, pp. 193–195.

Tölg, F.

1910   *Billaea pectinata* als Parasit von Cetoniden-und Cerambyciden-Larven. Zeits. wiss. Insektenbiol., vol. 6, pp. 208–211; 278–283; 331–336; 387–395; 426–430.

Townsend, L. H.

1935   The Mature Larva and Puparium of *Physocephala sagittaria*. Psyche, vol. 42, pp. 142–148.

Trägårdh, I.

1931   Zwei forstentomologisch wichtige Fliegen. Zeits. angew. Entom., vol. 18, pp. 672–690.

Wood, A. H.

1933   Notes on some Dipterous Parasites of *Schistocerca* and *Locusta* in the Sudan. Bull. Entom. Res., vol. 24, pp. 521–530.

Worthley, H. N.

1924   The Biology of *Trichopoda pennipes*, a Parasite of the Common Squash Bug. Psyche, vol. 31, pp. 7–16; 57–77.

COLEOPTERA, ETC.

Balduf, W. V.

1935   The Bionomics of Entomophagous Coleoptera. 220 pp., 108 figs. St. Louis, John S. Swift Co.

1939   The Bionomics of Entomophagous Insects, part 2, 384 pp., 228 figs. St. Louis, John S. Swift Co.

Bohart, R. M.

1936–   A Preliminary Study of the Genus *Stylops* in California. Pan-pacific. Entom.,
37   vol. 12, pp. 9–18; vol. 13, pp. 49–57.

Chapman, T. A.

1870   Some Facts toward a Life-history of *Rhipiphorus paradoxus*. Ann. Mag. Nat. Hist. (4), vol. 6, pp. 314–326, 1 pl.

Chobaut, A.

1891   Moeurs et métamorphoses de *Emmenada flabellata*, insecte Coléoptère de la famille des Rhipiphorides. Ann. Soc. Entom. France, vol. 60, pp. 447–456.

Cros, A.

1931   Biologie des Meloës. Ann. Sci. Nat.[Zool., vol. 14, pp. 189–227.

Dodd, A. P.

1912   Some Remarkable Ant-friend Lepidoptera of Queensland. Proc. Entom. Soc., London, 1911, pp. 577–590.

Hill, G. F.

1922   A New Species of *Mordellistena*, Parasitic on Termites. Proc. Linn. Soc. New South Wales, vol. 47, pp. 346–347.

Kemner, N. A.

1926   Zur Kenntniss der Staphyliniden Larven II. Die Lebensweise und die parasitische Entwicklung der echten Aleochariden. Entom. Tidskr., vol. 27, pp. 133–170.

Mank, H. G.

1923   The Biology of the Staphylinidae. Ann. Entom. Soc. America, vol. 16, pp. 220–237.

Murray, A.

1870   Conclusion of the Life History of the Wasp and *Rhipiphorus paradoxus*, with Description and Figure of the Grub of the Latter. Ann. Mag. Nat. Hist., vol. 6, pp. 204–213.

Parker, H. L., and H. D. Smith

1934   Further Notes on *Eoxenos laboulbenei* with a Description of the Male. Ann. Entom. Soc. America, vol. 27, pp. 468–477.

Pierce, W. D.

1910   A Monographic Revision of the Twisted Winged Insects Comprising the Order Strepsiptera. Bull. United States Nat. Mus., No. 66, xii + 232 pp., 15 pls., 3 figs., 1 map.

Pyle, R. W.

1940   Triungulins of a Rhipiphorid Beetle Borne by *Elis quinquecincta*. Entom. News, vol. 52, pp. 74–81, 1 fig.

Rouget, A.

1873   Sur les Coléoptères parasites des Vespides. Mém. Acad. Sci. Dijon, vol. 3, pp. 161–288.

Sabrowsky, C. W.

1934   Notes on the Larva and Larval Habit of *Isohydnocera curtipennis*. Journ. Kansas Entom. Soc., vol. 7, pp. 65–68.

Scott, H.

1920   Notes on the Parasitic Staphylinid, *Aleochara algarum* Fauvel, and its Hosts the Phycodromid Flies. Entom. Monthly Mag., vol. 56, pp. 148–157, 2 figs.

Silvestri, F.

1904   Contribuzione alla conoscenza della metamorfosi e dei costumi della *Lebia scapularis*. Redia, vol. 2, pp. 68–84.

1905   Descrizione di un nuovo genere di Rhipiphoridae. *Ibid.*, vol. 3, pp. 315–324.

1933   Descrizione della femmina e del maschio di una nuova specie di Mengenilla. Boll. Lab. Zool. Gen. Agrar. R. Scuola Agric. Portici, vol. 28, pp. 1–10.

Stamm, R. H.

1935–   A New Find of *Rhipidius pectinicornis* (*Symbius blattarum*). Entom. Meddel.,
36   vol. 19, pp. 286–288; 289–297.

Ulrich, W.

1930   Flächerflügler, Strepsiptera Kirby (1813). Tierwelt Mitteleuropas, vol. 5, 2 (XIII), pp. 1–26.

Voris, R.

1934   Biologic Investigations on the Staphylinidae. Trans. Acad. Sci. St. Louis, vol. 28, pp. 233–261.

Wadsworth, J. T.

1915   On the Life-history of *Aleochara bilineata*, a Staphylinid Parasite of *Chortophila brassicae*. Jour. Econ. Biol., vol. 10, pp. 1–27.

Zanon, V.

1922   La Larva triungulina di *Meloë cavensis* dannosa alle api in Cirenaica. L'Agric. Colon. Florence, vol. 16, pp. 345–354.

OVIPOSITION

Beling, I.

1934   Ueber den Ausflug der Schlupwespe *Nemeritis canescens*, und über die Bedeutung des Geruchsinns bei der Rückkehr zum Wirt. Biol. Zentralbl., vol. 54, pp. 147–169.

Bordas, L.

1917   Anatomie des glandes vénimeuses des Pimplinae. Bull. Soc. Entom. France, 1917, pp. 197–198, 1 fig.

Brues, C. T.

1917a Adult Hymenopterous Parasites attached to the Body of their Host. Proc. Nat. Acad. Sci., vol. 3, pp. 136–140, 1 fig.

1917b Notes on the Adult Habits of some Hymenopterous Egg-Parasites of Orthoptera and Mantoidea. Psyche, vol. 24, pp. 195–196.

Chopard, L.

1920 Observations sur la Mante réligieuse et ses parasites. C. R. Acad. Sci. Paris, vol. 170, pp. 140–143.

1923 Les parasites de la Mante réligieuse. Ann. soc. entom. France, vol. 91, pp. 249–274.

Culver, J. J.

1919 A Study of *Compsilura concinnata*, an Imported Tachinid Parasite of the Gipsy Moth and the Brown-tail Moth. Bull. United States Dept. Agric., No. 766, 27 pp., 10 figs.

Doten, S. B.

1911 Concerning the Relation of Food to Reproductive Activity and Longevity in Certain Hymenopterous Parasites. Tech. Bull. Nevada Agric. Expt. Sta., No. 78, pp. 1–30.

Eidmann, H.

1924 Die Eiablage von *Trioxys* nebst Bemerkungen über die wirtschaftliche Bedeutung dieses Blattlausparasiten. Zeit. f. angew. Entom., vol. 10, pp. 353–363.

Faure, J. C.

1924 Études sur la ponte et mode de nutrition chez *Pteromalus variabilis* Ratz. et *Eurytoma appendigaster* Dalm., Chalcidiens parasites de l'*Apanteles glomeratus* L. Rev. Zool. Agric. and Appl., vol. 23, pp. 225–233, 2 figs.

Flanders, S. E.

1934 The Secretion of the Colleterial Glands in the Parasitic Chalcids. Journ. Econ. Entom., vol. 27, pp. 861–862.

1935 An Apparent Correlation Between the Feeding Habits of Certain Pteromalids and the Condition of their Ovarian Follicles. Ann. Entom. Soc. America, vol. 28, pp. 438–444.

1942 Oösorbtion and Ovulation in Relation to Oviposition in Parasitic Hymenoptera. *Ibid.*, vol. 35, pp. 251–266, 3 pls.

Fulton, B. B.

1933 Notes on *Habrocytus cerealellae*, Parasite of the Angumois Grain Moth. Ann. Entom. Soc. America, vol. 26, pp. 536–552, 1 pl., 1 fig.

Graham, S. A.

1918 An Interesting Habit of a Wax Moth Parasite. *Ibid.*, vol. 11, pp. 175–180, 1 pl., 2 figs.

Griswold, G. H.

1926 Notes on Some Feeding Habits of two Chalcid Parasites. *Ibid.*, vol. 19, pp. 331–334, 1 fig.

Hase, A.

1922 Uber die eigentümliche Nahrungsgewinnung einer Schlüpfwespe (*Habrocytus*). Naturw. Wochenschr., vol. 21, pp. 110–111.

1924 Die Schlüpfwespen als Gifttiere. Biol. Zentralbl., vol. 44, pp. 209–243, 3 figs., 1 table.

Holloway, T. E.

1912 An Experiment on the Oviposition of a Hymenopterous Egg Parasite. Entom. News, vol. 23, pp. 329–330.

Howard, L. O.

1910 On the Habit with certain Chalcidoidea of Feeding at Puncture Holes made by the Ovipositor. Journ. Econ. Entom., vol. 3, pp. 257–260.

1923 A Curious Phase of Parasitism among the Parasitic Hymenoptera. Canadian Entom., vol. 55, pp. 223–224.

1925 Parasitic Hymenoptera Feeding by Indirect Suction. Entom. News, vol. 36, pp. 129–133.

Laing, J.

1937 Host-finding by Insect Parasites; Observations on *Alysia manducator, Monmoniella vitripennis* and *Trichogramma evanescens.* Journ. Anim. Ecol., vol. 6, pp. 298–317, 5 figs.

Landis, B. J.

1940 *Paradexodes epilachnae,* a Tachinid Parasite of the Mexican Bean Beetle. Tech. Bull. United States Dept. Agric., No. 721, 32 pp., 23 figs.

Loxinski, P.

1910 Ueber Anpassungserscheinungen bei Ichneumoniden. Zeits. wiss. Insektenbiol., vol. 6, pp. 298–300.

Marsh, Frank L.

1937 Biology of the Tachinid, *Winthemia datanae.* Psyche, vol. 44, pp. 138–140.

Maxwell, H.

1922 The Stinging of an Ichneumon Fly. Scottish Naturalist, 1922, pp. 17–18.

McClure, H. E.

1933 The Effectiveness of the Sting of *Aenoplex carpocapsae.* Entom. News, vol. 44, pp. 48–49.

Rabaud, E.

1922 Notes sur le comportement de *Rielia manticida* Proctotrypide parasite des oöthèques de Mantes. Bull. Soc. Zool. France, vol. 47, pp. 10–15.

Rockwood, L. P.

1917 An Aphis Parasite Feeding at Puncture Holes Made by the Ovipositor. Journ. Econ. Ent., vol. 10, p. 415.

Salt, G.

1937 The Sense Used by *Trichogramma* to Distinguish between Parasitized and Unparasitized Hosts. Proc. Roy. Soc. London, B, vol. 122, pp. 57–75.

Severin, H. H. P., H. C. Severin, and W. J. Hartung

1915 The Stimuli which Cause the Eggs of Leaf-ovipositing Tachinidae to Hatch. Psyche, vol. 22, pp. 132–137.

Smith, H. S.

1917 The Habit of Leaf-oviposition Among the Parasitic Hymenoptera. Psyche, vol. 24, pp. 63–68, 4 figs.

Townsend, C. H. T.

1908   A Record of Results from Rearings and Dissections of Tachinidae. Tech. Bull. United States Dept. Agric., No. 12, pt. 6, pp. 95–118, 6 figs.

1911   Review of Work by Pantel and Portschinsky on the Reproductive and Early Stage Characters of Muscoid Flies. Proc. Entom. Soc. Washington, vol. 13, pp. 151–170.

Voukassovitch, H., and P. Voukassovitch

1931   Sur la ponte des Hyménoptères parasites entomophages. C. R. Soc. Biol., Paris, vol. 106, pp. 695–697.

Whiting, A. R.

1940   Do *Habrobracon* Females Sting their Eggs? American Natural., vol. 74, pp. 468–471.

Williams, F. X.

1919   Some Observations on *Pipunculus*, flies which Parasitize the Cane Leafhopper, at Pahala, Hawaii. Proc. Hawaiian Entom. Soc., vol. 4, pp. 68–71.

LARVAL DEVELOPMENT AND FEEDING

Ayers, H.

1884   On the Development of *Oecanthus* and its Parasite, *Teleas*. Mem. Boston Soc. Nat. Hist., vol. 3, pp. 261–272.

Balfour-Browne, F.

1922   On the Life-history of *Melittobia acasta*, a Chalcid Parasite of Bees and Wasps. Parasitology, vol. 14, pp. 349–369, 1 pl.

Barber, H. S.

1915   *Macrosiagon flavipennis* in Cocoon of *Bembex spinolae*. Proc. Entom. Soc. Washington, vol. 17, pp. 187–188.

1939   A New Parasitic Beetle from California (Riphiphoridae). Bull. Brooklyn Entom. Soc., vol. 34, pp. 17–20.

Baume-Pluvinel, G. de la

1914   Évolution et formes larvaires d'un Braconide *Adelura gahani*, parasite interne d'un Phytomyzinae. Arch. de Zool. Expt. Gén., vol. 55, pp. 47–59.

Bess, H. A.

1936   The Biology of *Leschenaultia exul*, a Tachinid Parasite of *Malacosoma americanum* and *M. disstria*. Ann. Entom. Soc. America, vol. 29, pp. 593–613.

Bergold, G., and W. Ripper

1937   *Perilampus tristis* als Hyperparasit des Kieferntriebwicklers (*Rhyacionia buoliana*). Zeits. f. Parasitenk., vol. 9, pp. 394–417.

Brues, C. T.

1903   A Contribution to Our Knowledge of the Stylopidae. Zool. Jahrb. Abth. f. Anat., vol. 18, pp. 241–270.

Bugnion, E.

1891   Recherches sur le développement post-embryonnaire, l'anatomie et les moeurs de l'*Encyrtus fuscicollis*. Rec. Zool. Suisse, vol. 5, pp. 435–534, 2 pls.

Chrystal, R. N.

1930   Studies of the *Sirex* Parasites. The Biology and Postembryonic Development of *Ibalia leucospoides*. Oxford For. Mem., No. 11, 63 pp., 10 pls., 7 figs.

Clausen, C. P.

1923   The Biology of *Schizaspidia tenuicornis*, a Eucharid Parasite of *Camponotus*. Ann. Entom. Soc. America, vol. 16, pp. 195–217, 2 pls.

1928a The Manner of Oviposition and the Planidium of *Schizaspidia manipurensis*. Proc. Entom. Soc. Washington, vol. 30, pp. 80–86, 1 fig.

1928b *Hyperalonia oenomaus*, a Parasite of *Tiphia* Larvae.   Ann. Entom. Soc. America, vol. 21, pp. 642–658.

1929   Biological Studies on *Poecilogonalos thwaitesii*, Parasitic in the Cocoons of *Henicospilus*. *Ibid.*, vol. 31, pp. 67–79, 1 pl., 1 fig.

1931a Biological Notes on Trigonalidae. *Ibid.*, vol. 33, pp. 72–81, 2 figs.

1931b Biological Observations on *Agriotypus*. *Ibid.*, vol. 33, pp. 29–37.

1932   The Biology of *Encyrtus infidus*, a Parasite of *Lecanium kunoënsis*. Ann. Entom. Soc. America, vol. 25, pp. 670–687.

1939   The Effect of Host Size Upon the Sex Ratio of Hymenopterous Parasites and its Relation to Methods of Rearing and Colonization. Journ. New York Entom. Soc., vol. 47, pp. 1–9.

Cole, A. C., Jr.

1930   *Muscina stabulans* on *Archanara subcornea*. Entom. News, vol. 41, p. 112.

1931   *Typha* Insects and their Parasites. *Ibid.*, vol. 42, pp. 6–11; 35–39.

Daniel, D. M.

1932   *Macrocentrus ancylivorus*, a Polyembryonic Braconid Parasite of the Oriental Fruit Moth. Tech. Bull. New York State Agric. Expt. Sta., No. 187, 101 pp., 18 figs.

Embleton, A. L.

1904   On the Anatomy and Development of *Comys infelix*, a Hymenopterous Parasite of *Lecanium hemisphaericum*. Trans. Linn. Soc., London, Zool., vol. 9, pp. 231–254.

Fenton, F. A.

1918   The Parasites of Leaf-hoppers, with Special Reference to the Biology of the Anteoninae. Ohio Journ. Sci., vol. 28, pp. 177–212; 243–278; 285–296.

Flanders, S. E.

1936   Sexual Dimorphism of Hymenopterous Eggs and Larvae. Science, vol. 84, p. 85.

Gatenby, J. B.

1918   Polyembryony in Parasitic Hymenoptera. Quart. Jour. Micros. Sci., vol. 63, pp. 175–196.

Giard, A.

1896   Retard dans l'évolution déterminé par anhydrobiose chez un Hyménoptère Chalcidien. C. R. Soc. Biol., Paris, (10) vol. 3, pp. 837–839.

Grandori, R.

1911   Contributo alla embriologia e alla biologia dell' *Apanteles glomeratus*, Imenottero parassita del bruco di *Pieris brassicae*. Redia, vol. 7, pp. 363–428, 4 pls., 1 fig.

Hanson, H. S.

1939 Ecological Notes on *Sirex* Wood Wasps and their Parasites. Bull. Entom. Res., vol. 30, pp. 27–65, 6 pls.

Hill, C. C.

1923 *Platygaster vernalis*, an Important Parasite of the Hessian Fly. Jour. Agric. Res., vol. 25, pp. 31–42.

Hill, C. C., and W. T. Emery

1927 The Biology of *Platygaster herrickii*, a Parasite of Hessian Fly. *Ibid.*, vol. 55, pp. 199–213.

Hood, J. D.

1913 Notes on the Life History of *Rhopalosoma poeyi*. Proc. Ent. Soc. Washington, vol. 15, pp. 145–147, 1 fig.

Horsfall, W. R.

1941 Biology of the Black Blister Beetle. Ann. Entom. Soc. America, vol. 34, pp. 114–126, 1 pl.

Ingram, J. W., and W. A. Douglas

1932 Notes on the Life History of the Striped Blister Beetle in Southern Louisiana. Journ. Econ. Entom., vol. 25, pp. 71–74.

James, H. C.

1928 On the Life-history and Economic Status of Certain Cynipid Parasites of Dipterous Larvae, with Descriptions of some New Larval Forms. Rev. Appl. Biol., vol. 15, pp. 287–316.

Jarvis, E.

1922 Early Stages of *Macrasiagon cucullata*. Queensland Agric. Journ., vol. 17, p. 307, 1 pl.

Kearns, H. G. H.

1931 The Larval and Pupal Anatomy of *Stenomalus micans*, a Chalcid Endoparasite of the Gout-fly. Parasitology, vol. 23, pp. 380–395.

Keilin, D., and G. de la Baume-Pluvinel

1913 Formes larvaires et biologie d'un Cynipide entomophage, *Eucoila keilini*. Bull. Sci. France et Belgique, vol. 47, pp. 88–104. 2 pls., 6 figs.

Keilin, D., and G. Picado

1913 Évolution et formes larvaires du *Diachasma crawfordi* Braconide parasite d'une mouche des fruits. Bull. Sci. France et Belgique, vol. 47, pp. 203–214, 1 pl., 4 figs.

Keilin, D., and W. R. Thompson

1915 Sur le cycle évolutif des Dryinidae, Hyménoptères parasites des Hémiptères-Homoptera. C. R. Soc. Biol., Paris, vol. 78, pp. 83–87.

Leiby, R. W.

1922 The Polyembryonic Development of *Copidosoma gelechiae* with Notes on its Biology. Journ. Morphol., vol. 37, pp. 195–285, 18 pls.

1926 The Origin of Mixed Broods in Polyembryonic Hymenoptera. Ann. Entom. Soc. America, vol. 19, pp. 290–299.

1929 Polyembryony in Insects. Trans. Fourth Intern. Congr. Entom., 1928, Ithaca, vol. 2, pp. 873–877.

Leiby, R. W., and C. C. Hill

1923 The Twinning and Monembryonic Development of *Platygaster hiemalis*, a Parasite of the Hessian Fly. Journ. Agric. Res., vol. 25, pp. 337–350, 5 pls.

1924 The Polyembryonic Development of *Platygaster vernalis. Ibid.*, vol. 28, pp. 829–840, 8 pls.

McClure, H. E.

1933 Unusual Variation in the Life Cycle of the Male of *Aenoplex carpocapsae*, a Codling Moth Parasite. Ann. Entom. Soc. America, vol. 26, pp. 345–347.

MacGill, E. J.

1934 On the Biology of *Anagrus atomus*, an Egg Parasite of the Leaf-hopper *Erythroneura pallidifrons*. Parasitology, vol. 26, pp. 57–63.

Marchal, P.

1898 La dissociation de l'oeuf en un grand nombre d'individus distincts et le cycle évolutif chez l'*Encyrtus fuscicollis*. C. R. Acad. Sci. Paris, vol. 126, pp. 662–664.

1903 Le cycle évolutif du *Polygnotus minutus*. Bull. Soc. Entom. France, 1903, pp. 90–93.

1906 Recherches sur la biologie et le développement des Hyménoptères parasites. II. Les Platygasters. Arch. Zool., vol. 4, pp. 485–640.

Martin, F.

1914 Zur Entwicklungsgeschichte des Polyembryonalen Chalcidiers *Ageniaspis fuscicollis*. Zeits. f. Wiss. Zool., vol. 110, pp. 419–479.

Muir, F.

1918 Pipunculidae and Stylopidae in Homoptera. Entom. Monthly Mag., London, vol. 54, p. 137.

Munro, J. W.

1917 The Structure and Life-history of *Bracon* sp.: A Study in Parasitism. Proc. Roy. Soc. Edinburgh, vol. 36, pp. 313–333.

Noskiewicz, J., and G. Poluszynski

1924 Un nouveau cas de polyembryonie chez les insectes (Strepsiptères). C. R. Soc. Biol., Paris, vol. 90, pp. 896–898.

Oglobin, A. A.

1924 Le rôle du blastoderme extraembryonnaire du *Dinocampus terminalis* pendant l'état larvaire. Mém. Soc. Sci. Bohême. Cl. Sc. 1924, No. 3, 27 pp., 13 figs.

Paillot, A.

1937 Sur le développement polyembryonnaire d'*Amicroplus collaris* parasite des chenilles d'*Euxoa segetum*. C. R. Acad. Sci., Paris, vol. 204, pp. 810–812.

Pantel, J.

1898 Le *Thrixion halidyanum*. Essai monographique sur les charactères extérieures, la biologie et l'anatomie d'une larve parasite du groupe des Tachinaires. La Cellule, vol. 15, pp. 1–290, 6 pls.

1909 Notes de neuropathologie comparée. Ganglions d'insectes parasités par des larves d'insectes. Neuraxe, vol. 10, pp. 267–297, 14 figs.

1910 Recherches sur les Diptères à larves entomobies. I. La Cellule, vol. 26, pp. 27–216, 5 pls., 26 figs.

Parker, H. L.

1924 Recherches sur les formes postembryonnaires des Chalcidiens. Ann. Soc. Entom. France, vol. 93, pp. 361–379, 39 pls.

1931 *Macrocentrus figuensis*, a Polyembryonic Braconid Parasitic in the European Corn Borer. Tech. Bull. United States Dept. Agric., No. 230, 62 pp., 21 figs.

Parker, H. L., and W. R. Thompson

1925 Notes on the Larvae of the Chalcidoidea. Ann. Entom. Soc. America, vol. 18, pp. 384–395, 3 pls.

Patterson, J. T.

1915 Observations on the Development of *Copidosoma gelechiae*. Biol. Bull., vol. 29, pp. 333–373.

1919 Polyembryony and Sex. Journ. Hered., vol. 10, pp. 344–352.

1921 The Development of *Paracopidosomopsis*. Journ. Morph., vol. 36, pp. 1–69.

Perkins, R. C. L.

1905 Leaf-hoppers and their Natural Enemies (Dryinidae). Bull. Hawaiian Sugar Planters' Expt. Sta., No. 1, pt. 1, pp. 1–69.

1905 Leaf-hoppers and their Natural Enemies (Stylopidae). *Ibid.*, No. 1, pt. 3, pp. 90–111.

Phillips, W. J.

1927 *Eurytoma parva* and its Biology as a Parasite of the Wheat Jointworm, *Harmolita tritici*. Journ. Agric. Res., vol. 34, pp. 743–758.

Phillips, W. J., and F. W. Poos

1921 Life-history Studies of Three Jointworm Parasites. *Ibid.*, vol. 21, pp. 405–426.

Picard, F., and E. Rabaud

1914 Sur le parasitisme externe des Braconides. Bull. Entom. Soc. France, 1914, pp. 266–269.

Riley, C. V.

1878 On the Larval Characters and Habits of the Blister Beetles Belonging to the Genera *Macrobasis* and *Epicauta*, with Remarks on other Species of the Family Meloidae. Trans. Acad. Sci. St. Louis, vol. 3, pp. 544–562.

Schell, S. C.

1943 The Biology of *Hadronotus ajax*, a Parasite in the Eggs of the Squash-bug, *Anasa tristis*. Ann. Entom. Soc. America, vol. 36, pp. 625–635, 1 pl.

Silvestri, F.

1914 Prime fasi di sviluppo del *Copidosoma Buyssoni*, Imenottero Chalcidide. Anat. Anz., vol. 47, pp. 45–56.

1921 Sviluppo del *Platygaster dryomyiae*. Portici R. Scuola Super Agr. Lab. Zool. Gen. Agr. Bol., vol. 11, pp. 299–326.

1937 Insect Polyembryony and its General Biological Aspects. Bull. Mus. Comp. Zool., vol. 81, pp. 469–498.

Smith, H. S.

1912 The Chalcidoid Genus *Perilampus* and its Relations to the Problem of Parasite Introduction. Bull. United States Dept. Agric., Tech. Ser., No. 19, pt. 4, pp. 33–69, 31 figs.

Speicher, K. G., and B. R. Speicher

1938 Diploids from Unfertilized Eggs in *Habrobracon*. Biol. Bull., vol. 74, pp. 247–252.

Spencer, H.

1926 Biology of the Parasites and Hyperparasites of Aphids. Ann. Entom. Soc. America, vol. 19, pp. 119–153, 4 pls., 2 figs.

Thompson, W. R.

1924 Les larves primaires des Tachinaires à oeufs microtypes. Ann. Paras. Hum. et Comp., vol. 2, pp. 185–201; 279–306, 4 pls., 5 figs.

1926 Recherches sur les larves des Tachinaires *Sturmia*, *Winthemia*, *Carcelia* et *Exorista*. Ann. Parasit. Hum. et Comp., vol. 4, pp. 111–125; 207–227, 2 pls.

Thorpe, W. H.

1932a Experiments upon Respiration in the Larvae of Certain Parasitic Hymenoptera. Proc. Roy. Soc. London, B, vol. 109, pp. 450–471, 16 figs.

1932b The Primary Larvae of Three Ophionine Ichneumonids, Parasitic on *Rhyacionia buoliana*. Parasitology, vol. 24, pp. 107–110.

1934 The Biology and Development of *Cryptochaetum grandicorne* (Diptera), an Internal Parasite of *Guerinia serratulae* (Coccidae). Quart. Journ. Micro. Sci., vol. 77, pp. 273–304, 30 figs.

1936 On a New Type of Respiratory Interrelationship between an Insect (Chalcid) Parasite and its Host (Coccidae). Parasitology, vol. 25, pp. 517–539.

Vandel, A.

1932 Le sexe des parasites dépend-il du nombre d'individus renfermés dans le même hôte? Liv. Cent. Soc. Ent., France, pp. 245–252.

Voukassovitch, H., and P. Voukassovitch

1927 La lutte pour la possession de l'hôte chez les larves de Chalcidides (ectoparasites solitaires). Bull. Biol. France et Belgique, vol. 61, pp. 315–325, 2 figs.

1931 Sur la lutte pour la possession de l'hôte chez les larves d'ectoparasites solitaires. C. R. Soc. Biol., Paris, vol. 106, pp. 697–700.

Waterston, J.

1923 On an Internal Parasite of a Thrips from Trinidad. Bull. Entom. Res., vol. 13, pp. 453–455, 2 figs.

Wheeler, E. H.

1923 Some Braconids Parasitic on Aphids and their Life-history. Ann. Entom. Soc. America, vol. 16, pp. 1–29, 9 figs.

Wheeler, G. C., and E. H. Wheeler

1937 New Hymenopterous Parasites of Ants. Ann. Entom. Soc. America, vol. 30, pp. 163–175.

Withington, C. H.

1908– Notes on the Life-history and habits of *Lysiphlebus cerasaphis* and *Ephedrus*
09   *rosae*. Trans. Kansas. Acad. Sci., vol. 22, pp. 314–322.

EGG PARASITES

Bakkendorf, O.

1933– Biological Investigations on some Danish Hymenopterous Egg-parasites, es-
34   pecially in Homopterous Eggs, with Taxonomic Remarks and Descriptions of
       New Species. Entom. Meddel., vol. 19, pp. 1–96; 97–134.

Caffrey, D. J.

1921 Biology and Economic Importance of *Anastatus semiflavidus*, a Recently Described Egg Parasite of *Hemileuca oliviae*. Journ. Agr. Res., vol. 21, pp. 373–384.

Clausen, C. P.

1927 The Bionomics of *Anastatus albitarsis*, Parasitic in the Eggs of *Dicryoploca japonica*. Ann. Entom. Soc. America, vol. 20, pp. 461–473.

Crossman, S. S.

1925 Two Imported Egg Parasites of the Gipsy Moth, *Anastatus bifasciatus* and *Schedius kuvanae*. Jour. Agr. Res., vol. 30, pp. 643–675.

Girault, A. A.

1907– Hosts of Insect Egg-parasites in North and South America. Part I, Psyche,
11 vol. 14, pp. 27–39. Part II, Psyche, vol. 18, pp. 146–153.

1914 Hosts of Insect Egg-parasites in Europe, Asia, Africa, and Australasia, with a supplementary American list. Zeitschr. wiss. Insektenbiol., vol. 10, pp. 87–91.

Hase, A.

1925 Beiträge zur Lebensgeschichte der Schlüpwespe *Trichogramma evanescens*. Arb. Biol. Reichsanst. f. Land u. Forstw., vol. 14, pp. 171–224.

Kieffer, J. J.

1920 Sur les Hyménoptères parasites des oöthèques de Mantides. Bull. Soc. entom. France 1919, pp. 357–359.

Marchal, P.

1936 Recherches sur la biologie et le développement des Hyménoptères parasites. Les Trichogrammes. Ann. Epiphyt. et Phytogénét., vol. 2, pp. 447–567.

1941 Les modifications rythmiques du cycle annuel d'un parasite (*Trichogramma*) suivant l'hôte dans lequel il se développe. Verh. VIIten Internat. Entom. Kongr. Berlin, vol. 2, pp. 825–826.

Satterthwait, A. F.

1931 *Anaphoidea calendrae*, a Mymarid Parasite of Eggs of Weevils of the Genus *Calendra*. Journ. New York Entom. Soc., vol. 39, pp. 171–190.

Schulze, Hanna

1926 Uber die Fruchtbarkeit der Schlüpfwespe *Trichogramma evanescens*. Zeitschr. Morph. Ökol. Tiere, vol. 6, pp. 553–585, 2 figs.

Silvestri, F.

1920 Parassiti delle ova del grilletto canterino (*Oecanthus pellucens*). Bol. Lab. Zool. Gen. Agr. Portici, vol. 14, pp. 219–250.

Smith, L. M.

1930 *Macrorileya oecanthi*, a Hymenopterous Egg Parasite of Tree Crickets. Univ. California Pubs., Entom., vol. 5, pp. 165–172.

Williams, C. B.

1914 Notes on *Podagrion pachymerum*, a Chalcid Parasite of Mantid Eggs. Entomologist, vol. 47, pp. 262–266.

Zackvatkine, A. A.

1931 Parasites and Hyperparasites of the Egg-pods of Injurious Locusts (Acridoidea) of Turkestan. Bull. Entom. Res., vol. 22, pp. 385–391.

PARASITES OF PUPAE AND ADULTS

Balduf, W. V.

1926   The Bionomics of *Dinocampus coccinellae*. Ann. Entom. Soc. America, vol. 19, pp. 465–498.

Berry, P. A.

1928   *Tetrastichus brevistigma*, a Pupal Parasite of the Elm Leaf Beetle. Circ. U. S. Dept. Agr., No. 485, 11 pp.

1939   Biology and Habits of *Ephialtes turionellae*, a Pupal Parasite of the European Pine Shoot Moth. Journ. Econ. Entom., vol. 32, pp. 717–721.

Cousin, G.

1933   Étude biologique d'un chalcidien, *Mormoniella vitripennis*. Bull. Biol. France et Belgique, vol. 67, pp. 371–400.

Crandell, H. A.

1939   The Biology of *Pachycrepoideus dubius*, a Pteromalid Parasite of *Piophila casei*. Ann. Entom. Soc. America, vol. 32, pp. 632–654.

Diaz, B.

1923   Un Braconido parasito de insecto perfecto. Rev. Fitopathologia, vol. 1, pp. 108–110.

Gahan, A. B.

1922   A New Hymenopterous Parasite upon Adult Beetles. Ohio Journ. Sci., vol. 22, pp. 140–142.

Jackson, D. J.

1928   The Biology of *Dinocampus* (*Perilitus*) *rutilus*, a Braconid Parasite of *Sitona lineata*. Proc. Zool. Soc. London, 1928, pp. 597–630.

Knab, F.

1909   Notes on Tachinid Parasites of Chrysomelidae. Psyche, vol. 16, pp. 34–35.

Loiselle, A.

1908   Sur l'éclosion tardive de certains parasites. Bull. Soc. Entom. France, 1908, pp. 213–214.

Pinkus, H.

1913   The Life History and Habits of *Spalangia muscidarum*. Psyche, vol. 20, pp. 148–158.

Richardson, C. H.

1913   Studies on the Habits and Development of a Hymenopterous Parasite, *Spalangia muscidarum*. Journ. Morph., vol. 24, pp. 513–557.

Roubaud, E.

1917   Observations biologiques sur *Nasonia brevicornis*, Chalcidide parasite des pupes muscides. Bull. Sci. France et Belgique, vol. 50, pp. 425–439.

Schulze, P.

1910   Schlüpfen von Schmarotzern aus Imagines. Internat. Entom. Zeitschr., vol. 4, p. 10.

Séguy, E.

1929   Un nouveau parasite de l'abeille domestique. Encycl. Entom. Sér. B. Dipt., II, vol. 5, pp. 169–170, 4 figs.

Sherman, F., Jr.

1914 Rearing of Moths and Tachinid Flies from Larvae and Pupae of Army-worm in North Carolina in 1914. Journ. Econ. Entom., vol. 7, pp. 299–302.

Smith, H. D.

1932 *Phaeogenes nigridens*, an Important Ichneumonid Parasite of the Pupa of the European Corn Borer. Tech. Bull. United States Dept. Agr., No. 331, 45 pp.

Speyer, W.

1925 *Perilitus melanopus* als Imaginalparasit von *Ceuthorhynchus*. Zeits. f. angew. Entom., vol. 11, pp. 132–146.

Stich, R.

1929 Ueber Imaginalparasiten aus der Familie der Braconiden bei Käfern. Zeits. wiss. Insektenbiol., vol. 24, pp. 89–96.

HYPERPARASITISM, SUPERPARASITISM, ETC.

Chittenden, F. H.

1905 An Instance of Complete Parasitism of the Imported Cabbage Worm. Bull. United States Dept. Agric., Bur. Entom., No. 54, p. 59.

Dowden, P. B.

1935 *Brachymeria intermedia*, Primary Parasite and *B. compsilurae*, a Secondary Parasite of the Gipsy Moth. Journ. Agr. Res., vol. 50, pp. 495–523.

Faure, J. C.

1926 Contribution a l'étude d'un complexe biologique: la Piéride du chou (*Pieris brassicae*) et ses parasites hyménoptères. 222 pp., 7 pls. Lyon, Faculté des Sciences de l'Université.

1926 Sur la spécificité relative des insectes parasites polyphages. C. R. Acad. Sci., Paris, vol. 182, pp. 243–245.

Fiske, W. F.

1903 A Study of the Parasites of the American Tent Caterpillar. Tech. Bull. New Hampshire Expt. Sta., No. 6, pp. 185–203, 6 figs.

1910 Superparasitism, an Important Factor in the Natural Control of Insects. Journ. Econ. Entom., vol. 3, pp. 88–97.

Fiske, W. F., and A. F. Burgess

1910 The Natural Control of *Heterocampa guttivitta*. *Ibid.*, vol. 3, pp. 389–394.

Flanders, S. E.

1936a A Reproduction Phenomenon. Science, n.s., vol. 83, p. 499.

1936b A Biological Phenomenon Affecting the Establishment of Aphelinidae as Parasites. Ann. Entom. Soc. America, vol. 29, pp. 251–255.

1937 Ovipositional Instincts and Developmental Sex Differences in the Genus *Coccophagus*. Univ. California Pubs. Entom., vol. 6, pp. 401–422.

1939 Environmental Control of Sex in Hymenopterous Insects. Ann. Entom. Soc. America, vol. 32, pp. 11–26, 7 figs.

Frisch, J. A.

1936 *Perilampus*, a Secondary Parasite on Sarcophagids and Tachinids, Parasitic on Katydids and Long-horned Grasshoppers. Psyche, vol. 43, pp. 84–85.

Griswold, G. H.

1929 On the Bionomics of a Primary Parasite and of Two Hyperparasites of the Geranium Aphid. Ann. Entom. Soc. America, vol. 22, pp. 438–452.

Hase, A.

1924 Ueber den Stech- und Legeakt, sowie über den Wirtswechsel von *Lariophagus distinguindus*. Naturwissenschaften, vol. 12, pp. 377–384, 3 figs.

Haviland, M. D.

1921a Preliminary Note on a Cynipid Hyperparasite of Aphides. Proc. Cambridge Philos. Soc., vol. 20, pp. 235–238.

1921b On the Bionomics and Postembryonic Development of Certain Cynipid Hyperparasites of Aphides. Quart. Journ. Micro. Sci., n.s., vol. 65, pp. 451–478, 11 figs.

1922a On the Larval Development of *Dacnusa areolaris*, a Parasite of Phytomyzinae. Parasitology, vol. 14, pp. 167–173, 5 figs.

1922b On the Post-embryonic Development of certain Chalcids, Hyperparasites of Aphides, with Remarks on the Bionomics of Hymenopterous Parasites in General. Quart. Journ. Micr. Sci., n.s., vol. 66, pp. 321–338, 7 figs.

Kulagin, N.

1898 Beiträge zur Kenntnis der Entwicklungsgeschichte von *Platygaster*. Zeits. f. wiss. Zool., vol. 63, pp. 195–235.

Larrimer, W. H., and W. B. Noble

1926 Determination of the Percentage of Parasitism of the Hessian Fly, *Phytophaga destructor*. Journ. Agr. Res., vol. 32, pp. 1049–1051.

Mangan, J.

1910 Some Remarks on the Parasites of the Large Larch Sawfly. *Nematus erichsonii*. Journ. Econ. Biol., vol. 5, pp. 92–94.

Martelli, G.

1907 Contribuzioni alla biologia della *Pieris brassicae* e di alcuni suoi parassiti ed iperparassiti. Boll. Lab. Zool. Portici, vol. 1, pp. 170–224.

Morris, K. R. S.

1938 *Eupelmella vesicularis* as a Predator of Another Chalcid, *Microplectron fuscipennis*. Parasitology, vol. 30, pp. 20–32, 5 figs.

Muesebeck, C. F. W.

1931 *Monodontomerus aëreus*, both a Primary and Secondary Parasite of the Browntail Moth and the Gipsy Moth. Journ. Agric. Res., vol. 43, pp. 445–460, 3 figs.

Muesebeck, C. F. W., and S. M. Dohanian

1927 A Study in Hyperparasitism, with Particular Reference to the Parasites of *Apanteles melanoscelus*. Dept. Bull. United States Dept. Agric., No. 1487, 35 pp., 10 figs.

Parker, D. L.

1933 The Interrelations of two Hymenopterous Egg-Parasites of the Gipsy Moth. Journ. Agric. Res., vol. 46, pp. 23–34.

Proper, A. B.

1934 Hyperparasitism in the Case of some Introduced Lepidopterous Tree Defoliators. *Ibid.*, vol. 48, pp. 359–376.

Roubaud, E.

1924 Histoire des Anacamptomyies, mouches parasites des guêpes sociales d'Afrique. Contribution à l'étude du parasitisme chez les Muscides entomobies. Ann. Sci. nat. Zool., (10) vol. 7, pp. 197–248, 3 pls., 15 figs.

Salt, G.

1932 Superparasitism by *Collyria calcitrator*. Bull. Entom. Res., vol. 23, pp. 211–216.

1934 Experimental Studies in Insect Parasitism. Proc. Roy. Soc. London (B), vol. 114, pp. 450–476, 1 pl., 6 figs.

1936 The Effect of Superparasitism on Populations of *Trichogramma evanescens*. Journ. Exp. Biol., vol. 13, pp. 363–375.

Smith, H. S.

1929 Multiple parasitism: Its Relation to Biological Control of Insect Pests. Bull. Entom. Res., vol. 20, pp. 141–149.

Smits van Burgst, C. A. L.

1921 Hyperparasitisme bij primaire Parasieten van de gestreepte Dennewrups. Tijdschr. Plantenziekten, vol. 27, pp. 45–49.

Varsiliew, I.

1905 Beitrag zur Biologie der Gattung *Anthrax*. Zeits. wiss. Insektenbiol., vol. 1, pp. 174–175.

Voukassovitch, H., and P. Voukassovitch

1931 Sur la lutte pour possession de l'hôte chez les larves d'ectoparasites solitaires. C. R. Soc. Biol., Paris, vol. 106, pp. 697–700.

Worth, C. B.

1939 Observations on Parasitism and Superparasitism. Entom. News, vol. 50, pp. 137–141.

MUTUAL RELATIONS OF HOSTS AND PARASITES

Beard, R. L.

1940 Parasitic Castration of *Anasa tristis* by *Trichopoda pennipes* and its Effect on Reproduction. Journ. Econ. Entom., vol. 33, pp. 269–272.

Bess, H. A.

1939 The Resistance of Mealy-bugs to Parasitization by Internal Hymenopterous Parasites with Special Reference to Phagocytosis. Ann. Entom. Soc. America, vol. 32, pp. 189–226.

Bowen, M. F.

1936 A Biometrical Study of Two Morphologically Similar Species of *Trichogramma*. Ann. Entom. Soc. America, vol. 29, pp. 119–125, 4 figs.

Bradley, W. G., and K. D. Arbuthnot

1938 The Relation of Host Physiology to Development of the Braconid Parasite, *Chelonus annulipes*. Ann. Entom. Soc. America, vol. 21, pp. 359–365.

Brooks, C. C.

1930 Recovery from Parasitism. Nature, vol. 124, pp. 14–15.

Brues, C. T.

1908   The Correlation between Habits and Structural Characters among Parasitic Hymenoptera. Journ. Econ. Entom., vol. 1, pp. 123–128.

1910a   The Parasitic Hymenoptera of the Tertiary of Florissant, Colorado. Bull. Mus. Comp. Zool., vol. 54, pp. 1–125, 1 pl., 88 figs.

1910b   Some Notes on the Geological History of the Parasitic Hymenoptera. Bull. New York Entom. Soc., vol. 18, pp. 1–22, 5 figs.

1921   Correlation of Taxonomic Affinities with Food-habits in Hymenoptera, with Special Reference to Parasitism. American Natural., vol. 55, pp. 134–164.

1936   Aberrant Feeding Behavior among Insects and its Bearing on the Development of Specialized Food Habits. Quart. Rev. Biol., vol. 2, pp. 305–319.

1940   Is Ours the "Age of Insects"? Scientific Monthly, vol. 50, pp. 413–418.

Bukovski, V.

1933   Ecological Races of Braconidae Dependent Upon Different Hosts. [In Russian.] Rev. Entom. URSS., vol. 25, pp. 83–88. Summary in Rev. Appl. Entom., A, vol. 22, p. 372.

Chapman, R. N.

1931   Animal Ecology. New York, McGraw-Hill Book Co.

Cushman, R. A.

1926a   Location of Individual Hosts versus Systematic Relation of Host Species as a Determining Factor in Parasitic Attack. Proc. Entom. Soc. Washington, vol. 28, pp. 5–6.

1926b   Some Types of Parasitism among the Ichneumonidae. Ibid., vol. 28, pp. 29–51, 51 figs.

Dustan, A. G.

1923   A Histological Account of Three Parasites of the Fall Webworm (Hyphantria cunea). Proc. Acadian Entom. Soc., 1922, pp. 73–94.

Eckstein, F.

1931   Ueber Immunität bei Insekten. Anz. f. schädlingsk., vol. 7, pp. 49–55, 13 figs.

Faure, J. C.

1926   Sur la spécificité relative des insectes parasites polyphages. C. R. Acad. Sci., France, vol. 182, pp. 243–245.

Flanders, S. E.

1935   Host Influence on the Prolificacy and Size of Trichogramma. Pan-Pacific Entom., vol. 11, pp. 175–177.

1942   Abortive Development in Parasitic Hymenoptera Induced by the Foodplant of the Insect Host. Journ. Econ. Entom., vol. 35, pp. 834–835.

Fluiter, H. J. de

1933   Nogmaals iets over het uitkomen van de imago van Diplostichus tenthredinum uit den gesloten cocon van Diprion pini. Entom. Ber., vol. 8, pp. 487–493.

Gautier, C., and S. Bonnamour

1932   Quelques observations sur Praon abjectum. Bull. Soc. Entom. France, vol. 37, pp. 99–100.

Harland, S. C., and O. M. Atteck

1933   Breeding Experiments with Biological Races of Trichogramma minutum in the West Indies. Zeits. f. indukt. Abstamm.- u. Vererbungsl., vol. 44, pp. 54–76.

Hase, A.

1923   Ueber die Monophagie und Polyphagie der Schmarotzerwespen; ein Beitrag zur Kenntnis des Geruchssinnes der Insekten. Naturwissenschaften, vol. 11, pp. 801–806.

Hodson, A. C.

1939   *Sarcophaga aldrichi* as a Parasite of *Malacosoma disstria*. Journ. Econ. Entom., vol. 322, pp. 396–401, 4 figs.

Holdaway, F. G., and H. F. Smith

1932   A Relation between Size of Host Puparia and Sex Ratio of *Alysia manducator*. Australian Jour. Expt. Biol. and Med. Sci., vol. 10, pp. 247–259.

Jaynes, H. A., and T. R. Gardner

1924   Selective Parasitism by *Tiphia* sp. Journ. Econ. Entom., vol. 17, pp. 366–369.

Kornhauser, S. J.

1919   The Sexual Characteristics of the Membracid *Thelia bimaculata*. I. External Changes Induced by *Aphelopus theliae*. Journ. Morph., vol. 32, pp. 531–635.

Lartschenko, K.

1933   Die Unempfänglichkeit der Raupen von *Loxostege sticticalis* und *Pieris brassicae* gegen Parasiten. Z. Parasitenk., Berlin, vol. 5, pp. 679–707, 13 figs.

Lloyd, D. C.

1940   Host Selection by Hymenopterous Parasites of the Moth, *Plutella maculipennis*. Proc. Roy. Soc., London, Ser. B, vol. 128, pp. 451–484, 2 pls., 1 fig.

Marchal, P.

1907   Sur le *Lygellus epilachnae* (parasitisme; erreur de l'instinct; évolution). Bull. Soc. Entom. France, 1907, pp. 14–16.

1927   Contribution à l'étude génotypique et phénotypique des Trichogrammes. Les lignées naturelles de Trichogrammes. C. R. Acad. Sci. France, vol. 185, pp. 489–493; 521–523.

Meyer, N. F.

1926a Biologie von *Angitia fenestralis*, des Parasiten von *Plutella maculipennis*, und einige Worte über Immunität der Insekten. Zeits. angew. Entom., vol. 12, pp. 139–152, 10 figs.

1926b Ueber die Immunität einiger Raupen ihren Parasiten, den Schlüpfwespen, gegenüber. *Ibid.*, vol. 12, pp. 376–384.

Payne, N. M.

1933   The Differential Effect of Environmental Factors upon *Microbracon hebetor* and its Host, *Ephestia kühniella*. Biol. Bull., vol. 65, pp. 187–205.

Pictet, A.

1899   Hyménoptères et Diptères parasites de chenilles. Arch. Sci. Phys. Nat., (4) vol. 7, pp. 79–80.

Pierce, W. D.

1908   Factors Controlling Parasitism, with Special Reference to the Cotton Boll-weevil. Journ. Econ. Entom., vol. 1, pp. 315–323.

Salt, G.

1927 The Effect of Stylopization on Aculeate Hymenoptera. Journ. Expt. Zool., vol. 48, pp. 223-331.

1931 A Further Study of the Effect of Stylopization on Wasps. *Ibid.*, vol. 59, pp. 133-166.

1935 Host Selection (by *Trichogramma*). Proc. Roy. Soc. London, B, vol. 117, pp. 413-435, 1 fig.

1938a Experimental Studies in Insect Parasitism. — VI. Host Suitability. Bull. Entom. Res., vol. 29, pp. 223-246, 3 figs.

1938b Further Notes on *Trichogramma semblialis*. Parasitology, vol. 30, pp. 511-522, 3 figs.

Schmieder, R. G.

1933 The Polymorphic Forms of *Melittobia chalybii* and the Determining Factors Involved in their Production. Biol. Bull., vol. 65, pp. 338-354, 4 figs.

Seyrig, André

1935 Relations entre le sexe de certaines Ichneumonides et l'hôte aux dépens duquel ils sont vécu. Bull. Entom. Soc. France, 1935, pp. 67-70.

Silvestri, F.

1906 Biologia del *Litomastix truncatellus*. Ann. R. Scuola Sup. Agric. Portici, vol. 6, pp. 1-51, 4 pls., 13 figs.

Smith, H. S.

1916 An Attempt to Redefine the Host Relationships Exhibited by Entomophagous Insects. Journ. Econ. Entom., vol. 9, pp. 477-486.

Strickland, E. H.

1930 Phagocytosis of Internal Insect Parasites. Nature, vol. 126, p. 95.

Thompson, W. R.

1913 La Spécificité des Parasites entomophages. C. R. Soc. Biol., Paris, vol. 73, pp. 559-560.

1915a The Cuticula of Insects as a Means of Defense Against Parasites. Proc. Cambridge Philos. Soc., vol. 18, pp. 51-55.

1915b Les conditions de la résistance des insectes parasites internes dans l'organisme de leurs hôtes. C. R. Soc. Biol., Paris, vol. 77, pp. 562-564.

1923 Recherches sur la biologie des Diptères parasites. Bull. Biol. France et Belgique, vol. 57, pp. 174-237.

1930a Entomophagous Parasites and Phagocytes. Nature, vol. 125, p. 167.

1930b Reaction of the Phagocytes of Arthropods to their Internal Insect Parasites. *Ibid.*, vol. 125, pp. 565-566.

Thompson, W. R., and H. L. Parker

1927a Études sur la biologie des insectes parasites: La vie parasitaire et la notion morphologique de l'adaptation. Ann. Entom. Soc. France, vol. 96, pp. 113-146.

1927b The Problem of Host Relations with Special Reference to Entomophagous Parasites. Parasitology, vol. 19, pp. 1-34.

Thorpe, W. H., and F. G. W. Jones

1937 Olfactory Conditioning in a Parasitic Insect and its Relation to the Problem of Host Selection. Proc. Roy. London, Ser. B, vol. 124, pp. 58-61, 1 fig.

Timberlake, P. H.

1912 Experimental Parasitism: A Study of the Biology of *Linnerium validum.* Bull. United States Dept. Agric. Bur. Entom. Tech. Ser., No. 19, pt. 5, pp. 71–92.

1916 Note on an Interesting Case of Two Generations of a Parasite Reared from the same Individual Host. Canadian Entom., vol. 48, pp. 89–91.

Timon-David, J.

1931 Contribution à l'étude de la spécificité biochimique des parasites. Huile d'*Exeristes roborator.* C. R. Soc. Biol., Paris, vol. 106, pp. 829–831.

Tower, D. G.

1916 Comparative Study of the Amount of Food Eaten by Parasitized and Non-parasitized Larvae of *Cirphis unipuncta.* Journ. Agric. Res., vol. 6, pp. 455–458.

Varley, G. C., and C. G. Butler

1933 The Acceleration of Development of Insects by Parasitism. Parasitology, vol. 25, pp. 263–268.

Webber, R. T., and J. V. Schaffner

1926 Host Relations of *Compsilura concinnata,* an Important Tachinid Parasite of the Gipsy Moth and the Brown-tail Moth. Bull. United States Dept. Agric., No. 1363, 31 pp.

Weese, A. O.

1930 Differential Reactions to Environment of a Host and its Parasite. Proc. Oklahoma Acad. Sci., vol. 10, pp. 18–19.

West, L. S.

1923 Immunity to Parasitism in *Samia cecropia.* Entom. News, vol. 34, pp. 23–25.

Wheeler, W. M.

1907 The Polymorphism of Ants, with an Account of Some Singular Abnormalities due to Parasitism. Bull. American Mus. Nat. Hist., vol. 23, pp. 1–93.

1910 Effects of Parasitic and Other Kinds of Castration Upon Insects. Journ. Expt. Zool., vol. 8, pp. 377–438.

PARASITIC FAUNAE

Allen, H. W.

1925 Biology of the Red-tailed Tachina-fly, *Winthemia quadripustulata.* Tech. Bull. Mississippi Agric. Expt. Sta., No. 12, 32 pp., 9 figs.

Bruneteau, J.

1937 Recherches sur les ennemis naturels du Doryphore en Amérique. Ann. Epiphyt. Phytogénét., vol. 3, pp. 113–135.

Burgess, A. F., and S. S. Crossman

1929 Imported Insect Enemies of the Gipsy Moth and the Brown-tail Moth. Tech. Bull. United States Dept. Agr., No. 86, 147 pp.

Davis, J. J.

1919 Contributions to a Knowledge of the Natural Enemies of *Phyllophaga.* Bull. Illinois Nat. Hist. Surv., vol. 13, pp. 53–133, 13 pls., 46 figs.

Fluke, C. L.

1929　The Known Predacious and Parasitic Enemies of the Pea Aphid in North America. Res. Bull. Wisconsin Agric. Expt. Sta., No. 93, 47 pp., 50 figs.

Handschin, E.

1934　Studien an *Lyperosia exigua* und Ihren Parasiten. Rev. Suiss Zool., vol. 41, pp. 267–297.

Hardy, J. E.

1933　The Natural Control of the Cabbage Caterpillars, *Pieris* spp. Journ. Anim. Ecol., vol. 2, pp. 210–231.

Howard, L. O.

1897　A Study in Insect Parasitism: A Consideration of the Parasites of the White-marked Tussock Moth, with an Account of their Interrelations, and with Descriptions of New Species. Bull. United States Dept. Agric., Tech. Ser., No. 5, 57 pp., 24 figs.

Hunter, S. J.

1909　The Green Bug and its Natural Enemies — A Study in Insect Parasitism. Bull. Univ. Kansas, vol. 9, pp. 1–163, 3 pls., 48 figs.

Imms, A. D.

1916–　Observations on the Insect Parasites of Some Coccidae. Quart. Journ. Micr.
18　Sci., vol. 61, pp. 217–274, 2 pls.; vol. 63, pp. 293–374, 34 figs.

Marsh, F. L.

1937　Ecological Observations upon Enemies of *Cecropia* with Particular Reference to its Hymenopterous Parasites. Ecology, vol. 18, pp. 106–112, 2 figs.

Melander, A. L., and C. T. Brues

1903　Guests and Parasites of the Burrowing Bee, *Halictus*. Biol. Bull., vol. 5, pp. 1–27.

Morris, K. R. S., E. Cameron, and W. F. Jepson

1937　The Insect Parasites of the Spruce Sawfly (*Diprion polytomum*) in Europe. Bull. Entom. Res., vol. 28, pp. 341–393.

Pemberton, C. E., and H. F. Willard

1918　A Contribution to the Biology of Fruit-fly Parasites in Hawaii. Journ. Agr. Res., vol. 15, pp. 419–465.

Picard, F.

1922　Contribution à l'étude des parasites de *Pieris brassicae*. Bull. Biol. France et Belgique, vol. 56, pp. 54–130.

Pierce, W. D.

1908　A List of Parasites Known to Attack American *Rhyncophora*. Journ. Econ. Entom., vol. 1, pp. 380–396.

1912　The Insect Enemies of the Cotton Boll-weevil. Bull. Unites States Dept. Agric. Bur. Entom., No. 100, 99 pp., 26 figs.

Richards, O. W., and W. S. Thomson

1932　The Genera *Ephestia*, *Strymax* and *Plodia*, with Notes on Parasites of the Larvae. Trans. Entom. Soc. London, vol. 80, pp. 169–250, 8 pp.

PLATE XIX

A. Larva of *Dermatobia hominis*, a bot-fly that frequently causes myiasis in Man. Photograph by D. H. Linder.

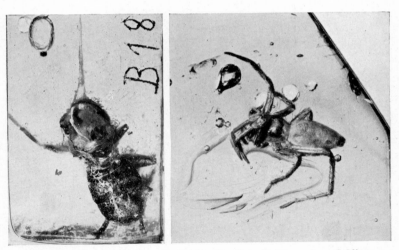

B. "The Spider and the Fly." Two fossils in Baltic amber of Oligocene age. The fly, at left, gives evidence of having been partly devoured, perhaps by a spider, before fossilization in the resin some 60 million years ago. Original photographs.

PLATE XX

Fruiting stage of a fungus (*Cordyceps melolonthae*) parasitic on the grub of a South American melolonthid beetle. Photograph by D. H. Linder.

Roberts, R. A.

1935 Some North American Parasites of Blowflies. Journ. Agri. Res., vol. 50, pp. 479–494.

Sachtleben, H.

1927 Beiträge zur Naturgeschichte der Forleule, *Panolis flammea*, und ihre Parasiten. Arb. biol. Reichanst., vol. 15, pp. 437–536, 3 pls., 3 figs.

Salt, G.

1931 Parasites of the Wheat-stem Sawfly, *Cephus pygmaeus*, in England. Bull. Entom. Res., vol. 22, pp. 479–545.

1932 The Natural Control of the Sheep Blowfly, *Lucilia sericata*. *Ibid.*, vol. 23, pp. 235–245.

Schaffner, J. V., and C. L. Griswold

1934 Macrolepidoptera and their Parasites Reared from Field Collections in the Northeastern Part of the United States. Misc. Pub. United States Dept. Agr., No. 188, 160 pp.

Siltala, A. J., and J. C. Nielsen

1906 Zur Kenntnis der Parasiten der Trichopteren. Zeits. f. wiss. Insektenbiol., vol. 2, pp. 382–386.

Silvestri, F.

1910– Contribuzioni alla conoscenza degli insetti dannosi e dei loro simbionti. I,
11 *Galerucella* dell'olmo. Bol. Lab. Zool. gen. Agr., Portici, vol. 4, pp. 246–289. II. *Plusia gamma.* Vol. 10, 35 pp., 26 figs.

Skaife, S. H.

1921 Some Factors in the Natural Control of the Wattle Bagworm. South African Journ. Sci., vol. 17, pp. 291–301.

Smith, H. S., and H. Compère

1928 A Preliminary Report on the Insect Parasites of the Black Scale, *Saissetia oleae.* Univ. California Pubs., Ent., vol. 4, pp. 231–334.

Taylor, R. L.

1929– The Biology of the White Pine Weevil (*Pissodes strobi*) and a Study of the
30 Insect Parasites from an Economic Standpoint. Entom. Americana, vol. 9, pp. 167–246; vol. 10, pp. 1–86.

Taylor, T. H. C.

1937 The Biological Control of an Insect in Fiji. An Account of the Coconut Leaf-mining Beetle and its Parasite Complex. London, 239 pp.

Tothill, J. D.

1922 The Natural Control of the Fall Webworm (*Hyphantria cunea*) in Canada. Tech. Bull. Dept. Agr. Canada, No. 3, 107 pp.

Tothill, J. D., T. H. C. Taylor, and R. W. Paine

1933 The Coconut Moth in Fiji, a History of its Control by Means of Parasites. Imp. Inst. Entom., London, 269 pp.

Trouvelot, B.

1932 Recherches sur les parasites et prédateurs attaquant le Doryphore en Amérique du Nord. Ann. des Epiphyt., 17ᵉ Ann., pp. 408–445.

Waterston, J.

1921 Report on Parasitic Hymenoptera, Bred from Stored Grain. Repts. Grain Pests Comm. Royal Soc. London, No. 9, pp. 8–32.

Willard, H. F.

1927 Parasites of the Pink Boll-worm in Hawaii. Tech. Bull. United States Dept. Agr., No. 19, 15 pp.

PARASITES OF OTHER INVERTEBRATES

Barnard, K. H.

1911 Chironomid-larvae and Water Snails. Entom. Monthly Mag., vol. 47, pp. 76–78.

Bequaert, J.

1918 The Arthropod Enemies of Molluscs, with Description of a New Dipterous Parasite from Brazil. Parasitology, vol. 11, pp. 201–212, 1 fig.

Bishopp, F. C.

1934 Records of Hymenopterous Parasites of Ticks in the United States. Proc. Entom. Soc. Washington, vol. 36, pp. 87–88.

Bovey, P.

1936 Sur la ponte et la larve primare d'*Oncodes pallipes*. Bull. Soc. Vaud. Sci. Nat. vol. 59, pp. 171–176.

Brauer, F.

1869a Beitrag zur Biologie der Acroceriden. Verh. zool.-bot. Gesell. Wien, vol. 19, pp. 737–740.
1869b Beschreibung der Verwandlungsgeschichte der *Mantispa styriaca*. Ibid., vol. 19, pp. 831–840.

Bristowe, W. S.

1932 *Mantispa*, a Spider Parasite. Entom. Monthly Mag., vol. 68, pp. 222–224.

Brues, C. T.

1903 A Dexiid Parasite of the Sow bug. Entom. News, vol. 14, p. 291.

Buysson, R. du

1912 Un hyménoptère parasite d'*Ixodes*. Arch. Parasitol., vol. 15, pp. 246–247, 1 fig.

Cole, F. R.

1919 The Dipterous Family Cyrtidae in North America. Trans. American Entom. Soc., vol. 45, pp. 1–69.

Cooley, R. A.

1929 Tick Parasites. 7th Ann. Rept. Montana State Board Entom., 1927–28, pp. 10–31.

Cooley, R. A., and G. M. Kohls

1933 A Summary on Tick Parasites. Proc. 5th Pacific Sci. Cong., Vancouver, Canada, vol. 5, pp. 3375–3381.

Gahan, A. B.

1934 On the Identities of Chalcidoid Tick Parasites. Proc. Entom. Soc. Washington, vol. 36, pp. 89–97.

Gillette, C. P.

1924 A Peculiar Egg-laying Experience with the Spider Parasite, *Oncodes costatus.* Circ. Colorado State Entom., No. 43, pp. 49–51.

Hungerford, H. B.

1936 The Mantispidae of the Douglas Lake, Michigan Region, with some Biological Observations. Entom. News, vol. 47, pp. 69–72; 85–88.

Kaston, B. J.

1937 Notes on Dipterous Parasites of Spiders. Journ. New York Entom. Soc., vol. 45, pp. 415–420.

Keilin, D.

1909 Sur le parasitisme de la larve de *Pollenia rudis* dans *Allolobophora chlorotica.* C. R. Soc. Biol., Paris, vol. 67, pp. 201–203.

1915 Recherches sur les larves de Diptères cyclorraphes (cycle évolutif de *Pollenia rudis* parasite de *Allolobophora chlorotica*). Bull. Soc. Sci. France et Belgique, vol. 49, pp. 15–196, 16 pls., 27 figs.

1919 On the Life-history and Larval Anatomy of *Melinda cognata*, Parasitic on the Snail, *Helicella virgata.* Parasitology, vol. 11, pp. 430–455.

1921 On some Dipterous Larvae Infesting the Branchial Chambers of Land-crabs. Ann. Nat. Hist. London, (9) vol. 8, pp. 601–608.

King, J. L.

1916 Observations on the Life-history of *Pterodontia flavipes.* Ann. Entom. Soc. America, vol. 9, pp. 309–321, 2 pls.

König, A.

1894 Ueber die Larve von *Oncodes.* Verh. zool.-bot. Gesell. Wien, vol. 44, pp. 163–166.

Lundbeck, W.

1923 Some Remarks on the Biology of the Sciomyzidae. Vidensk. Medd. Dansk. Naturh. Foren., vol. 76, pp. 101–109.

Mathias, P., and L. Boulle

1933 Sur une larve de Chironomide parasite d'un Mollusque. C. R. Acad. Sci., Paris, vol. 196, pp. 1744–1746.

Melander, A. L.

1902 Notes on the Acroceridae. Entom. News, vol. 13, pp. 178–182.

Millot, J.

1938 Le développement et la biologie larvaire des Oncodides, Diptères parasites d'araignées. Bull. Soc. Zool. France, vol. 63, pp. 162–181; 183–197.

Myers, J. G.

1934 Aggressive Parasitism of a Millipede by a Phorid. Proc. Roy. Entom. Soc. London, vol. 9, pp. 62–63.

Picard, F.

1930 Sur le parasitisme d'une phoride (*Megaselia cuspidata*) aux dépens d'un myriapode. Bull. Soc. Zool. France, vol. 55, pp. 180–183.

Pierce, W. D.

1938 The Black Widow Spider and its Parasites. Bull. South California Acad. Sci., vol. 37, pp. 101–104.

Rostand, J.

1920 Sur la biologie de *Sarcophaga filia*. Bull. Soc. Entom. France, 1920, pp. 215–216.

Schmitz, H.

1917 Biologische Beziehungen zwischen Dipteren und Schnecken. Biol. Zentralbl., vol. 37, pp. 24–43.

Séguy, E.

1921 Les Diptères qui vivent aux dépens des escargots. Bull. Soc. Ent. France, 1921, pp. 238–239.

1928 Étude sur le *Pollenia hasei*. Zeits. angew. Entom., vol. 14, pp. 369–375, 10 figs.

Thompson, W. R.

1917 Sur un Diptère parasite des Isopodes terrestres. C. R. Soc. Biol., Paris, vol. 80, pp. 785–788, 7 figs.

1923 Recherches sur la biologie des diptères parasites. Bull. Biol. France et Belgique, vol. 57, pp. 174–237.

1934 The Tachinid Parasites of Wood-lice. Parasitology, vol. 26, pp. 378–448, 8 pls., 5 figs.

Webb, J. L., and R. H. Hutchinson

1916 A Preliminary Note on the Bionomics of *Pollenia rudis* in America. Proc. Washington Entom. Soc., vol. 18, pp. 197–199.

PARASITES OF VERTEBRATES

Aldrich, J. M.

1915 The Deer Bot flies (*Cephenomyia*). Journ. New York Entom. Soc., vol. 23, pp. 145–150.

1923a A New Genus and Species of Fly Reared from the Hoof of the Carabao. Philippine Journ. Sci., vol. 22, pp. 141–142.

1923b The Genus *Philornis*, a Bird-infesting Group of Anthomyiidae. Ann. Entom. Soc. America, vol. 16, pp. 304–309.

Baer, W. S.

1931 The Treatment of Chronic Osteomyelitis with the Maggot. Journ. Bone and Joint Surgery, vol. 13, pp. 438–475.

Belschner, H. G.

1937 A Review of the Sheep Blowfly Problem in New South Wales. Dept. Agric., New South Wales, 60 pp.

Bequaert, J.

1916 Parasitic Muscid Larvae collected from the African Elephant and the White Rhinoceros. Bull. American Mus. Nat. Hist., vol. 35, pp. 377–387.

Bezzi, M.

1922 On the Dipterous Genera *Passeromyia* and *Ornithomusca*, with Notes and Bibliography on the Nonpupiparous *Myiodaria* Parasitic on Birds. Parasitology, vol. 14, pp. 29–46.

Bishopp, F. C.

1915 Flies which Cause Myiasis in Man and Animals. Journ. Econ. Entom., vol. 8, pp. 317–329.

Bishopp, F. C., and W. E. Dove

1926 The Horse Bots and Their Control. Farmers' Bull., U. S. Dept. Agric., No. 1503.

Bishopp, F. C., E. W. Laake, and R. W. Wells

1926 The Cattle Grubs or Ox-warbles, their Biologies and Suggestions for Control. Dept. Bull., U. S. Dept. Agric., No. 1369, 119 pp.

1929 Cattle Grubs or Heel Flies, with Suggestions for their Control. Farmers' Bull., U. S. Dept. Agric., No. 1596, 22 pp.

Blacklock, B., and M. G. Thompson

1923 A Study of the Tumbu-fly, *Cordylobia anthropophaga* in Sierra Leone. Ann. Trop. Med. Parasitol., vol. 17, pp. 443–501.

Brauer, F.

1863 Mongraphie der Oestriden. vi + 292 pp. K. K. zool.-bot. Ges., Wien.

Chevrel, R.

1909 Sur la myiase des voies urinaires. Arch. Parasitol., vol. 12, pp. 369–450.

Cushing, E. C., and D. G. Hall

1937 Morphological Differences between the Screw worm Fly and other Similar Species in North America. Proc. Entom. Soc. Washington, vol. 39, pp. 195–198.

Cushing, E. C., and W. S. Patton

1933 Studies on the Higher Diptera of Medical and Veterinary Importance. *Cochliomyia americana.* Ann. Trop. Med. Parasitol., vol. 27, pp. 539–551.

Dove, W. E.

1918 Some Biological and Control Studies of *Gastrophilus haemorrhoidalis* and Other Bots of Horses. Bull. U. S. Dept. Agric., No. 597, 51 pp.

1935 Screwworms in the Southeastern States. Journ. Econ. Entom., vol. 28, pp. 765–772.

Dunn, L. H.

1930 Rearing the Larvae of *Dermatobia hominis* in Man. Psyche, vol. 37, pp. 327–342.

Fülleborn, F.

1908 Untersuchungen über den Sandfloh. Arch. f. Schiffs- und Tropen-Hygiene, vol. 12, pp. 269–275.

Hall, M. C., and J. T. Muir

1913 A Critical Study of a Case of Myiasis due to *Eristalis.* Arch. Intern. Med., vol. 2, pp. 193–203.

Hadwen, S.

1915 Warble Flies — A Further Contribution on the Biology of *Hypoderma lineatum* and *H. bovis.* Parasitology, vol. 7, pp. 331–338.

1916 Observations on the Migration of Warble Larvae through the Tissues. Bull. Sci. Ser., Canada Dept. Agric., Health of Animals Branch., No. 22, 14 pp.

1917 Anaphylaxis in Cattle and Sheep Produced by the Larvae of *Hypoderma bovis, H. lineatum* and *Oestrus ovis.* Journ. American Vet. Med. Assoc., vol. 51, pp. 16–44.

Knipling, E. F.

1937 The Biology of *Sarcophaga cistudinis*, a Species Parasitic on Turtles. Proc. Entom. Soc. Washington, vol. 39, pp. 91–101, 2 pls.

1939 A Key for Blowfly Larvae concerned in Wound and Cutaneous Myiases. Ann. Entom. Soc. America, vol. 32, pp. 376–383.

Laake, E. W.

1936 Economic Studies of Screwworm Flies. Iowa State Coll. Jour. Sci., vol. 10, pp. 345–359.

Laake, E. W., E. C. Cushing, and H. E. Parish

1936 Biology of the Primary Screwworm, *Cochliomyia americana.* Tech. Bull., U. S. Dept. Agric., No. 500, 24 pp., 1 pl., 14 figs.

Meleney, H. E., and P. D. Harwood

1935 Human Intestinal Myiasis due to the Larvae of the Soldier Fly, *Hermetia illucens.* American Journ. Trop. Med., vol. 15, pp. 45–49.

Muir, F.

1912 Two New Species of *Ascodipteron.* Bull. Mus. Comp. Zool., Harvard, vol. 54, pp. 351–366.

Munro, H. K.

1922 The Sheep Blow-fly in South Africa. Journ. Dept. Agric., Union South Africa, Reprint No. 44, 10 pp.

Phibbs, G.

1922 The Larval Mouth-hooks of *Hypoderma.* Irish Naturalist, vol. 31, pp. 25–30.

Robinson, W.

1933a Surgical Maggots in the Treatment of Infected Wounds. Journ. Lab. Clin. Med., vol. 18, pp. 406–412.

1933b The Use of Blowfly Larvae in the Treatment of Infected Wounds. Ann. Entom. Soc. America, vol. 26, pp. 270–276.

Rodhain, J., and J. Bequaert

1916a Les Myiases cutanées produites par *Staisia (Cordylobia) rodhaini.* Bull. Sci. France et Belgique, vol. 49, pp. 262–289.

1916b Révision des Oestrinae du continent africain. Bull. Sci .France et Belgique, (7) vol. 50, pp. 53–164, 2 pls., 30 figs.

1920 Oestrides d'antilopes et de zèbres en Afrique orientale. Rev. Zool. Africaine, vol. 8, pp. 169–228, 3 figs.

Roubaud, E.

1914 Études sur la faune parasitaire de l'Afrique occidentale française. Pp. 114–169. Paris, Larose. (Contains complete account of *Cordylobia anthropophaga*).

Sambon, L. W.

1922 Tropical and Subtropical Diseases. Journ. Trop. Med. Hygiene, vol. 25, pp. 170–185.

Townsend, C. H. T.

1895 The Grass-quit Bot, an Anthomyiid Parasite of Nestling Birds in Jamaica. Journ. Inst. Jamaica, vol. 2, pp. 173–174.

Warburton, C.

1922 The Warble Flies of Cattle, *Hypoderma bovis* and *H. lineatum.* Parasitology, vol. 14, pp. 322–341.

Wells, R. W., and E. F. Knipling
1938 Recent Studies on Species of *Gastrophilus* Occurring in Horses in the United States. Iowa State College Journ. Sci., vol. 12, pp. 181–203.

Woodworth, H. E., and J. B. Ashcraft
1923 The Foot Maggot, *Boöponus intonsus*, a New Myiasis-producing Fly. Philippine Journ. Sci., vol. 22, pp. 143–156.

ECONOMIC APPLICATIONS

Berlese, A.
1916 Entomophagous Insects and their Practical Employment in Agriculture. Internat. Rev. Sci. and Practic. Agric., vol. 7, pp. 321–332.

Clausen, C. P.
1936 Insect Parasitism and Biological Control. Ann. Entom. Soc. America, vol. 29, pp. 201–223.

Collins, C. W.
1931 Parasite Introductions for the Gipsy Moth and Some other Forest and Shade Trees Insects. Proc. 7th Nat. Shade Tree Conf., pp. 86–87. Yonkers, N. Y.

Crossman, S. S., and R. T. Webber
1924 Recent European Investigations of Parasites of the Gipsy Moth, *Porthetria dispar* and the Brown-tail Moth, *Euproctis chrysorrhoea*. Journ. Econ. Entom., vol. 17, pp. 67–76.

Fiske, W. F.
1903 A Study of the Parasites of the American Tent Caterpillar. Tech. Bull. New Hampshire Agric. Expt. Sta., No. 6, pp. 183–230, 6 figs.
1910 Parasites of the Gipsy and Brown-tail Moths, Introduced into Massachusetts. 56 pp., 22 figs., 3 diagrams. Office of the State Forester, Boston, Mass.

Friedrichs, K.
1932 Kieferspanner und Parasiten nach der Gradation. Zeits. angew. Entom., vol. 19, pp. 130–143.

Fullaway, D. D.
1920 Natural Control of Scale Insects in Hawaii. Proc. Hawaiian Entom. Soc., vol. 4, pp. 237–246.

Howard, L. O.
1917 The Practical Use of the Insect Enemies of Injurious Insects. Yearbook United States Dept. Agric. for 1916, pp. 273–288.
1922 A Side Line in the Importation of Insect Parasites of Injurious Insects from One Country to Another. Proc. Nat. Acad. Sci., vol. 8, pp. 133–139.
1924 Insect Parasites of Insects. Proc. Entom. Soc. Washington, vol. 26, pp. 27–46.
1926 The Parasite Element of Natural Control of Injurious Insects and its Control by Man. Journ. Econ. Entom., vol. 19, pp. 271–282. Ann. Rept. Smithsonian Inst. for 1926, pp. 411–420 (1927).

Howard, L. O., and W. F. Fiske
1911 The Importation into the United States of the Parasites of the Gipsy Moth and the Brown-tail Moth. Bull. United States Dept. Agric., Bur. Entom., No. 91, 312 pp., 74 figs., 28 pls., 1 map.

Marchal, P.

1907   Utilisation des insectes auxiliaires entomophages dans la lutte contre les insectes nuisibles à l'agriculture. Ann. Inst. Nat. Agron., ser. 2, vol. 6, pp. 281–354, 26 figs.

Morrill, A. W.

1931   A Discussion of Smith and Flanders *Trichogramma* Fad Query. Journ. Econ. Entom., vol. 24, pp. 1264–1273.

Myers, J. G.

1929   The Principles of Biological Control. Tropical Agric., vol. 6, pp. 163–165.

Smith, H. S.

1923   What may we Expect from Biological Control? Journ. Econ. Entom., vol. 16, pp. 506–511.

1929   On Some Phases of Preventive Entomology. Scientific Monthly, vol. 29, pp. 177–184.

Sweetman, H. L.

1936   The Biological Control of Insects. Ithaca, N.Y., Comstock Pub. Co. 461 pp.

Thompson, W. R.

1922   Théorie de l'action des parasites entomophages. Accroissement de la proportion d'hôte parasites dans le parasitisme cyclique. C. R. Acad. Sci., France, vol. 175, pp. 65–68.

1923   A Criticism of the "Sequence" Theory of Parasitic Control. Ann. Entom. Soc. America, vol. 16, pp. 115–128, 2 figs.

1927   A Method for the Approximate Calculation of the Progress of Introduced Parasites of Insect Pests. Bull. Entom. Res., vol. 17, pp. 273–277.

1928   A Contribution to the Study of Biological Control and Parasite Introduction in Continental Areas. Parasitology, vol. 20, pp. 90–112.

1929   On the Relative Value of Parasites and Predators in the Biological Control of Insect Pests. Bull. Entom. Res., vol. 19, pp. 343–350.

1930   The Principles of Biological Control. Ann. Appl. Biol., vol. 17, pp. 306–338.

1939   Biological Control and the Theories of the Interactions of Populations. Parasitology, vol. 31, pp. 299–388, 1 fig.

Thompson, W. R., and H. L. Parker

1928   The European Corn Borer and its Controlling Factors in Europe. Tech. Bull. United States Dept. Agric., No. 59, 62 pp., 3 figs.

Timberlake, P. H.

1927   Biological Control of Insect Pests in the Hawaiian Islands. Proc. Hawaiian Entom. Soc., vol. 6, pp. 529–556, 5 pls., 6 figs.

Trouvelot, B.

1924   Recherches de biologie appliquée sur la teigne des pommes de terre et ses parasites et considérations générales sur l'utilisation des insectes entomo-phages en agriculture. Ann. Epiphyties, vol. 10, 132 pp., 4 pls., 32 figs., 4 diagrams.

Voukassovitch, P.

1932   Sur l'importance des insectes parasites entomophages dans les biocénoses des insectes. C. R. Soc. Biol., Paris, vol. 110, pp. 499–501.

CHAPTER X

INSECTS AS FOOD FOR MAN AND
OTHER ORGANISMS

Αὐτὸς δὲ ὁ ᾽Ιωάννης εἶχε τὸ ἔνδυμα αὐτοῦ ἀπὸ τριχῶν
καμήλου, καὶ ζώνην δερματίνην περὶ τὴν ὀσφὺν αὐτοῦ· ἡ
δὲ τροφὴ αὐτοῦ ἦν ἀκρίδες καὶ μέλι ἄγριον.[1]

Now THAT we have detailed at such length the great success of the
insects in developing food habits that exploit practically everything
in their living environment, it may be of interest to view the matter
from another angle. Their food relationships with other living organisms are by no means restricted to the categories so far mentioned, for many other animals and even some plants have discovered
in the insects themselves a large and nutritious, if not wholly savory,
food supply. We have already devoted one earlier chapter to predatory insects and another to parasitic ones and found these to subsist
almost entirely on other insects. Many other groups of carnivorous
animals feed, at least to some extent, on insects both as predators
and true parasites. Even man himself does not completely exclude
insects from his dietary although the consumption of insects by
members of our western civilizations is at the present time entirely
unintentional on their part. Such is not true of other peoples, and
if space permits we may return to this as it serves royally to bolster
up the feeling of race superiority, nordic or otherwise.

Indirectly, insects are of great importance to the food supply of
man the world over, as they form, let us say, the cruder materials
that are transformed into the bodies of our various food animals,
especially fishes and birds whose flesh later finds its place in culinary
art. They may be twice or thrice removed, but are as much an inte-
gral part of the chain as the corn or garbage that enables our bacon
and pork roast to develop its appetizing flavor.

Many species of birds are insectivorous, wholly or partly, and a

[1] "John's clothing was of camels' hair, with a leathern girdle, and his food consisted
of locusts and wild honey."—Synoptic Gospel of Matthew.

vast volume of literature has been written about their food habits.
The fact that such birds may be useful to agriculture by eating insect pests destructive to cultivated crops has been seized upon as
an argument against the wanton destruction of our native birds.
One cannot doubt the good intentions of those who have tried to
impress upon the uninitiated the value of birds in this respect, but
it seems hardly likely that such efforts will be of much avail in influencing the behavior of persons unwilling to preserve the birds
for their own sake. Moreover, the part played by insectivorous
birds in regulating the abundance of insects has led to much controversy and has often been grossly exaggerated, due partly no
doubt to the assumption that the end justifies the means.

Nevertheless some species of birds may at times become a prime
factor in lessening the depredations of insect pests on growing vegetables and fruits and they always play a part in regulating the
abundance of insects in natural associations. One of our pioneer
American ecologists, Professor S. A. Forbes ('80) early called attention to the economic value of birds as destroyers of insects in Illinois, based on an examination of the stomach contents of a number
of common species. He continued this work during the later years
of his life and also published data on the relative numerical abundance of birds and insects in the same region. Extensive material
collected by naturalists in many parts of the world has since added
innumerable data almost as hard to digest as the chitinous bodies
of the insects themselves, and it is difficult to evaluate the real
status of the feathered vertebrates in respect to insect control. Many
isolated observations on specific insects show their importance.
McAtee ('24) found that the fall webworm fell prey to the red-
eyed vireo to the extent of 11 to 89% during a period of eight years
in Canada. Barber ('25) estimated that 61% of the overwintering
larvae of the European corn-borer in New England were eaten by
birds. Baird ('23) found that birds destroyed 10% of the brood of
the larch sawfly and about one-quarter of the caterpillars of the
larch case-bearer in New Brunswick. In South Africa, Skaife ('21)
found that about 1% of the wattle bagworm, a psychid caterpillar,
were eaten by birds and rats, among over 99% that failed for various reasons to survive the vicissitudes of larval life.

A most important agency in collating data on the food of North
American birds is the Bureau of Biological Survey in Washington.
After years of painstaking work, the gentlemen in this organization
can speak more authoritatively than any others concerning the

actual food of insectivorous birds as well as those that share our strawberries and cherries or fishing worms, according to their tastes. Much of this material has been summarized by McAtee ('32) in a paper dealing primarily with the protective adaptations of insects against birds and other predatory animals, but his data have an equally direct bearing on the question we are considering. They show in a general way that birds are not partial to any particular kinds of insects, but that they select them mainly on the basis of abundance and availability at the moment, changing their diet to suit the exigencies of the occasion. This is at once in great contrast to the fixed diets of nearly all predatory insects. On account of their comparatively large size and ravenous appetites birds may consume *per capita* very large numbers of insects within a short space of time. He cites 200 to 300 ants at a meal as an ordinary number and has found flickers that had in their stomachs as many as 2,000, or in one case over 5,000 ants. Grasshoppers are naturally a favorite food for birds, no doubt on account of their large size and succulent bodies. The swarms of migratory locusts are especially fine feeding grounds for birds in regions where such periodic invasions occur.

Actual estimates of Bryant ('12) attribute a daily toll of over 120,000 grasshoppers per square mile during an outbreak of these insects in California. While large in the aggregate this would be less than 200 per acre and thus involve only a small part of the hopper population. The bird population of any given area is hard to estimate as most birds are conspicuous animals, even at considerable distances. Forbes and Gross ('23) have summarized extensive field observations made in Illinois which show that the number of birds per square mile varies from an average of 520 in the winter to 644 in the summer, or approximately one bird per acre at the time of their maximum and 0.812 bird at their minimum abundance. However, as the absurdly mathematical bird-fraction must needs be seminivorous at this inhospitable season, its importance to the insect population is greatly reduced. When an acre of woodland may support a population of 10,000,000 ants of a single species, an acre of cornfield 35,000 corn earworms or 1,000,000 corn-borers, and an acre of potatoes 5,000,000 leaf-hoppers in addition to similarly stupendous numbers of potato beetles and sundry other potato-eating pests, it can be seen that exceptional conditions are required to make much impression on the abundance of insects. Swarms of locusts are of course too erratic in occurrence to deal with in this connection, but the case mentioned in a previous chapter of a non-migratory grasshopper (*Camnula pellucida*) represented by over 20,000 hoppers per acre is quite within the range of abundance of other common species.

Perhaps we should pause at this point to consider in a more generalized way the place that this insect food, consumed by various vertebrates, occupies in the scheme of Nature, or more properly, how it affects the balance of Nature. From what has been presented in the several preceding chapters no one will question the complexity of this problem. Before the advent of civilized man a balance of Nature actually did exist in a reasonably stable form, although undergoing gradual progressive changes, and undoubtedly experiencing many sudden, perhaps almost catastrophic, adjustments of greatly limited extent.

The quite recent discovery of specific, ultrapowerful insecticides that may be readily applied in wholesale fashion to destroy insects noxious to man, to his food products, or to his other pet possessions has been a great boon to uncontrolled human multiplication and has served to mitigate at least to some extent the menace of insects to human welfare. It is quite obvious that there are serious drawbacks to the full utilization of all that has been learned concerning the efficacy of these control measures, with the sole exception of the so-called natural control of insect pests by specific parasitic and predatory enemies. Chief among these disadvantages are the factors whose long-term action on the insect fauna as a whole cannot be formulated precisely in advance. Like the future of any biological association or biocoenosis, they are not predictable even within wide limits.

The uncertainty of this complicated situation has been suddenly thrust upon us as a very practical problem by the recent discovery that 2, 2-bis (*p*-chlorophenyl) 1, 1, 1, trichloroethane, familiarly nicknamed DDT, is almost unbelievably poisonous to insects although very much less so to man and the higher vertebrates. Used extensively during the war to destroy mosquitoes, lice and other disease-bearing insects, DDT has been applied to the destruction of many other forms of insect life. We are now confronted by the question whether to utilize this material indiscriminately in a full-scale campaign to exterminate insect life over large areas. Such might include, for example, the New England forests overrun by the imported gipsy moth, vast areas of coastal marshes that serve to breed mosquitoes and sand-flies, innumerable rushing brooks that support the larvae of the pestiferous black-flies, or thousands of square miles devoted to the cultivation of cotton, maize, rice, or the like. Many have been quick to see in DDT the panacea for all ills, annoyances and losses due to insects, without further inquiry into the far-reaching biological problems involved. The effects of such wholesale destruction may be gauged to some extent by what has been learned from the wide use of other infinitely less efficient insecticides such as the arsenicals, nicotine, rotenone, pyrethrum, etc., but the matter must now be viewed in a new light.

Extensive experimental investigations at present under way by federal and other agencies to determine the effects of DDT on terrestrial and aquatic insects as well as on other forms of animal life should make possible a rational program for its use. This will necessarily take into consideration the food relations of insects to the biota as a whole, as well as the immediate decimation of the insect fauna. As we have seen, these relations are manifold since they involve also the predatory and entomophagous insects. In addition, we must consider also the insectivorous birds, fishes and other animals dealt with in the present chapter, and finally the earthworms on which the continued natural fertility of soils depends in great measure.

At the present moment of writing no final or adequate analysis of these matters can be presented, but we may rest assured that it will be forthcoming, based on careful experimental investigation.

Meanwhile we may cheerfully dust the family cat and dog, using "Flee-away" or some other barbarically named powder spiked with DDT. Equally efficacious are similar preparations or sprays for the destruction of the several semidomesticated flies, clothes-moths, cockroaches, carpet beetles and the like. This deadly material may even be incorporated in the paint on walls or screens to kill flies, mosquitoes and other interlopers.

Insects in great variety form the major food supply of many lizards although we know much less concerning the details than is the case with insectivorous birds. In warm climates and especially in dry regions where lizards are abundant these small vertebrates are an actual factor in regulating the abundance of insects. It is generally believed that at least many lizards restrict their feeding to certain kinds of insects and it has been frequently stated with some basis of observation and experimentation that the selection of some insects and the rejection of others is dependent upon their flavors. Hence, selection by lizards may be regarded as actively associated with the development of aposematic colors and as related to other moot questions concerning protective coloration and mimicry among insects.

Observations and dissections by Knowlton ('38) show that several lizards common in Utah (*Sceleporus, Uta, Cnemidophorus*, etc.) are especially fond of grasshoppers, termites and of many Homoptera and Hemiptera although they eat all sorts of insects, including aculeate Hymenoptera. Similarly, Burt ('28) found that the common lizards of Kansas consume insects in great variety, concentrating their attention upon grasshoppers which form about one-half of their insect food. Curiously enough, he found them to be very fond of Lepidoptera despite the dusty covering of scales which must be swallowed with the edible portions. The same is true of the nocturnal geckos as I have myself frequently observed in both the American and East Indian tropics. These semi-domesticated lizards appear at nightfall when lights are lit on porches and lie in wait for the insects that will soon put in their appearance. On such occasions they pounce upon moths by preference although indicating a comical disgust as they work the prey into the mouth and swallow it.

The Texas "horned toads," lizards of the genus *Phrynosoma*, specialize in the large agricultural ant (*Pogonomyrmex*) which abounds in that region. I have often observed one of these horned toads, raised high on its short legs near the entrance to a nest snapping in rapid succession at the ants, gulping them down after crushing the body and rendering the powerful sting inactive. How they can survive stings which they often receive from this ant is a mystery. Their food includes also, according to Winton ('15) a considerable proportion of stink bugs, a further indication of their rank as first class lacertilian yogis.

On the other hand Pack ('23) found in the stomachs of a common species of *Callisaurus* an extremely miscellaneous assortment of insects and spiders which appears to be typical of the rather generalized dietaries of most lizards. Likewise an examination of a number of common Philippine lizards by Villadolid ('34) reveals their great fondness for insects of diverse sorts as these make up the bulk of their stomach contents.

Many amphibians are carnivorous, especially after they reach maturity and insects form the bulk of their animal food, although snails, earthworms and small crustaceans are eaten also. Unfortunately most of the available data on the food habits of these animals are restricted to those of the common garden toad and its relatives, to a few frogs and a small number of salamanders.

The garden toad is *par excellence* insectivorous and the same appears to be true of the other members of the almost cosmopolitan genus, *Bufo*. Hamilton ('30) has given a very concise statement on the food of our common *Bufo americanus*. He found a great variety of insects in their stomachs, including a surprisingly large number of dipterous larvae. On account of its crepuscular habits the toad feeds extensively on those insects that sally forth at nightfall. They seem to lack entirely any distaste for the most vile-smelling prey, as I have seen them feeding on *Calosoma* ground-beetles, emitting only an apparent smile of satisfaction as they heave the abdomen to settle the beetle in its final resting place. Although the toad is lauded as the savior of the family flower garden, its diet of insects is by no means restricted to obnoxious species, but it doubtless does more good than harm. Its grotesque appearance and nonchalant air of domesticity certainly make it an object of interest and one, let us hope, that will never disappear as the result of senseless destruction.

One other species of toad deserves mention. This is the giant *Bufo marinus* of the American tropics, which has been introduced and naturalized on a number of West Indian islands and elsewhere. In Trinidad and British Guiana, according to Weber ('38), ants form its main article of diet. It was introduced into Puerto Rico some years ago and its fondness for the white grubs destructive to sugar cane has greatly reduced the damage by these pests. Observations by Wolcott ('37) appear to indicate that it selects these white grubs quite consistently in spite of their decreased abundance. Other species of toads have similar tastes. Power ('31) found a South African toad (*Bufo regularis*) to eat insects almost exclusively and observed one individual that ate 147 lamellicorn beetles, members of the same family as the white grubs just mentioned.

Frogs are also eminently insectivorous, but as most of their feeding is done about the borders of ponds they consume many lacustrine insects or those that fall into the water. Records show that they eat various kinds of insect larvae as well as adult beetles and grasshoppers in numbers. Our common leopard frog which ventures further from water than most species is known to be particularly fond of grasshoppers. The bullfrog, *Rana catesbiana*, has been seen by Mallonee ('16) eating one of our abundant yellow swallow-tail butterflies, *Papilio turnus* and it is insisted by many proponents of color-mimicry among butterflies that this is a general habit among other kinds of frogs.

Tree-frogs are probably entirely insectivorous although their abundance never rises to a point where they can have any appreciable effect on the insect population.

The larval tadpole stage of both toads and frogs is in great part herbivorous, but in pools where plant material is scarce they often feed extensively on aquatic insect larvae. In some of our Arizona sandstone canyons the pools left after violent floodings serve as breeding places for mosquitoes and toads. Wherever the tadpoles become numerous the mosquito larvae are absent and although I have not observed the tragedy of engulfment there can be little doubt that the mosquito larvae serve as food for the hungry little tadpoles.

The tailed amphibians or Urodela, commonly known as salamanders are carnivorous and consume many insects as well as other small invertebrates. The terrestrial forms like *Ambystoma* and *Plethodon* include centipedes, sowbugs, earthworms and snails in their diet while the larval forms and purely aquatic adults of others like *Triturus* consume many aquatic insect larvae as well as varied insects and other invertebrates of suitable size. Although they have been supplanted by small fish in the control of mosquitoes, salamanders are useful aids in reducing the numbers of mosquito larvae and their propagation has been advocated (Garfolini '24).

For many fishes, the insects are an important source of food, especially for the finer game-fish like trout. The catholic taste of these fish is attested by the avidity with which they snap up the contraptions of feathers successfully used by anglers. No self-respecting entomologist nor any other insectivorous animal having previous experience would confuse these "flies" with Nature's product, but curiosity in this case is their undoing. Their reaction to actual food is less impulsive but surer, provided, of course, that they are hungry at the moment. At such times they take a great variety of food. That trout actually consume great quantities of aquatic insect larvae is attested by the fact that the naturalization of these fish in New Zealand has led to the threatened destruction of the native insect fauna (Fig. 61) and the consequent ruin of the enterprise unless the fish are netted in great quantities sufficient to keep them within bounds.

In the mountain streams of our northern Rockies where trout are abundant, the larvae and adults of various stone-flies form the principal food of these fish, together with may-flies in lesser numbers. Muttkowski ('25) states that the food of trout in Yellowstone Park, especially of the cutthroat trout (*Salmo clarkii*) consists mainly of

several Plecoptera, especially the large *Pteronarcys californica* together with species of *Acroneuria*, *Perla* and *Alloperla* which often
form as much as 90% of the stomach contents. May-flies are also
favored as food, but in much lesser numbers. In New Zealand,
Tillyard ('21) found the food of the trout there to be essentially
similar and to include the larvae of caddis flies in great numbers,
and the same is true in our own country. Many other fishes consume
insects to a lesser extent, but often in considerable amount, although

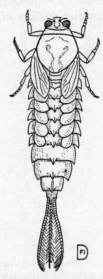

Fɪɢ. 61. Nymph of a New Zealand mayfly (*Oniscogaster distans*),
threatened with extinction due to the introduction of American
trout into that region. Redrawn from Tillyard.

any analysis of their food would needs be dealt with at great length
to be of specific interest. A number of relevant publications are cited
in the bibliography.

Of the invertebrate enemies of insects, aside from other predatory members of their own class, the majority fall into the category of parasites rather than predators. Notable among the latter
are the spiders whose proverbial addiction to flies (and other miscellaneous insects) has not escaped the attention even of poets
(Pl. XIX B). Curiously enough there seem to be few serious
studies by competent observers. Bilsing ('20) has given an interesting account of the food of a score of North American spiders be-

longing to several families, including both the hunting and web-spinning types and those that lie in wait for their prey. Most of the web-spinning species will spin up any insect that blunders into their webs. Grasshoppers are frequently caught by the large garden spider, *Agelena naevia*, and the bulk of the prey of *Acacesia foliata* nesting on a house was found to consist of houseflies. Bilsing found that the related *Araneus trifolium*, nesting in pastures and waste land, had among other insects caught many honey bees. It would appear that the trapped insects are quite random catches of species locally numerous. The jumping spiders of the family Attidae usu-

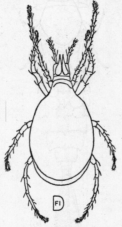

Fig. 62. A parasitic mite (*Macrocheles carolinensis*). This and many related forms are external parasites of beetles and other insects. Redrawn from Banks.

ally capture small insects, often very active flies, but appear to avoid hard-bodied beetles entirely. The little yellow crab spider, *Misumena*, hides on flowers like the common field daisy and captures small bees and flies that visit the flowers for food. It may be said in general that spiders do not exercise any noticeable choice in selecting their prey but that insects are the chief victims. Many other small Arachnida, commonly known as mites and forming a part of the order Acarina, prey on insects. On account of their small size they live as external parasites on the bodies of beetles, flies, ants and other diverse insects where they secure their nourishment by piercing the integumental membranes with their sharp mouthparts and extracting the body juices. Many of these mites are of a brilliant red color and ornament the bodies of their luckless hosts as with gaudy coral beads (Fig. 62). One very minute form, *Pediculoides*, is an important enemy of certain pests of grain and of the larvae of bark beetles, and may on occasion turn its attention to man, producing severe urticaria.

Among the lower invertebrates, many nematode worms are parasitic on insects either wholly or during a part of their life-cycle. The well known Filaria that causes filariasis in man passes a part of its developmental stages in the thoracic muscles of mosquitoes which acquire the parasites by feeding on infected persons. In this case the mosquito harbors the parasite during only a part of its develop-

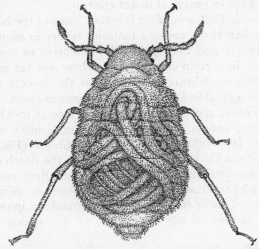

FIG. 63. An aphid (*Anoecia*) infested with a parasitic nematode worm. After Davis.

mental period and the same is true of another human parasite that is transmitted by certain black-flies (*Simulium*).

Other "nemas" that have no alternate hosts are important parasites of a great variety of insects (Glaser and Wilcox; Christie '37). Among those frequently noticed are members of the family Mermithidae, some of which occur in grasshoppers. The young hoppers ingest the eggs which are laid on plants and the larval worms perforate the alimentary canal to develop in the body cavity of the host whence they emerge at maturity as long, thread-like adults. Others occur similarly in the abdomen of ants, aphids (Fig. 63), beetles and other insects. As the parasites are commonly of large size the bodies of their hosts are often hypertrophied and among ants thus affected peculiar inter-caste forms known as mermi-thergates are sometimes produced as a result of a castration of the host by the contained parasite.

Nematodes are undoubtedly of great importance as enemies of insects, but unfortunately they have been sadly neglected by zoologists, no doubt on account of their drab appearance and the technical difficulties of classifying them accurately. At least one species, *Neoplectana glaseri* Steiner is a useful parasite of the Japanese beetle. Glaser ('31, '32) succeeded in cultivating this species on artificial media during its entire life cycle, thus opening up an entirely new field in practical insect control.

Many parasitic Protozoa affect insects. Some of the better known are Sporozoa like those causing malarial fevers in man and other higher vertebrates and undergoing the definitive or sexual part of their life cycle in certain mosquitoes. These are far more serious enemies of their vertebrate hosts than of the insects which commonly survive such attacks without fatal consequences. Such is not the case with other sporozoan infections of insects in which no secondary hosts are known. Larvae of black-flies often succumb in great numbers from infection by species of *Glugea* (Fig. 64) which form large cysts in the body cavity leading to the death of the host. Similar diseases commonly known as pébrine affect various caterpillars. Indeed the classical studies of Pasteur on pébrine in the silkworm, caused by *Nosema bombycis*, formed an important step in the early history of our knowledge of contagious diseases in general.

Many flagellate Protozoa are parasitic in insects and some of these have alternate hosts among the vertebrates, including man. Thus the trypanosomes causing African sleeping sickness and Chagas' fever find their definitive hosts in insects. As a consequence the tsetse flies (*Glossina*) and certain large reduviid bugs (*Conorhinus*) act respectively as the vectors of these human diseases and at once become objects of great human concern. How great a part protozoan diseases play in the regulation of insect abundance cannot be stated on the basis of present fragmentary knowledge as this field has not been thoroughly explored.

In considering the status of insects as food for plants, it will be convenient to reverse the order just followed with animals and to begin with the lower plants since the relation of these to insects is much similar to that of the lower invertebrates like the nematodes and protozoans.

Many bacteria live in the bodies of insects causing a great variety of diseases about which we know as yet very little. There appears, however, to be no great novelty in these as they readily fall into the patterns established in the case of similar diseases of man and we may anticipate that the future will show their number to be legion. So far, most of the work on such insect diseases has been limited to maladies of the honey-bee, of certain caterpillars and a few other

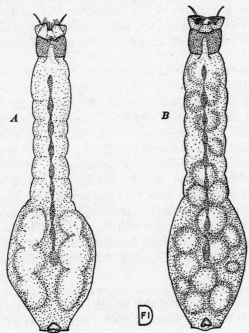

FIG. 64. Larvae of black-flies infected with protozoan parasites. *A, Glugea bracteata* in *Simulium bracteatum; B, Glugea multispora* in *Simulium vittatum.* Redrawn from Strickland.

miscellaneous insects, and it is already obvious that they involve difficulties similar to those met with by students of human disease.

Contrary to expectations based on human pathology, the rôle of fungi in causing disease among insects is highly important. The number of such plants that are parasitic on insects is very large; the manifestations of their occurrence are extremely diverse and frequently assume epidemic proportions. Such diseases, known as mycoses are better known and more easily recognized and differentiated than those due to bacteria.

Perhaps the most conspicuous of these insect mycoses are the "vegetable caterpillars," known as awetos in New Zealand where they are especially abundant. They arise through the infection of subterranean caterpillars or other insects by a fungus of the genus *Cordyceps* (Pl. XX) or of some related form. After the mycelium has consumed its host, a stalk emerges from the mummified body, pushes up through the soil and produces the aërial fruiting body which is often of striking shape and color. These fungi are widespread, affecting diverse insects, particularly those which live in the soil, although scale insects are commonly affected by smaller species.

The sudden decimation of our housefly population in the autumn is due to another type of fungous parasite (*Empusa muscae* and related species) which invades the bodies of adult flies. Here again the mycelium fills the body and breaks through the intersegmental membranes to produce spores from innumerable spore-bearing stalks or conidia. These spores are thrown off to an appreciable distance at maturity and often form a whitish ring or halo on the substratum where the luckless fly has found its last resting place. Other members of this group (Entomophthorales) affect caterpillars, aphids and Hemiptera. One economically important species is parasitic on the destructive chinch-bug, *Blissus leucopterus* and another (*Entomopthora aulicae*) has been the most important check to the imported brown-tail moth in America, parasitizing the larvae of this species.

Some fungi actually catch minute animals by means of mesh-like or loop-like modifications of the hyphae that form their mycelium. Numerous species are known that trap protozoans, nematode worms and crustaceans, but very recently Drechsler has described *Arthrobotys entomophaga* that entraps spring-tails (Collembola) by means of a net of intersecting hyphae, exuding droplets of adhesive liquid.

This account of fungi might be drawn out at great length, but one other peculiar group must be mentioned. These are the Laboulbeniales, minute forms of dubious relationships which develop as spiny outgrowths on the chitinous integument of many insects (Fig. 65). Many bizarre types have been described and beautifully figured by Thaxter ('96–'26) who devoted many years of his life to their study.

Although the diseases of insects due to fungi are better known than those caused by other organisms, it may be stated with reasonable assurance that they are far and away the most important in the toll which they exact from the insect population the world over.

Among the flowering plants there occur a number of truly insectivorous forms and curiously enough these belong to several quite diverse groups, each of which has obviously acquired the habit independently.

One of these includes the bladderworts, members of a widely distributed genus of aquatic plants, *Utricularia*. Each of the finely divided leaves bears one or several small, bladder-like vesicles. Each bladder is furnished with a tightly fitting valve-like lid that readily

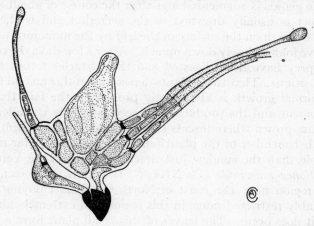

FIG. 65. *Dimeromyces formicicola*, a laboulbeniaceous fungus parasitic on an ant (*Prenolepis silvestri*). Female, with male attached. After Thaxter.

opens inwards and forms a sort of trap-door through which small insects and crustaceans may easily enter but which prevents their escape. Imprisoned in the bladders, which are filled with water, the prey dies, and after undergoing decomposition supplies nitrogenous substances that are absorbed by the plant and furnish it with a part of its nourishment.

Quite different methods are pursued by terrestrial plants that also depend to a great extent upon animal food. Commonest among these are the sundews of the genus *Drosera* (Pl. XXI; *B*). These small plants often grow abundantly in sand along the shores of ponds or in swampy places, frequently in sphagnum bogs. In this case the leaves are greatly modified in form by the development of numerous long tentacle-like processes. Each tentacle is surmounted at the tip by a gland that secretes a minute sphere of very viscous, sticky material. These innumerable specks give the appearance of dew on the leaves which has led to the name of sundew that is highly appropriate as they glisten in the sun. The tentacles on the disk of the small leaves are short, but those at the margin are much longer, equalling fully half the width of the leaf. Small insects, usually tiny flies alighting on the leaves of the *Drosera* are caught by the sticky secretion and as soon as they come into contact with the surface of the glands the surrounding tentacles initiate a bending toward the intruder. This flexion continues as the secretion from the glands is augmented and after the course of about an hour the insect is usually drowned in the secretion and is then later pressed down upon the surface of the leaf by the numerous tentacles that have folded closely down upon it. After a few days the contents of the prey have been digested and the tentacles return to their erect positions. That these plants depend upon this animal food for their normal growth is abundantly proved by the fact that their development and the production of seed is grossly curtailed if the plants are grown where insects cannot have access to them.

Another member of the plant family Droseraceae is far more remarkable than the sundew just described. This is the Venus' flytrap, *Dionea muscipula* (Pl. XXI; *C, D*), often to be seen in the sandy region near the coast of North Carolina, enjoying a most remarkably restricted range in this region and extremely abundant where it does occur. The leaves of this small plant form a rosette, with the body of each leaf expanded and fleshy and sharply divided into two lobes by a median groove on the upper side. Each lateral margin is fringed by a series of about fifteen sharp spines or spikes and the sides of the leaf are capable of folding upwards, bringing the upper surface of the two halves together. When this closure occurs the spines interdigitate forming a series of cross bars that effectually entrap any insect that may have been on the leaf. Each half of the leaf is provided with several fine, hair-like filaments arising vertically from its upper surface. These are the so-called trigger hairs, for when they are disturbed by contact with a fluttering or crawling insect the two halves of the leaf suddenly fold up like the shell of a bivalve mollusk. As this folding is accomplished within the unbelievably short space of less than half a second, the insect is imprisoned unless it be small enough to escape between the spines which bar its exit. Once the prey is caught, the leaf slowly continues to fold more tightly until the insect is pressed between its two surfaces. Glands on the leaf then exude a copious digestive liquid, and the victim disintegrates to furnish nitrogenous food that is absorbed by the plant. After a few days the leaf slowly reopens and is ready to repeat the process. This performance in itself should be sufficient to excite our wonder and admiration, but there is still another anatomical feature which adds to the efficiency of the device. Insects are attracted to the leaves by a series of nectaries distributed along the margin and the trap is ordinarily sprung only by those large enough to touch the trigger hairs while feeding. Thus, very small insects may visit the leaves without evoking any movement on the part of the trap mechanism, while those of sufficient size to be worth while snap the trap. *Dionea*, therefore, actually sorts out the larger of its visitors for capture, while as we have already seen, it allows the smaller fry to escape even though they may spring the trap by venturing away from the zone of nectaries. How many times a leaf may go through the feeding process is not definitely known, but undoubtedly repetition is limited. As with the sundew it is definitely known that insect food is essential to the proper growth and fruiting of the Venus' fly-trap.

Among the insectivorous flowering plants the pitcher plants of the genus *Sarracenia* are of great interest. They form the typical genus of a small American family and some half a dozen species are locally abundant in the eastern United States, especially from the Carolinas southward. The leaves, including their petioles, are modified to form hollow tubes, expanded toward their tips (Pl. XXI; *A*). Although the leaf or pitcher is open at the top, in most species it is erect and provided with an appendage that partially or almost completely covers it and serves to prevent or delay the filling of the pitchers during rains. The inner surface bears extra-floral nectaries and insects attracted into the leaves frequently tumble into the tubes where their downfall is aided by a smooth surface beset with downwardly directed hairs that likewise render escape more difficult. Many kinds of insects, especially small flies, find their way into these traps, and an accumulation of partially disintegrated insects is to be found in the narrowly tubular base of the leaf. A variable amount of liquid is also present and this contains an active proteolytic enzyme that actually digests the captures. That this liquid is a direct secretion of the plant seems certain, although as might be expected, proteolytic bacteria are present also. Increased captures stimulate the secretion of liquid and the products of digestion are absorbed by the plant. There is considerable variation in the form of the leaves in the several species of *Sarracenia*, especially in the shape of the lid or hood. In one species, our common *S. purpurea* of the Northeastern states the leaves are decumbent, not protected by an umbrella and these are usually partly filled with water which greatly dilutes the digestive secretion. In spite of the deadly nature of the traps certain highly adapted insect larvae regularly develop in the water or muck of the pitchers. Several flesh

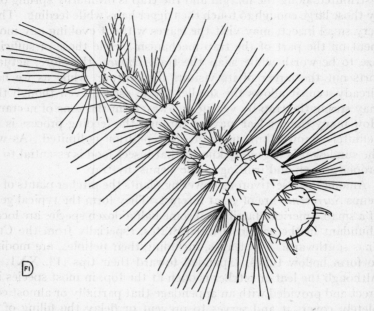

Fig. 66. Larva of *Wyeomyia smithii*, a mosquito in the water that accumulates in the leaves of the pitcher plant, *Sarracenia purpurea*. Redrawn from Knab.

flies of the genus *Sarcophaga*, especially *S. sarraceniae*, occur frequently in the mass of disintegrating insects. In North Carolina where the gorgeous yellow *Sarracenia flava* is locally very abundant I have on several occasions found a *Sarcophaga* larva in almost every leaf feeding in gluttonous comfort on the accumulation of insect chop-suey and not at all affected by the bath of digestive fluid in which it is immersed. Likewise, in the larger volume of water contained in the pitchers of *S. purpurea*, the peculiar larva of a mosquito, *Wyeomyia smithii* (Fig. 66) is commonly to be found and this particular mosquito is restricted to this very unusual habitat. A few small Diptera of several other families are also known to develop in *Sarracenia* leaves.

Of particular interest are the caterpillars of three species of *Exyra*, noctuid moths that feed within the pitchers, although upon the tissues of the plant itself. These larvae have capitalized the unusual form of the Sarracenia pitchers to serve as retreats during their period of growth and hibernation and since the structure of the pitchers is extremely diverse in the several species these moths have adopted different methods in feeding and in the construction of their hibernacula. An adequate account of these would demand more space than could be allotted here, but the reader is referred to a most illuminating description by Jones ('21) who has made extensive studies on pitcher plant insects.

A closely related Californian plant, *Darlingtonia*, with great hooded leaves is even more remarkable in appearance. The leaves also digest insects which enter the hood, and enticed by transparent window-like spots on its upper surface, seek escape, only to catapult into the depths of the hollow leaf below where they are drowned and digested.

Some not very distant relatives of our pitcher plants are those of the genus *Nepenthes*, native to the Malayan region and often cultivated in greenhouses throughout the world. These are trailing or climbing plants that have the tip of the leaf extended and formed into a pendant cup or pitcher. Water accumulates in the latter, and supplemented by a digestive secretion and acid derived from the plant, insects drowned by falling into the pitcher furnish food which is absorbed. Although *Nepenthes* may be grown without animal food they flourish best when this is available, especially in the very sterile soils on which some of the species develop. Like *Sarracenia*, *Nepenthes* harbors in its pitchers various aquatic insects that live as commensals feeding on the trapped food and living unharmed by the digestive ferments. All together the insectivorous plants include some 450 species belonging to fifteen genera distributed among six of the natural families. Of these, six genera occur in the United States. The traps are of several distinct types and have been classified by Lloyd ('42) as "pitfalls" represented by the leaves of the pitcher plants, as "lobster pots" in the tropical *Genlisia*, as "fly-paper traps" in *Drosera*, as a "steel-trap" in *Dionaea* and as "mouse-traps" in *Utricularia*.

It was promised at the beginning of the present chapter that something might be said concerning the eating of insects by the human species if space would permit after considering the insectivo-rous behavior of the more lowly organisms. To do this with the proper biological background it is necessary to refer first to certain other mammals, including the monkeys and apes. As is well known, biologists firmly insist that man is more closely related to these latter quadrupeds than to the mythical angels, devils and deities even though the latter are pictured as close morphological mimics of *Homo sapiens*.

The application of the name Insectivora to a group of mammals clearly indicates an early recognition of the fact that some of its members, including the moles and shrews, of Europe and North America, eat insects extensively. Whether they are statistically important in this respect cannot be stated. Many bats, members of another primitive order of mammals, consume insects in large quantities, particularly small flying forms whose activity at dusk suits them to the crepuscular feeding habits of the bats. Among Carnivora, skunks are fond of certain insects, even including bumble-bees which they first damage with their paws to avoid the ready stings of their prey. The American black bear is notoriously fond of sweets and will submit to the stings of wild honeybees to satisfy its appetite. They are also known to eat ants, noted for their acid taste. Reported instances of these bears eating hornets might even lead one to believe that the stings of aculeate Hymenop-tera may lend a piquancy to their food that is actually relished.

The most notable insect eating mammals are of course the ant-eaters. The South American ant-bear, *Myrmecophagus*, is a large edentate which captures ants and termites by means of its promi-nent, sticky, prehensile tongue and this animal is anatomically unfit to consume other sorts of food. Other related edentates with similar habits include forms in the tropics of the old world especially the pangolins of Africa and the Indomalayan region. A similarly modi-fied large African type represents another order, Tubulidentata. These, known as aard-varks (*Orycteropus*) burrow into the ground in search of their food.

Many of our closer relatives among the Primates consume varied insects, mainly as tid-bits, but a notable exception is formed by the tiny spectral lemurs of certain Malayan islands that subsist exclusively on insects.

Likewise, insects fall mainly into the class of delicacies so far as human consumption is concerned and necessarily so as the supply of these tiny creatures can at best only on rare occasions go far toward satisfying the Gargantuan requirements of such ponderous social organisms as ourselves. Nevertheless certain insects do on occasions form a real source of food, and several such instances relate to our own American Indians. A common brine fly, *Ephydra hians*, which breeds in countless numbers in Mono Lake, California, is collected by the Indians in the pupal stage which is dried and furnishes a nutritious, fatty, though hardly appetizing, food. Other highly alkaline and saline lakes in this same region so teem with the developmental stages of this fly that they are capable of supplying sizeable quantities of this material, known as koochabe. Other aquatic insects, this time the eggs of several waterbugs of the families Corixidae and Notonectidae are utilized in Mexico for the production of an edible meal known as ahuatle. This is prepared by drying and is sold in the markets as are also dried cakes containing the insects themselves. According to Ancona ('33) a kilo of the insects and eggs sells for a dollar or less in Mexican currency in the towns near Lake Texcoco which is a center for this queer ahuatle-fishing industry.

Another mass utilization of insects for food by the Californian Indians has been described by Aldrich ('12). In this instance the material consists of adult flies belonging to the genus *Atherix* of the family Leptidae, much larger than the *Ephydra* previously mentioned. The flies gather along streams during the month of May to lay their eggs and are gathered in the early morning when still numbed by the cold. The flies are mashed and kneaded like dough preparatory to being made into loaves and baked in a primitive pit oven heated by stones. Thus cooked the mess may be sliced and eaten. I am unable to find any description of its consistency or flavor, but we may surmise that the chitinous integuments and coagulated ripe eggs would form a peaty omelet and that it might initiate the removal of undesirable materials from the alimentary tract quite as successfully as the bran custards or other similar "nature foods" recommended by more civilized food enthusiasts.

Naturally, grasshoppers have also been used for food on account of their large size and frequent availability in large quantities. Essig ('34) describes the process of chasing the hoppers (also by Californian Indians) into pits containing hot stones which serve to kill and roast the catch simultaneously.

It must not be assumed that these southwestern Indians are different from other aboriginal tribes the world over, but they do seem to have utilized the insects for food on a larger scale than many others. Large caterpillars are also included in their dietary, particularly an abundant giant silk-worm, *Coloradia pandora,* that feeds on the foliage of pines. These and also the pupae are dried and roasted or stewed. This procedure is still practiced (Pl. XXII) but the recent influx of wide-eyed and hilariously mirthful tourists has naturally made the Indians bashful in disclosing their more interesting and unusual customs.

In the old world grasshoppers have for at least many centuries been eaten by man, especially the migratory locusts and these are still of importance, particularly to the African negroes as they supply

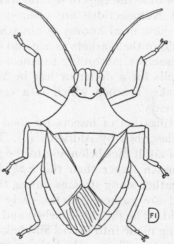

FIG. 67. A stink-bug of the family Pentatomidae. Redrawn from Hickman.

animal proteins otherwise difficult to obtain. Bequaert ('21) gives an account of the collection of winged termites in the same region and refers to the sale of baskets of dried termites, in this case the wingless soldier caste, at a native market in the Belgian Congo. Termites are also eaten in the Oriental tropics, but although they are also abundant in the warmer parts of the new world they seem to have been generally omitted from the diet of the American aborigines.

When one comes to consider insects as delicacies or entomological *hors d'oeuvres,* he is overwhelmed by a wealth of material that could be catalogued only by a bibliophile. Such tidbits range in size from the large egg-filled females of termites that rival a shad-roe in size to small pentatomid stink-bugs (Fig. 67) that are used to fortify the piquancy of sauces or other condiments. The termite queens are roasted or fried in fat by the African natives and the same method is used to prepare the larvae of various large beetles. One of the most widespread of these is that of the palm weevil, *Rhyncophorus* (Fig. 68). This is practically tropicopolitan and natives of many lands extract the large, plump grubs from the palms to roast, fry or broil them. The beetle larvae are most generally, however, those of lamellicorns or longicorns on account of their comparatively great bulk and plump, fatty bodies. Bequaert mentions the larva of the African *Goliathus*, the largest living beetle, as a delicacy enjoyed by natives of the African forest. Related beetles of the genus *Oryctes*

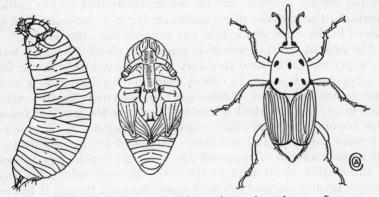

FIG. 68. The palm weevil, *Rhyncophorus ferrugineus*. Larva, pupa and adult beetle. Redrawn from Bainbrigge Fletcher.

that develop in dung are prized by the Laos natives of Siam according to Bristowe ('32) who states that the adult beetles are roasted or fried, but that the larvae and pupae are given a preliminary soaking before they are roasted. He gives a very remarkable account of the insects eaten by these people, including a long series belonging to several orders. One striking example is the giant water-bug, *Lethocerus*. This insect, which is some two inches long is steamed thoroughly and then picked like a lobster. Bristowe tells us that the meat has a strong flavor, reminiscent of Gorgonzola cheese. The bugs are also cooked, dried and pulverized to add zest for curries or sauces. Combined with shrimps, lime juice, garlic and pepper they form a popular native sauce known as namphla. A similar use in Mexico of stink-bugs of the genus *Euschistus* has been described by Ancona ('33). These are toasted, ground up with chile or pimiento pepper, dashed with tomato sauce and used as a substitute for maple syrup on the native pancakes or tortillas. A stronger flavor may be secured if the scent glands are extracted from the more innocuous contents of the body by pinching off the tip of the abdomen to which they are attached. The delicacy of this insect is familiar to any of my readers who may have unexpectedly crushed one of these bugs in a mouthful of fresh raspberries, a fruit which they occasionally frequent.

Although the preceding account has been drawn out to considerable length, many matters that are worthy of attention have been completely ignored or given only scant notice. Such inconsistencies are not readily avoided, and the writer trusts that he has neither overtaxed the patience of his audience, nor neglected any field involved in the main theme that has engaged our attention.

The present trend in biological research is clearly leaning toward the experimental method to clarify our ideas, not only in the field of physiology, but also in others where its application is a much more recent innovation. If this aspect has been more lightly touched upon in our text, such lack of emphasis is due to the fact that knowledge accumulated from other sources over many decades is far more extensive, concrete and stabilized. As noted in the introduction, the newer methods of approach are so far applicable only to some of the matters dealt with in the text. In such fields they have already greatly clarified our understanding even though it has not yet been possible to apply them too indiscriminately — a goal anticipated by some workers.

It is hoped that some of the comparisons between human behavior and that of the insects may not offend those who hold that the intellect of man transcends utterly the bare animal activities we have described. As a matter of fact, the attainment of his present civilized syndrome of behavior patterns by the human species has entailed the acquisition of taboos and inhibitions almost as varied as those we have described for the hungry insects. Culled from superstition and religious dogmas, from the accumulated lore born of trial and error, from the intuitive processes and from varied sources — by no means least as a compensation for physical inferiority — these mores are so hopelessly confused that we might expect them to lead any other animal to racial suicide.

No such dire eventuality is apparent among the insects.

Plate XXI

Insectivorous plants. *A*, a pitcher plant, *Sarracenia flava*, which traps small insects in its hollow tubular leaves; *B*, sundew, *Drosera rotundifolia*, with an ant that has just been caught in the sticky secretion produced by the leaves; *C*, rosette of leaves of the Venus' fly-trap, *Dionaea muscipula*; *D*, two leaves of the same plant, one open and the other with the trap closed. Photographs by the author.

PLATE XXII

Basket made and used by the Piute Indians in collecting and drying caterpillars for food. Original photograph.

REFERENCES

MAN

Aldrich, J. M.

1912a Larvae of a Saturniid Moth Used as Food by California Indians. Journ. New York Entom. Soc., vol. 20, pp. 28–31, 1 pl.

1912b Flies of the Leptid Genus *Atherix* used as Food by California Indians. Entom. News, vol. 33, pp. 159–163.

Ancona, Leopoldo H.

1933 Los jumiles de cuautla *Euschistus zopilotensis*. An. Inst. Biol., Mexico, vol. 4, pp. 103–108.

1933 El ahuautle de Texcoco. *Ibid.*, vol. 4, pp. 51–69, 17 figs.

Arndt, W.

1923 Bemerkungen ueber die Rolle der Insekten im Arzneischatz der alten Kulturvölker. Deuts. Entom. Zeits., p. 553.

Banks, N.

1912 The Structure of Certain Dipterous Larvae with Particular Reference to those in Human Foods. Bull. U. S. Dept. Agr., Bur. Entom., Tech. Ser., vol. 22, 44 pp.

Bequaert, J.

1921 Insects as Food. How they have Augmented the Food Supply of Mankind in Early and Recent Times. Journ. American Mus. Nat. Hist., vol. 21, pp. 191–200, 8 figs.

Berensberg, H. von P.

1907 The Uses of Insects as Food Delicacies, Medicines, or in Manufactures. Natal Journ. and Min. Rec., vol. 10, pp. 757–762, 1 pl.

Bristowe, W. S.

1932 Insects and other Invertebrates for Human Consumption in Siam. Trans. Entom. Soc., London, vol. 80, pp. 387–404.

Engelhardt, G. P.

1924 The Saturniid Moth, *Coloradia pandora*, a Menace to Pine Forests and a Source of Food to Indians in Eastern Oregon. Bull. Brooklyn Entom. Soc., vol. 19, pp. 35–37.

Essig, E. O.

1934The Value of Insects to the California Indians. Scientific Monthly, vol. 38, pp. 181–186.

Fagan, M. M.

1918The Uses of Insect Galls. American Natural., vol. 52, pp. 155–176.

Horikawa, Y.

1929Animals used as Medicine in Formosa. [In Japanese.] Contrib. Extra. Dept. Hyg., Res. Inst. Formosa. 84 pp.

Netolitzky, F.

1920Käfer als Nahrungs- und Heilmittel. Kol. Rundschau, vol. 8, pp. 21–26; 47–60.

Nguyen-Cong-Tieu

1928Notes sur les insectes comestibles au Tonkin. Bull. écon. Indochine, vol. 31, pp. 735–744, 4 pls., 3 figs.

Pierce, W. D.

1915The Uses of Certain Weevils and Weevil Products in Food and Medicine. Proc. Ent. Soc. Washington, vol. 17, pp. 151–154.

Reum, W.

1924Insekten und andere Tracheaten als menschliche Nahrungsmittel. Entom. Jahrb., vol. 33–34, pp. 89–92.

MAMMALS

Bailey, V., and C. C. Sperry

1929Life History and Habits of Grasshopper Mice, Genus *Onychomys*. Techn. Bull. 145, United States Dept. Agric., 19 pp., 4 pls.

Bigelow, N. K.

1922Insect Food of the Black Bear (*Ursus americanus*). Canadian Entom., vol. 54, pp. 49–50.

Bodkin, G. E., and L. D. Cleare

1919An Invasion of British Guiana by Locusts in 1917. Bull. Entom. Res., vol. 9, pp. 341–357, 8 figs.

Fisher, A. K.

1909The Economic Value of Predaceous Birds and Mammals. Yearb. United States Dept. Agric., 1908, pp. 187–194, 3 pls.

Froggatt, W. W.

1911Bag-shelter Caterpillars of the Family Liparidae that are Reputed to Kill Stock. Agric. Gaz., New South Wales, vol. 22, pp. 443–447.

Graham, S. A.

1928The Influence of Small Mammals and Other Factors Upon the Larch Sawfly Survival. Journ. Econ. Entom., vol. 21, pp. 301–310.

Hamilton, W. J., Jr.

1930The Food of the Soricidae. Journ. Mamm., vol. 11, pp. 26–39.

1931Skunks as Grasshopper Destroyers. Journ. Econ. Entom., vol. 24, p. 918.

Howard, L. O.

1916 *Lachnosterna* Larvae as a Possible Food Supply. *Ibid.*, vol. 9, pp. 390–392.

Jack, R. W.

1924 Locusts as Food for Stock. Bull. Minist. Agric. and Lands, S. Rhodesia, No. 518, 4 pp.

Künckel D'Herculais, J.

1899 Empleo de las langostas como abono. An. Soc. Cient. Argentina, vol. 48, pp. 368–377.

Lantz, D. E.

1923 Economic Value of North American Skunks. Farmer's Bull. 587, United States Dept. Agric., 24 pp., 10 figs.

Nelson, E. W.

1926 Bats in Relation to the Production of Guano and the Destruction of Insects. Bull. United States Dept. Agric., No. 1395, 12 pp., 4 figs.

Poulton, E. B.

1929 British Insectivorous Bats and their Prey. Proc. Zool. Soc. London, pp. 277–303.

Riley, C. V., A. S. Packard, and C. Thomas

1878 Uses to which Locusts may be Put. 3rd Rept. United States Entom. Comm., pp. 437–443.

Sheffer, T. H.

1914 The Common Mole of the Eastern United States. Farmer's Bull. United States Dept. Agric., No. 583, 10 pp.

Sim, R. J.

1934 Small Mammals as Predators on Japanese Beetle Grubs. Journ. Econ. Entom., vol. 27, pp. 482–485.

BIRDS

Baer, W.

1913 Die Bedeutung der insektenfressenden Vögel für die Forstwirtschaft. Aus der Natur, Leipsig, vol. 9, pp. 659–679.

Baird, A. B.

1923 Some Notes on the Natural Control of the Larch Sawfly and Larch Case Bearer in New Brunswick in 1922. Proc. Acadian Entom. Soc., vol. 8, pp. 158–171.

Barber, G. W.

1925 The Efficiency of Birds in Destroying Over-wintering Larvae of the European Corn Borer in New England. Psyche, vol. 32, pp. 30–46.

Beal, F. E. L.

1909 The Relation Between Birds and Insects. Yearb. United States Dept. Agric., 1908, pp. 343–350.

Beal, F. E. L., and W. L. McAtee

1912 Food of Some Well Known Birds of the Forest, Farm and Garden. United States Dept. Agric. Farmer's Bull., No. 506, 35 pp., 16 figs.

Bigglestone, H. C.

1913   A Study of the Nesting Behavior of the Yellow Warbler.   Wilson Bull., vol. 25, pp. 49–67.

Brower, A. E.

1934   Predatory Checks, Especially Birds, on the Birch Leaf-mining Sawfly.   Journ. Econ. Entom., vol. 27, pp. 342–344.

Bryant, H. C.

1912a   Birds in Relation to a Grasshopper Outbreak in California.   Publ. Univ. California, Zool., vol. 11, pp. 1–20.

1912b   The Economic Status of the Meadow Lark in California.   Month. Bull. Comm. Hort. California, vol. 1, pp. 226–231, 2 figs.

Dalke, P. D., W. K. Clark, and L. J. Korschgen

1942   Food Trends of the Wild Turkey in Missouri.   Journ. Wildlife Management, vol. 6, pp. 237–243.

Florence, Laura

1912   The Food of Birds.   Trans. Highland Agric. Soc. Scotland (5), vol. 24, pp. 180–219.

Forbes, S. A.

1880   The Food of Birds.   Bull. Illinois State Lab. Nat. Hist., vol. 1, 2nd ed., pp. 86–159.

1883   The Regulative Action of Birds Upon Insect Oscillations.   *Ibid.*, vol. 1, pp. 1–32.

1907   An Ornithological Cross-section of Illinois in Autumn.   *Ibid.*, vol. 7, pp. 305–335.

1908   The Midsummer Bird Life of Illinois: a Statistical Study.   American Nat., vol. 42, pp. 505–519.

Forbes, S. A., and A. O. Gross

1921   The Orchard Birds of an Illinois Summer.   Bull. Illinois Nat. Hist. Surv., vol. 14, pp. 1–8, 6 pls.

Jones, F. M.

1932   Insect Coloration and the Relative Acceptability of Insects to Birds.   Trans. Entom. Soc. London, vol. 80, pp. 345–386.

Judd, S. D.

1899   The Efficiency of Some Protective Adaptations in Securing Insects from Birds.   American Nat., vol. 33, pp. 461–484.

1901   The Food of Nestling Birds.   Yearb. United States Dept. Agric., 1900, pp. 411–436, 5 pls., 9 figs.

Kalmbach, E. R.

1914   Birds in Relation to the Alfalfa Weevil.   Bull. 107, United States Dept. Agric., 64 pp., 5 pls.

Lea, A. M., and J. T. Gray

1935   On the Food of Australian Birds.   Emu, vol. 34, pp. 275–292; vol. 35, pp. 65–98; 145–178; 251–280; 335–346.

Little, A.

1924   Locusts and Locust Meal as Poultry Feed.   Rhodesia Agric. Journ., vol. 21, p. 334.

Mason, C. W., and H. Maxwell-Lefroy

1912 The Food of Birds of India. Mem. Dept. Agric. India, Entom. Ser., vol. 3, 371 pp.

McAtee, W. L.

1907 Birds that Eat Scale Insects. Yearbook United States Dept. Agric., 1906, pp. 189–198.

1911a Bird Enemies of the Codling Moth. *Ibid.*, 1911, pp. 237–246, 2 pls.

1911b Woodpeckers in Relation to Trees and Wood Products. Bull. United States Dept. Agric., Bur. Biol. Survey, No. 39, 99 pp., 12 pls., 44 figs.

1918 Bird Enemies of Brine Shrimps and Alkali Flies. Auk, vol. 35, p. 372.

1924 Birds as Factors in the Control of the Fall Webworm. *Ibid.*, vol. 41, p. 372.

1932 Effectiveness in Nature of the So-called Protective Adaptations in the Animal Kingdom, Chiefly as Illustrated by the Food Habits of Nearctic Birds. Smithsonian Misc. Coll., vol. 85, No. 7, 201 pp.

1934 Protective Resemblances in Insects, Experiment and Theory. Science, vol. 79, pp. 361–363.

McAtee, W. L., and F. E. L. Beal

1912 Some Common Game, Aquatic and Rapacious Birds in Relation to Man. United States Dept. Agric., Farmer's Bull., No. 497, 30 pp., 14 figs.

Pitman, C. R. S.

1929 The Economic Importance of Birds in Uganda and Parts of Kenya Colony from the Point of View of Locust Destruction. Bull. Soc. Roy. Entom. Egypte, Fasc. 1–3, pp. 93–103.

Skaife, S. H.

1921 Some Factors in the Natural Control of the Wattle Bagworm. South African Journ. Sci., vol. 17, pp. 291–301.

Terres, J. K.

1940 Birds Eating Tent Caterpillars. Auk, vol. 57, p. 422.

LIZARDS

Anon.

1920 Nuestro amigo el Anolis. Rev. Agric. Puerto Rico, San Juan, vol. 4, pp. 11–21.

Burt, C. E.

1928 Insect Food of Kansas Lizards with Notes on Feeding Habits. Journ. Kansas Entom. Soc., vol. 1, pp. 50–68.

Cole, A. C.

1932 Analyses of the Stomach Contents of Two Species of Idaho Lizards, with Special Reference to the Formicidae. Ann. Entom. Soc. America, vol. 25, pp. 638–640.

Eltringham, H.

1909 An Account of Some Experiments on the Edibility of Certain Lepidopterous Larvae. Trans. Ent. Soc. London, 1909, pp. 471–478.

Hunter, W. D.

1912 Two Destructive Texas Ants. United States Dept. Agric., Bur. Entom. Circ. 148, 6 pp.

Knowlton, G. F.

1934 Lizards as a Factor in the Control of Range Insects. Journ. Econ. Entom., vol. 27, pp. 998–1004.

1938 Lizards and Insect Control. Ohio Journ. Sci., vol. 38, pp. 235–238.

Knowlton, G. F., W. D. Fronk, and D. R. Maddock

1942 Seasonal Insect Food of the Brown shouldered Uta. Journ. Econ. Entom., vol. 35, pp. 942–945.

Knowlton, G. F., and M. J. Janes

1932 Studies of the Food Habits of Utah Lizards. Ohio Journ. Sci., vol. 32, pp. 467–470.

1933 Lizards as Predators of the Beet Leafhopper. Journ. Econ. Entom., vol. 26, pp. 1011–1016.

Pack, H. J.

1923 Food Habits of *Callisaurus ventralis*. Proc. Biol. Soc. Washington, vol. 36, pp. 79–81.

Villadolid, D. V.

1934 Food Habits of Six Common Lizards Found in Los Baños, Laguna, Philippine Islands. Philippine Journ. Sci., vol. 55, pp. 61–67.

Winton, W. M.

1915 A Preliminary Note on the Food Habits and Distribution of the Texas Horned Lizards. Science, n.s., vol. 41, pp. 797–798.

Wolcott, G. N.

1924 The Food of Porto Rican Lizards. Journ. Dept. Agric. Porto Rico, vol. 7, pp. 5–37.

AMPHIBIANS

Frost, S. W.

1924 Frogs as Insect Collectors. Journ. New York Entom. Soc., vol. 32, pp. 174–185, 1 pl.

Garofolini, L.

1924 L'utilizzazione del *Triton cristatus* par la distruzione delle larve di anofele. Atti R. Accad. Naz. Lincei., Rendi., vol. 33, pp. 129–131.

Hamilton, W. J., Jr.

1930 Notes on the Food of the American Toad. Copeia, 1930, p. 45.

Illingworth, J. F.

1941 Feeding Habits of *Bufo marinus*. Hawaiian Entom. Soc. Proc., vol. 11, p. 51.

Klugh, A. B.

1922 The Economic Value of the Leopard Frog. Copeia, 1922, pp. 14–15.

Lamborn, W. A.

1912 Butterflies a Natural Food of Monkeys. Trans. Entom. Soc. London, 1912, p. iv.

Leonard, M. D.

1933 Notes on the Giant Toad, *Bufo marinus* in Puerto Rico. Journ. Econ. Entom., vol. 26, pp. 67–72.

Mallonee, A. M.

1916 Frogs Catching Butterflies. Science, n.s., vol. 43, pp. 386–387.

Pack, H. J.

1922 Toads in Regulating Insect Outbreaks. Copeia, 1922, pp. 46–47.

Power, J. H.

1931 The Genus *Bufo* as an Economic Asset. South African Journ. Sci., vol. 28, pp. 376–377.

Smallwood, W. M.

1928 Notes on the Food of Some Onondaga Urodela. Copeia, 1928, pp. 89–98.

Van Volkenberg, H. L.

1935 Biological Control of an Insect Pest by a Toad. Science, vol. 82, pp. 278–279.

Weber, N. A.

1938 The Food of the Giant Toad, *Bufo marinus* in Trinidad and British Guiana. Ann. Entom. Soc. America, vol. 21, pp. 499–503.

Wolcott, G. N.

1937 What the Giant Surinam Toad is Eating now in Puerto Rico. Journ. Agric. Univ. Puerto Rico, vol. 21, pp. 79–84.

FISHES

Anon.

1923 Use of Fish for Mosquito Control. Rockefeller Foundation, Internat. Health Bd., 10th Ann. Rept., 1923, pp. 80–93, 10 figs.

Dimick, R. E., and D. C. Mote

1934 A Preliminary Survey of the Food of Oregon Trout. Bull. Oregon Agric. Expt. Sta., No. 223, pp. 23, 15 figs.

Forbes, S. A.

1880 The Food of Fishes. Bull. Illinois State Lab. Nat. Hist., vol. 1, pp. 19–85.

1890 On the Food Relations of Fresh-water Fishes. *Ibid.*, vol. 2, pp. 475–538.

Hildebrand, S. F.

1919 Fishes in Relation to Mosquito Control in Ponds. Rept. United States Comm. Fisheries, 1918, App. 19, 15 pp., 18 figs.

Hildebrand, S. F., and I. L. Towers

1927 Food of Trout in Fish Lake, Utah. Ecology, vol. 8, pp. 389–397.

Kendall, W. C., and W. A. Dence

1927 A Trout Survey of the Allegheny State Park in 1922. Roosevelt Wild Life Bull., vol. 4, pp. 291–482, 33 figs.

Morofsky, W. F.

1940 A Comparative Study of Insect Food of Trout. Journ. Expt. Biol., vol. 17, pp. 295–306, 7 figs.

Morris, D.

1910 Destruction of Mosquitoes by Small Fish in the West Indies. Proc. 1st Intern. Entom. Congress, vol. 2, pp. 171–172.

Muttkowski, R. A.

1925 The Food of Trout in Yellowstone National Park. Roosevelt Wild Life Bull., vol. 2, pp. 470–490, 21 figs.

Needham, J. G.

1920 Burrowing Mayflies of our Larger Lakes and Streams. Bull. United States Bur. Fisheries, vol. 36, (1917–18), pp. 269–292.

Pearse, A. S.

1921 Distribution and Food of the Fishes of Green Lake, Wisconsin, in Summer. *Ibid.*, vol. 37, (1919–1920), pp. 255–272, 1 map.

Pearse, A. S., and Henrietta Achtenberg

1917 Habits of Yellow Perch in Wisconsin Lakes. *Ibid.*, vol. 36, pp. 297–366, 1 pl., 35 figs.

Russell, P. F., and V. P. Jacob

1939 Experiments in the Use of Fish to Control *Anopheles* Breeding in Casuarina-pits. Journ. Malaria Inst. India, vol. 2, pp. 273–291, 3 pls.

Sibley, C. K.

1929 The Food of Certain Fishes of the Lake Erie Drainage Basin. Suppl. 18th Ann. Rept. New York Conserv. Dept. 1928, pp. 180–188.

Tillyard, R. J.

1921 Neuropteroid Insects of the Hot Springs Region, New Zealand, in Relation to the Problem of Trout-food. New Zealand Journ. Technol., vol. 3, pp. 271–279.

MOLLUSCA

Bishop, S. C., and R. C. Hart

1931 Notes on some Natural Enemies of the Mosquito in Colorado. Journ. N. Y. Entom. Soc., vol. 39, pp. 151–157.

ARACHNIDA

Anderson, J.

1930 Isle of Wight Disease in Bees. Bee World, vol. 11, pp. 37–42; 50–53.

Bilsing, S. W.

1920 Quantitative Studies in Food of Spiders. Ohio Journ. Sci., vol. 20, pp. 215–260.

Bullamore, G. W.

1922 *Nosema apis* and *Acarapis woodi* in Relation to Isle of Wight Bee Disease. Parasitology, vol. 14, pp. 53–62.

Börchert, A.

1932 Untersuchungen an der Acarapismilbe. Zeits. f. Parasitenk., vol. 4, pp. 331–368, 10 figs.

1934 On the Breeding Places of the External Mite (*Acarapis*) on the Honey Bee. Bee World, vol. 15, pp. 43–44.

Wehrle, L. P., and P. S. Welch

1925 The Occurrence of Mites in the Tracheal System of Certain Orthoptera. Ann. Entom. Soc. America, vol. 18, pp. 35–44.

NEMATODES

Bovien, P.

1937 Some Types of Association Between Nematodes and Insects. Vidensk. Medd. Dansk Naturk. Förening, vol. 101, pp. 1–114, 31 figs.

Buckley, J. J. C.

1938 On Culicoides as a Vector of *Onchocerca gibsoni*. Journ. Helminthol., vol. 16, pp. 121–158, 5 pls., 15 figs.

Chitwood, B. G.

1932 A Synopsis of the Nematodes Parasitic in Insects of the Family Blattidae. Zeits. Parasitenk., vol. 5, pp. 14–50, 59 figs.

Christie, J. R.

1931 Some Nemic Parasites (Oxyuridae) of Coleopterous Larvae. Journ. Agric. Res., vol. 42, pp. 463–482, 14 figs.

1937 *Mermis subnigrescens*, a Nematode Parasite of Grasshoppers. *Ibid.*, vol. 55, pp. 353–364.

Cobb, N. A.

1926 The Species of *Mermis*. A Group of Very Remarkable Nemas Infesting Insects. Journ. Parasitology, vol. 12, pp. 66–72, 1 pl., 1 fig.

Cobb, N. A., G. Steiner, and J. R. Christie

1925 The Nemic Parasites of Grasshoppers. Official Rec. United States Dept. Agric., vol. 4, p. 5.

Davis, J. J.

1916 A Nematode Parasite of Root Aphids. Psyche, vol. 23, pp. 39–40, 1 fig.

Glaser, R. W.

1931 The Cultivation of a Nematode Parasite of an Insect. Science, vol. 82, pp. 614–615.

1932 Studies of *Neoplectana glaseri*, a Nematode Parasite of the Japanese Beetle, *Popillia japonica*. Circ. New Jersey Dept. Agric., No. 211, 34 pp., 3 pls.

1940 Continued Culture of a Nematode Parasitic in the Japanese Beetle. Journ. Expt. Zool., vol. 84, No. 1, pp. 1–12.

Glaser, R. W., and H. Fox

1930 A Nematode Parasite of the Japanese Beetle (*Popillia japonica* Newm.). Science, vol. 71, pp. 16–17.

Glaser, R. W., E. E. McCoy, and H. B. Girth

1940 The Biology and Economic Importance of a Nematode Parasitic in Insects. Journ. Parasitol., vol. 26, pp. 479–495.

Glaser, R. W., and A. M. Wilcox

1918 On the Occurrence of a *Mermis* Epidemic Amongst Grasshoppers. Psyche, vol. 25, pp. 12–15.

Hall, M. G.

1929 Arthropods as Intermediate Hosts of Helminths. Smithsonian Misc. Coll., vol. 81, No. 15, 77 pp.

Oldham, J. N.

1933 Helminths in the Biological Control of Insect Pests. Notes Memor. Imp. Bur. Agric. Parasit., No. 9, 6 pp.

Pemberton, C. E.

1928 Nematodes Associated with Termites in Hawaii, Borneo and Celebes. Proc. Hawaiian Entom. Soc., vol. 7, pp. 148–150.

Thorne, G.

1940 The Hairworm, *Gordius*, as a Parasite of the Mormon Cricket. Journ. Washington Acad. Sci., vol. 30, pp. 219–231, 7 figs.

Van Zwaluwenburg, R. H.

1928 The Interrelationships of Insects and Roundworms. Bull. Expt. Sta. Hawaiian Sugar Planters' Assoc., Ent. Ser., vol. 20, 68 pp.

PROTOZOA

Drbolah, J. J.

1925 The Relation of Insect Flagellates to Leishmaniasis. American Journ. Hygiene, vol. 5, pp. 580–621.

Hertig, M.

1923 The Normal and Pathological Histology of the Ventriculus of the Honeybee, with Special Reference to Infection with *Nosema apis*. Journ. Parasitology, vol. 9, pp. 109–140, 3 pls.

King, R. L., and A. B. Taylor

1936 *Malpighamoeba locustae* n.sp. (Amoebidae), a Protozoan Parasitic in the Malpighian Tubes of Grasshoppers. Trans. American Micros. Soc., vol. 55, pp. 6–10, 8 figs.

Kowalczyk, S. A.

1938 Report on the Intestinal Protozoa of the Larva of the Japanese Beetle (*Popillia japonica*). Ibid., vol. 57, pp. 229–244.

Krassiltschik, J.

1905 Sur une affection parasitaire des Lépidoptères par un sporozoaire nouveau (*Microklassia prima*). C. R. Soc. Biol., Paris, vol. 58, p. 656.

Kudo, R. R., and J. D. de Coursey

1940 Experimental Infection of *Hyphantria cunea* with *Nosema bombycis*. Journ. Parasitol., vol. 26, pp. 123–125.

Paillot, A.

1928 Sur le cycle évolutif de *Nosema bombycis*, parasité de la pébrine du ver à soie. C. R. Soc. Biol., Paris, vol. 99, pp. 81–83, 16 figs.

1933 L'infection chez les insectes. 535 pp., 279 figs. Imp. Trevoux, Paris.

Payne, N. M.

1933 A Parasitic Hymenopteron as a Vector of an Insect Disease. Entom. News, vol. 44, p. 22.

Prell, H.

1926 Beiträge zur Kenntnis einer Amöbenseuche der Honigbiene. Zeits. angew. Entom., vol. 12, pp. 163–168.

Semans, F. M.

1933 Protozoan Parasites of the Orthoptera, with Special Reference to those of Central and Southeastern Ohio. Journ. Parasitol., vol. 20, pp. 125–126.

1939– Protozoan Parasites of the Orthoptera. Ohio Journ. Sci., vol. 39, pp. 157–
40 179; vol. 41, pp. 457–464.

Strickland, E. H.

1911 Some Parasites of *Simulium* Larvae and their Effects on the Development of the Host. Biol. Bull., vol. 21, pp. 302–338, 5 pls.

Zwölfer, W.

1927 Die Pebrine des Schwammspinners und Goldafters, eine neue wirtschaftlich bedeutungsvolle Infektionskrankheit. Zeits. angew. Entom., vol. 12, pp. 498–500.

PLANTS

Brown, F. M.

1930 Bacterial Wilt Disease. Journ. Econ. Entom., vol. 23, pp. 145–146.

Chorine, V.

1929 New Bacteria Pathogenic to the Larvae of *Pyrausta nubilalis*. Internat. Corn Borer Invest. Sci. Rept., vol. 2, pp. 39–53, 6 figs.

1930 On the Use of Bacteria in the Fight Against the Corn Borer. *Ibid.*, vol. 3, pp. 94–98.

1931 Sur l'utilisation des microbes dans la lutte contre la pyrale du maïs. Ann. Inst. Pasteur, vol. 46, pp. 326–336.

De Bach, P. H., and W. A. McOmie

1939 New Diseases of Termites Caused by Bacteria. Ann. Entom. Soc. America, vol. 32, pp. 137–146.

Duggar, B. M.

1897 On a Bacterial Disease of the Squash-bug (*Anasa tristis* DeG.). Bull. Illinois State Lab. Nat. Hist., vol. 4, pp. 340–379, 2 pls.

Dutky, S. R.

1940 Two New Spore-forming Bacteria Causing Milky Diseases of Japanese Beetle Larvae. Journ. Agric. Res., vol. 41, pp. 57–68, 6 figs.

Glaser, R. W.

1914 The Bacterial Diseases of Caterpillars. Psyche, vol. 21, pp. 184–190.

1924a A Bacterial Disease of Adult House Flies. American Journ. Hyg., vol. 4, pp. 411–415, 1 pl.

1924b A Bacterial Disease of Silkworms. Journ. Bacter., vol. 9, pp. 339–355.

1925 Specificity in Bacterial Disease with Special Reference to Silkworms and Tent Caterpillars. Journ. Econ. Entom., vol. 18, pp. 769–771.

1926 Further Experiments on a Bacterial Disease of Adult Flies with a Revision of the Etiological Agent. Ann. Entom. Soc. America, vol. 19, pp. 193–198.

d'Hérelle, F.

1914 Le coccobacille des Sauterelles. Ann. Inst. Pasteur, Paris, vol. 28, pp. 280–328; 387–407.

Metalnikov, S.

1933 Rôle des microörganismes dans la déstruction des insectes nuisibles. Congr. Int. Entom. Paris, 1932, 5, 2 Trav., pp. 611–616.

Metalnikov, S., and V. Chorine

1929   On the Infection of the Gypsy Moth and certain other Insects with *Bacterium thuringiensis*. Internat. Corn Borer Invest. Sci. Rept., vol. 2, pp. 60–61.

Metalnikov, S., B. Hergula, and D. M. Strail

1931   Utilisation des microbes dans la lutte contre la pyrale du maïs. Ann. Inst. Pasteur, vol. 46, pp. 320–325.

Metalnikov, S., and S. Metalnikov, Jr.

1932   Maladies des vers du coton (*Gelechia gossypiella* et *Prodenia litura*). C. R. Acad. Agric. France, vol. 18, pp. 203–207.

1941   Utilisation des microbes dans la lutte contre les insectes nuisibles. Proc. 6th Entom. Congr., Madrid, vol. 2, pp. 555–566.

Paillot, A.

1926   Rôle des microbes sporulés dans la flachérie du ver à soie. C. R. Acad. Sci., Paris, vol. 183, pp. 704–707.

1927   Les affections du tube digestif chez le ver à soie. Rev. Zool. Agric. and Appl., vol. 26, pp. 37–41.

1928   Les maladies du ver à soie. Grassérie et dysentéries. 324 pp., 32 pls. Lyon.

1933   L'infection chez les insectes. 535 pp., 279 figs. Imp. Trevoux, Paris.

Sato, R.

1928   Ueber bakterielle Krankheiten der Seidenraupen, insbesondere über Septikämie. Centralbl. f. Bakt., Orig., vol. 107, pp. 234–278.

Steinhaus, E. A.

1942   Catalogue of Bacteria Associated Extracellularly with Insects and Ticks. iii + 206 pp. Minneapolis, Burgess Pub. Co.

White, G. F.

1923a   Hornworm Septicemia. Journ. Agric. Res., *Ibid.*, vol. 26, pp. 477–486, 1 pl., 2 figs.

1923b   Cutworm Septicemia. Vol. 26, pp. 487–496, 2 pls., 2 figs.

FUNGI

Arnaud, M.

1927   Recherches preliminaires sur les champignons entomophytes. Ann. Epiphyties, vol. 13, pp. 1–30, 16 figs.

Burnside, C. E.

1930   Fungus Diseases of the Honeybee. Tech. Bull. United States Dept. Agric., No. 149, 42 pp., 6 pls., 5 figs.

Cépède, C.

1914   Etude des Laboulbeniacées européennes. Arch. Parasitol., vol. 16, pp. 373–403.

Cooke, M. C.

1892   Vegetable Wasps and Plant Worms, A Popular History of Entomogenous Fungi, or Fungi Parasitic upon Insects. x + 364 pp., 4 pls. London.

Cunningham, G. H., and J. G. Myers

1921   The Genus *Cordyceps* in New Zealand. Trans. Proc. New Zealand Inst., vol. 53, pp. 372–382, 4 pls., 8 figs.

Dieuzeide, R.

1926   Le *Beauveria effusa* Vuillemin, parasite du Doryphore de la pomme de terre. Rev. Zool. Agric. and Appl., vol. 25, pp. 129–134; 145–154.

Dustan, A. G.

1920   Entomogenous Fungi. Proc. Ent. Soc. Nova Scotia, 1920, pp. 36–45.

Forbes, S. A.

1895. Experiments with the Muscardine Disease of the Chinch Bug. Bull. Illinois Agric. Expt. Sta., No. 38, pp. 25–86.

1896   On Contagious Diseases in the Chinch Bug. 19th Ann. Rept. State Entom. Illinois, pp. 16–176.

Glaser, R. W.

1914   The Economic Status of the Fungus Diseases of Insects. Journ. Econ. Entom., vol. 7, pp. 1–4.

1926   The Green Muscardine Disease in Silkworms and its Control. Ann. Entom. Soc. America, vol. 19, pp. 180–192.

Goldstein, B.

1927   An *Empusa* Disease of *Drosophila*. Mycologia, vol. 19, pp. 97–109, 3 pls.

Gosswald, K.

1939   Ueber den Insektentötenden Pilz, *Beauveria bassiana*. Arb. Biol. Reichanst. Land u. Forstwirtsch., Berlin, vol. 22, pp. 399–452.

Lakon, G.

1919   Die Insektenfeinde aus der Familie der Entomophthoreen. Zeits. angew. Ent. Berlin, vol. 5, pp. 161–216.

Lefebvre, C. L.

1934   Penetration and Development of the Fungus, *Beauveria bassiana* in the Tissues of the Corn Borer. Ann. Bot., vol. 48, pp. 441–452, 1 pl., 2 figs.

Massee, G.

1895   A Revision of the Genus *Cordyceps*. Ann. Bot., vol. 9, pp. 1–44, 2 pls.

1898– Revision du genre *Cordyceps*. Revue Mycologique, vol. 20, pp. 49–57; 85–99   94; vol. 21, pp. 1–16.

Metalnikov, S., and K. Toumanoff

1928   Recherches expérimentales sur l'infection de *Pyrausta nubilalis* par des champignons entomophytes. C. R. Soc. Biol., Paris, vol. 98, pp. 583–584.

Molliard, M.

1924   Epidémie d'*Empusa muscae* sur *Melanostoma* (*Syrphus*) *mellinum* L. Feuille Natural. Ann., 45, pp. 81–82.

Petch, T.

1925   Entomogenous Fungi and their Use in Controlling Insect Pests. Bull. Ceylon Dept. Agric., Paradeniya, No. 71, 40 pp., 2 pls.

1932   A List of the Entomogenous Fungi of Great Britain. Trans. Brit. Mycol. Soc., vol. 17, pp. 170–178.

Picard, F.

1913   Contribution à l'étude des Laboulbeniacées d'Europe et du Nord de l'Afrique. Bull. Soc. Mycol., France, vol. 29, pp. 503–571, 19 pls.

Reichensperger, A.

  1923 Neue eigenartige Parasiten von Termiten.   Bull. Soc. Fribourg. Sci. Nat.,
        vol. 26, pp. 103–114.

Roubaud, E.

  1926 Sur un champignon entomophyte parasite des fourmis en Afrique équatorielle.
        Bull. Soc. Path. Exot., vol. 19, pp. 815–819.

Sawyer, W. H., Jr.

  1931 Studies on the Morphology and Development of an Insect-destroying Fungus,
        Entomophthora sphaerosperma.   Mycologia, vol. 23, pp. 411–432, 2 pls.
  1933 The Development of *Entomophthora sphaerosperma* upon *Rhopobota vaccini-*
        ana.   Ann. Bot., vol. 47, pp. 799–809, 2 pls., 1 fig.

Seymour, A. B.

  1929 Host Index of the Fungi of North America.   Cambridge, Mass., Harvard
        Univ. Press.

Skaife, S. H.

  1925a The Locust Fungus, *Empusa grylli*, and its Effects on its Host.   South
         African Journ. Sci., vol. 22, pp. 298–308, 1 pl.
  1925b The Fungus Disease of Locusts.   Reports on a Preliminary Investigation
         in Southwest Africa.   Journ. Dept. Agric. Pretoria, vol. 11, pp. 179–185, 4 figs.

Snow, F. H.

  1891 Experiments for the Destruction of Chinch Bugs in the Field by the Artificial
        Introduction of Contagious Diseases.   Insect Life, vol. 3, pp. 279–285.
  1892 Experiments for the Destruction of Chinch Bugs by Infection.   Psyche, vol.
        6, pp. 222–233.
  1894 Contagious Diseases of Chinch Bug.   3rd Rept. Director Kansas Agric. Expt.
        Sta. for 1893, 247 pp.

Speare, A. T.

  1920 On Certain Entomogenous Fungi.   Mycologia, vol. 12, pp. 62–76, 2 pls.
  1922 Natural Control of the Citrus Mealybug in Florida.   Bull. United States
        Dept. Agric., No. 1117, 18 pp., 1 pl., 2 figs.

Swingle, H. S., and J. L. Seal

  1931 Some Fungus and Bacterial Diseases of Pecan Weevil Larvae.   Journ. Econ.
        Entom., vol. 24, p. 917.

Thaxter, Roland

  1888 The Entomophthoreae of the United States.   Mem. Boston Soc. Nat. Hist.,
        vol. 4, pp. 133–201.
  1896– Contribution Towards a Monograph of the Laboulbeniaceae.   Mem. Ameri-
  1926 can Acad. Arts Sci., 4 pts., vols. 12–15.
  1914 On Certain Peculiar Fungus-parasites of Living Insects.   Bot. Gaz., vol. 58,
        pp. 235–253, 4 pls.
  1920 Second Note on Certain Peculiar Fungus-Parasites on Living Insects.   Bot.
        Gaz., vol. 69, pp. 1–27, 5 pls.

Vincens, F.

  1923 Sur l'aspergillomycose des abeilles.   C. R. Acad. Sci., Paris, vol. 177, pp. 540–
        542.

Wallengren, H., and R. Johannson

1929 On the Infection of *Pyrausta nubilalis* by *Metarrhizium anisopliae*. Internat. Corn Borer Invest., Sci. Rept., vol. 2, pp. 131–145, 15 figs.

FLOWERING PLANTS

Danser, B. H.

1928 The Nepenthaceae of the Netherlands Indies. Bull. Jard. Bot. Buitenzorg, vol. (3) 9, pp. 249–438.

Darwin, Charles

1875 Insectivorous Plants. Second Edition, revised by Francis Darwin, 1893. xiv + 377 pp., 29 figs. London, John Murray.

Dover, C.

1928 Fauna of Pitcher Plants. Journ. Malayan Branch, Roy. Asiatic Soc., vol. 6, pp. 1–27.

França, C.

1922 Recherches sur les plantes carnivores II. *Utricularia vulgaris*. Bul. Soc. Brotériana, (2) vol. 1, pp. 11–37, 9 figs.

Gillette, C. P.

1898 An Insect-catching Plant. Entom. News, vol. 9, pp. 169–170.

Goodnight, C. J.

1940 Insects taken by the Southern Pitcher Plant. Trans. Illinois Acad. Sci., vol. 33, p. 213.

Guenther, K.

1913 Die lebenden Bewohner der Kannen der insektenfressenden Pflanze *Nepenthes distillatoria* auf Ceylon. Zeits. wiss. Insektenbiol., vol. 9, pp. 90–95; 122–130; 156–160; 198–207; 259–270, 14 figs.

Hepburn, J. S.

1918 Biochemical Studies of the Pitcher Liquor of *Nepenthes*. Proc. American Philos. Soc., vol. 57, pp. 112–129.

Hepburn, J. S., et al.

1927 Biochemical Studies of North American Sarraceniaceae. Trans. Wagner Free Inst., Philadelphia, vol. 11, pp. 1–95.

Hepburn, J. S., and F. M. Jones

1919 Occurrence of Anti-proteases in the Larvae of *Sarcophaga* Associates of *Sarracenia flava*. Contrib. Bot. Lab. Univ. Pennsylvania, vol. 4, pp. 460–463.

Jones, F. M.

1920 Another Pitcher Plant Insect, *Neosciara macfarlanei*. Entom. News, vol. 31, pp. 91–94.

1921 Pitcher Plants and Their Moths. The Influence of Insect-trapping Plants on Their Insect Associates. Journ. American Mus. Nat. Hist., vol. 21, pp. 297–316, 21 figs.

1923 The Most Wonderful Plant in the World. Nat. Hist., vol. 23, pp. 589–596, 7 figs.

Knab, F.

1905   A Chironomid Inhabitant of *Sarracenia purpurea*, *Metriocnemus knabi* Coq. Journ. New York Entom. Soc., vol. 13, pp. 69–73, 1 pl.

Lloyd, F. E.

1935   *Utricularia.*   Biol. Rev., vol. 10, pp. 72–110.
1942   The Carnivorous Plants.   xv + 352 pp., 38 pls.   Waltham, Mass., Chronica Botanica Co.

Meijere, J. C. H. de, and H. Jensen

1910   Nepenthes-Tiere, Systematik und biologische Notizen.   Ann. Jard. Botan. Buitenzorg, vol. 3, pp. 917–946, 4 pls.

Menzel, R.

1922–   Beiträge zur Kenntnis der Mikrofauna von Nederlandisch-Ost-Indien.
23   Treubia, Batavia, vol. 3, pp. 116–126; 189–196, 5 figs.

Rabaud, E.

1923   La capture des insectes par les plantes.   Bull. Soc. Entom. France, 1923, pp. 122–123.

Ricome, H.

1919   Une plante dangéreuse pur les insectes qui en assurent la pollinisation.   C. R. Soc. Biol., Paris, vol. 82, pp. 1045–1047.

Riley, C. V.

1875   On the Insects More Particularly Associated with *Sarracenia*.   Proc. American Assoc. Adv. Sci., vol. 23, pp. 17–25.

Romeo, A.

1933   Contributo alla Biologia fiorale dell' *Araujia sericifera* Brot.   Ann. Inst. Agr. Portici, (3) vol. 6, pp. 78–97.

Thienemann, A.

1932   Die Tierwelt der Nepenthes-Kannen.   Tropische Binnengewässer, vol. 3, pp. 1–54.   (Suppl. vol. 11, Archiv f. Hydrobiol.)

Wray, D. L., and C. S. Brimley

1943   The Insect Inquilines and Victims of Pitcher Plants in North Carolina.   Ann. Entom. Soc. America, vol. 36, pp. 128–137.

Zeeuw, J. de

1934   Versuche über die Verdauung in Nepentheskannen.   Biochem. Zeits., vol. 269, pp. 187–195.

VIRUS DISEASES

Acqua, C.

1931   Contributo alla conoscenza della natura degli ultravirus.   Ricerche sulla malattia della poliedria negli insetti (*Bombyx mori* L.).   Boll. R. Staz. sperim. Gelsic., vol. 10, pp. 159–201.

Atanasoff, D.

1934   Virus Diseases of Plants: A Bibliography.   Houdojnik Ptg. Co., Sofia, 1934, pp. iv + 219.

Cook, M. T.

1937 Insect Transmission of Virus Diseases of Plants. Sci. Monthly, vol. 44, pp. 174–177.

Gildermeister, E., et al.

1939 Handbuch der Viruskrankheiten. 2 vols., 1420 pp. Jena, Gustav Fischer.

Glaser, R. W.

1927 Studies on the Polyhedral Diseases Due to Filterable Viruses. Ann. Entom. Soc. America, vol. 20, pp. 319–342, 1 pl.

Komarek, J., and V. Breindl

1924 Die Wipfelkrankheit der Nonne und der Erreger deselben. Zeits. angew. Entom., vol. 10, pp. 99–162, 2 pls., 1 fig.

Lotmar, R.

1941 Die Polyederkrankheit der Kleidermotte, *Tineola*. Mitt. Schweitz. Entom. Ges., vol. 18, pp. 372–373.

Otero, J. I., and M. T. Cook

1934– Partial Bibliography of Virus Diseases of Plants. Journ. Agric. Univ.
35 Puerto Rico, vol. 18, pp. 5–410; *ibid.*, vol. 20, pp. 129–313.

Prell, H.

1926 Die Polyederkrankheiten. Verh. IIIten Entom. Kongr., Zürich, 1925, vol. 2, pp. 145–168.

Storey, H. H.

1939 Transmission of Plant Viruses by Insects. Bot. Rev., vol. 5, pp. 240–272.

INDEX OF AUTHORS

Page numbers in italics refer to bibliographical references.

INDEX OF SUBJECTS

Date Due

DEC 4 1976			
FEB 8 1983			
			UML 735